SSHRC

Violence and Racial Prejudice in the Context of Peacekeeping

Violence and Racial Prejudice in the Context of Peacekeeping

a study prepared for
the Commission
of Inquiry into
the Deployment of
Canadian Forces
to Somalia

Jean-Paul Brodeur

Printed and bound in Canada

Available in Canada through
your local bookseller or by mail from
Canadian Government Publishing
Ottawa, Canada K1A 0S9

Catalogue No. CP32-64/3-1997E
ISBN 0-660-16870-7

Canadian Cataloguing in Publication Data

Brodeur, Jean-Paul

Violence and racial prejudice in the context of
peacekeeping

Issued also in French under title: Violence et préjugés
raciaux dans les missions de maintien de la paix.

ISBN 0-660-16870-7
Cat. no. CP32-64/3-1997E

1. Somalia affair, 1992-
2. Canada — Armed Forces — Somalia.
3. Canada. Canadian Armed Forces. Canadian Airborne
Regiment Battle Group.
4. Security, International.
I. Commission of Inquiry into the Deployment of
Canadian Forces to Somalia.
II. Title.
III. Title: The Canadian Forces in Somalia.

FC603.B76 1997 355'.00971 C97-980054-4
F1028.B76 1997

Q - General:
"Yeah, did you ever hear anybody use the term nigger for example?"

A - A Trooper deployed to Somalia:
"Oh, yes, I did hear that."

Q - General:
"And in what context did you hear that term used?"

A - Same Trooper:
"Only as a descriptive term."

(Transcript from the hearings of the de Faye Board of Inquiry).

When there is peace, the warlike man attacks himself.

Friedrich Nietzsche

Contents

Abbreviations and Acronyms

ADM (Per)	Assistant Deputy Minister (Personnel)
AETHRP	Awareness Education and Training for Harassment and Racism Prevention (a CF program to be developed and implemented)
ASIO	Australian Security Intelligence Organization
ASIS	Australian Security Intelligence Service
BOI	Board of Inquiry (appointed by the CF to investigate the incidents that took place in Somalia; headed by Major-General T.F. de Faye)
CAR	Canadian Airborne Regiment
CARBG	Canadian Airborne Regiment Battle Group
CAR-TP	Canadian Airborne Regiment, Operation Cordon, Training Plan
CBC	Canadian Broadcasting Corporation
CCMA	Chain of Command and Military Accountability (CF Brief for the CIDCFS)
CCSI	Cabinet Committee on Security and Intelligence
CDS	Chief of the Defence Staff
CF	Canadian Forces
CFAO	Canadian Forces Administrative Order. The letters "CFAO" followed by a number indicate an official policy of the CF.
CFB	Canadian Forces Base
CFDS	Canadian Forces Diversity Study
CIA	Central Intelligence Agency
CIDA	Canadian International Development Agency

CIDCFS	Commission of Inquiry into the Deployment of the Canadian Forces to Somalia; chaired by Judge G. Létourneau.
CJFS	Canadian Joint Force Command (in Somalia)
CMSF	Canadian Mobile Striking Force
CNN	Cable News Network
CO	Commanding Officer
COI, REP	Report of the Commission of Inquiry established pursuant to UN Security Council 884 (1993) to investigate armed attacks on UNOSOM II personnel which led to casualties among them. Chaired by the Hon. Matthew N.S.W. Ngulube.
CSE	Communications Security Establishment
CSIS	Canadian Security Intelligence Service
CWO	Chief Warrant Officer
DG Secur	Director General Security
DM	Deputy Minister (for DND, in this report)
DND	Department of National Defence
DPIS	Demographic Profile of Department of National Defence Personnel Information System
EE	CF Employment Equity data base
FBI	Federal Bureau of Investigation
FMC	Force Mobile Command
FNG	"fucking new guy" (acronym used in the CAR to designate newcomers)
ICRC	International Committee of the Red Cross
ICSI	Interdepartmental Committee on Security and Intelligence
IG	Inspector General
JAG	Judge Advocate General (CF Military Justice)
KKK	Ku Klux Klan
MND	Minister of National Defence
MPIS	Military Personnel Information System
MVA	Motor Vehicle Accidents
MWO	Master Warrant Officer
NCM(s)	Non-commissioned member(s) of the CF
NCO	Non-commissioned Officer
NDA	*National Defence Act*

NDCC-SC	National Defence Consultative Committee on Social Change
NDHQ	National Defence Headquarters
OC	Officer in Command
PCIP	Lester B. Pearson Canadian International Peacekeeping Centre
PCO	Privy Council Office
PPCLI	Princess Patricia's Canadian Light Infantry (one of the three parent regiments of the CAR)
Q & As	Questions and Answers
QRF	Quick Reaction Force (U.S. Forces in Somalia)
RCMP	Royal Canadian Mounted Police
ROE	Rules of Engagement
RTQ	Response(s) to Query(ies)
SHARP	This is a project otherwise known as Harassment Awareness Training Project (HATP) for the CF and the DND
SILT	Somalia Inquiry Liaison Team
SIRC	Security Intelligence Review Committee
SIROS	An internal CF project designed to identify, among others, right-wing elements in CF personnel. The documents consulted do not reveal the precise meaning of the acronym "SIROS"
SOP	Standard Operating Procedure
ST-PSVE	Study Team on Public Service Values and Ethics
UN	United Nations
UNEF	United Nations Emergency Force (1956)
UNFICYP	United Nations Force of Intervention in Cyprus
UNITAF	United Task Force
UNOSOM I	First United Nations Operation in Somalia
UNOSOM II	Second United Nations Operation in Somalia
UNPROFOR	United Nations Protection Force
YOS	Years of Service

Introduction

In presenting this report my aim is not to make an unfocussed scholarly contribution on the subjects enunciated in the terms of reference or related to them, although all the canons of scholarly work will be, to the extent possible, respected. First, my aim is to present the results of research — both in the available literature and in my own analysis of relevant documents, field work and interviews — which are pertinent to the terms of reference of the Commission of Inquiry into the Deployment of the Canadian Forces to Somalia.

Second, I also believe it is important to answer the questions raised by the Canadian public and by the Canadian media regarding the topics covered by the Commission's terms of reference.

Finally, I am writing this report while the Inquiry is still proceeding. Hence, I am submitting it without the benefit of appraising all the factual results of the inquiry. This most important consideration determines my basic aim: it cannot be to formulate definitive conclusions with respect to an inquiry which is still ongoing at the writing of this report, but to provide the Somalia Inquiry with the main theoretical research options it might exercise in framing the issues to be discussed in its final report and in formulating its recommendations, when its investigations have been completed. Consequently, I devote a larger part of this report to conceptual analysis than was initially projected. Needless to say, the conceptual instruments I try to develop are far from exhausting the Somalia Inquiry's agenda and are only pursuant to my own limited terms of reference.

THE TERMS OF REFERENCE

There is a distinction between the terms of reference themselves and the interpretation given to them in the Commission's statement on 3 August 1995.

20 March 1995: The Terms of Reference

Without excluding any part of the terms of reference, I will be addressing in more detail the following articles:

Pre-deployment (prior to 10 January 1993)
(a) the suitability of the Canadian Airborne Regiment for service in Somalia;
(b) the mission and tasks assigned to the Canadian Airborne Regiment Battle Group (CARBG) and the suitability and its composition and organization for the mission and tasks assigned;
(d) the adequacy of the selection and screening of officers and non-commissioned members for the Somalia deployment;
(e) the appropriateness of the training objectives and standards used to prepare for deployment of the Airborne Regiment.

In-theatre (10 January 1993 to 10 June 1993)
(l) the extent, if any, to which cultural differences affected the conduct of operations;
(m) the attitude of all rank levels towards the lawful conduct of operations, including the treatment of detainees;
(n) the appropriateness of professional values and attitudes in the Task Force and the impact of deployment in Somalia on those values and attitudes;
(o) the extent to which the Task Force Rules of Engagement were effectively interpreted, understood and applied at all levels of the Canadian Forces chain of command;
(r) the effectiveness of the decisions and actions taken by leadership at all levels of National Defence Headquarters in response to the operational, disciplinary and administrative problems related to the Somalia deployment.

Post-deployment (11 June 1993 to 28 November 1994)
(s) the manner in which the chain of command of the Canadian Forces responded to the operational, disciplinary and administrative problems related to the Somalia deployment.

The Statement about the Terms of Reference (3 August 1995)

The CIDCFS has issued a statement on its own interpretation of its terms of reference. The purport was to draw a balance between a general inquiry into the Canadian Forces (CF) and an inquiry that focussed on the incidents that occurred in relation to the CF deployment in Somalia. In essence, the CIDCFS ruling stated that the inquiry would be limited to the incidents that actually occurred in Somalia, but that the Commission would not shirk investigating the systemic ramifications of these incidents. Indeed, it is difficult to imagine how the CIDCFS could make enlightened recommendations without investigating these systemic ramifications. There are several relevant excerpts from the 3 August statement for our purposes:

In our view, the Defence Minister's decision to disband the Airborne Regiment and restructure it differently is a decision...that falls outside the scope of our inquiry... (p.8)

The terms "cultural differences" are broader in scope than the issue of racism or human rights itself. They refer to all those sociological, anthropological, political, economic, intellectual and human characteristics which define a culture and serve to differentiate it from another. (p. 12)

In order to properly assess the impact of cultural differences on the conduct of operations, the Commission has to look at the appropriateness of the training objectives and standards of the Canadian Airborne Regiment Battle Group with respect to the Somalian culture and environment and the proper treatment of civilians and detainees, the adequacy of selection and screening of its officers and the state of discipline of its members with a view to determining the extent to which Somalian and human rights values, and the need to protect and respect them, have been properly taught, understood and respected in the context of a peacemaking mission in a fundamentally different cultural environment. (p. 12)

...

The inquiry as a whole is bounded by the necessity to investigate the deployment of Canadian troops to Somalia. The inquiry is not an investigation into the deployment of Canadian troops on peacekeeping missions in general and by the same token, the aspect of racism that is to be investigated does not call for a general investigation of racism in the Canadian military as a whole. (p. 13)

...

In investigating racism to the extent that our terms allow, the Commission will of necessity be required to investigate aspects of military operations possessing systemic dimensions and implications. Issues such as screening and training involve factual inquiries that lead beyond the narrow confines of any single regiment or unit and may require our analyzing various operations, procedures, manuals, training courses and curricula that may have system wide application. A concentration on the Somalia operation may hold meaning for the military as a whole. Although the Commission is not in a position to embark on an exploration of the state of racism and human rights violations in the Canadian Forces in general, it is quite prepared to call and examine all relevant evidence for the purpose of doing justice to such issues as validly fall within its Terms of Reference. (p. 13)

...

Indeed, the admissibility of evidence of racism in the military as a whole presupposes the existence and availability of such evidence while an inquiry into this very issue is an inquiry both to uncover evidence of racism and collect evidence of racism and determine the existence and extent of such racism. (p. 14)

...

The Commission will not investigate and report on the disbandment of the Canadian Airborne Regiment and the state of human rights violations and racism in the Canadian military as a whole although it will be looking, in the context of the Somalia operation, at issues of systemic dimensions and implications. (p. 14)

The statement narrows down the scope of possible inquiries. (1) It clearly appears that the Commission wants to limit its main inquiry to the case at hand, which is the deployment of the Canadian Forces in Somalia. Consequently, the reasons for the disbandment of the Canadian Airborne Regiment are seen as lying outside the Commission's terms of reference. This being so, there is all the more reason to assert that the Commission will not investigate the deployment of Canadian troops on peacekeeping missions in general, nor will it generally investigate allegations of racism with respect to the whole Canadian Forces. (2) However, the Commission will allow itself to go beyond its main object of inquiry when issues, such as screening, selection and training have, by their very nature, systemic ramifications. (3) The Commission will consider evidence collateral to the Somalia operation — e.g., the training of the Airborne Regiment for the Western Sahara Operation — as admissible on the basis of its existence and availability. However, it will not embark on an investigation to uncover and collect new evidence of racism. In addition to these two extensions of the limits set to the terms of reference, i.e., systemic issues and availability of existing evidence, there are two further instances in which a cautious broadening of the scope of the inquiry appears to be reasonably implied by the terms of reference: (4) it will on occasion be necessary to address some issues in a general way to provide context for the inquiry into the operation in Somalia and (5) providing context will be especially crucial in making recommendations that will address the shortcomings of the operation in Somalia.

I shall thus conduct my own research within the framework provided by the terms of reference of the Commission and the five rules for their interpretation that I have just enunciated on the basis of the statement by the Commission on the scope of its terms of reference.

The Issues

The issues I shall address may pertain to both substance and process. The three main substantial issues are ethnicity, the use of force and accountability. Ethnicity encompasses being exposed and reacting to ethnic and cultural differences as well as the various reactions to this exposure — acceptation, positive valuation, xenophobia and racism. All these reactions and attitudes will have many variants. There is also the presence, in the Canadian Airborne Regiment Battle Group, of individuals with affiliations with organizations and groups belonging to the extreme right. The use of force will be discussed within the context of the kind of mission that was given to the Battle Group and its rules of engagement (ROE). Accountability refers to the action taken at the different levels of command of the Canadian Forces in reacting to the situation that had developed in Somalia, as it was revealed in particular by the beating death of Shidane Arone. This required shedding as much light as possible on what was happening, applying sanctions to those who deserved them and employing remedies that would avoid the repetition of a similar situation in the future.

The issues that relate to the process are recruitment, selection, screening, training, cross-cultural sensitization and, most important, the enhancement of accountability through the development of mechanisms that promote transparency and monitor the application of the policies adopted to remedy what was found to be wanting in the operation in Somalia. Although how these issues are treated and dealt with in police organizations is not seen by us as mandatory with regard to every discussion undertaken in this report, I will make frequent use of my experience and knowledge in the field of policing. The methods that I will use, including documentary analysis and interviews with key informants, are the standard tools of the social sciences.

There is one last point that needs to be made. Although the unfortunate incidents that occurred during the deployment of the Battle Group in Somalia are few, they have a formidable depth, requiring many layers of analysis. To give one example, the fate suffered by Shidane Arone at the hands of members of an Airborne Regiment commando sent to Somalia on a peacekeeping/peacemaking operation despite a history of problems with internal discipline, and the presence of Aboriginal and other ethnic minority members in the regiment, raises (1) the issues of the nature and variants of international policing operations and what is permissible within their framework, (2) the wisdom of choosing to send such a unit to Somalia, (3) the ability of the leadership to impose discipline in the field of operation, (4) the particular subculture that pervades airborne military units (5) with their hazing rituals, (6) the behaviour of soldiers in a physical, ethnic and cultural set-up that was very different from their usual environment and at times hostile to them, (7) the pressures ethnic minorities, such as Aboriginal or Black soldiers may be subjected to in this kind of unit, (8) the effect of the presence of right wing organization sympathizers in the unit, (9) the measures taken to remedy the situation and (10) to bring to military justice those who might have committed serious crimes. This list is not exhaustive and every researcher could extend it, depending on his or her specialization. What is striking is that all these issues are specific in varying degrees, and they have been the object of an impressive body of research. Hazing ritual is much too specific an issue to warrant the assumption that social scientists are generally familiar with it; in order to gain this familiarity, one has to review a considerable quantity of material. In the case of less specific issues, such as xenophobia and racism, the available data are overwhelming. The upshot of these remarks is that no one researcher could ever exhaustively cover the issue of the beating death of Shidane Arone. The only research strategy that can be pursued in coming to terms with incidents that involve so many layers of analysis has to be selective. It is to be hoped that the Commission will be able to bring the different pieces of the numerous puzzles it is confronting into a comprehensive and consistent whole.

CHAPTER ONE

Conceptual Analysis

This chapter is devoted to conceptual analysis. The three concepts to be discussed are peacekeeping, racism and accountability. The first pertains to the mandate of the Canadian Forces sent to Somalia and applies principally to the pre-deployment phase of the Commission's terms of reference; the second concept will be discussed in the subsequent chapters of this report with respect to the performance of the Canadian Airborne Regiment Battle Group (CARBG) in-theatre. Accountability is mostly relevant to post-deployment. Although the meaning of these different concepts is not exhausted by making them correspond to the phases of the operation, there is justification for discussing them in the same order as these three phases (pre-deployment, in-theatre, post-deployment).

Not only are these concepts fairly broad notions, but there is scant consensus about their use. Since all three concepts play a major role in my report, I shall endeavour to clarify their meaning and to identify their different, and often ambiguous, usages.

PEACEKEEPING AND RELATED CONCEPTS

The Canadian troops deployed in Somalia did not wear blue helmets or berets, the hallmark of a United Nations (UN) peacekeeping mission. They were originally sent to Somalia as part of the UN Operation in Somalia (UNOSOM, later UNOSOM I) authorized by Security Council Resolution 751, dated 24 April 1992. However, Resolution 794 of the UN Security Council (4 December 1992), acting under Chapter VII of the UN Charter, called for an international coalition led by the United States. Its mandate was to establish, if necessary through the use of military force, a secure environment for preventing obstruction to humanitarian relief operations in Somalia. This mandate was to be carried out by a multinational military force, of which Canada was a member, that operated under the name of the United Task Force (UNITAF).[1]

Although the UNITAF was only sponsored by the UN, and not an official peacekeeping operation such as UNOSOM I or II, it was carried under the mantle of the UN and was associated with UN peacekeeping operations by all interested parties. I shall treat it as if it were a UN peacekeeping operation and then discuss the ambivalence generated by the change of status from UNOSOM I to UNITAF (and then to UNOSOM II).

A discussion of the concept of peacekeeping is important for this report in two respects. First, it sheds some light on the problems that the Canadian troops deployed in Somalia were to face in the circumstances in which it was legitimate to use force. Second, no recommendation for future deployment can be made without confronting the numerous ambiguities that surround the theory and practice of international peacekeeping.

Peacekeeping in Historical Context:The New Schizophrenic World Order

Before addressing peacekeeping and its related concepts as such, it is worth noting that the idea of peacekeeping is, at least in part, the result of a historical process fraught with fundamental ambiguity. Following the influential work of German sociologist Norbert Elias (1982a, b and c; see also Elias, 1983 and 1994), the civilizing process was defined as an attempt to eliminate violence from daily life and to make humans behave more peacefully. According to Elias (1982: 259), this internal pacification of Western society began in the 11th century and reached its conclusion in the 17th and 18th centuries. It is generally assumed by authors writing on policing that this civilizing process has continued up until the present day, its main result being the state's monopoly on the use of legitimate force (see e.g., Bittner, 1990).

Elias's thesis has not escaped criticism. In a century like ours, which has witnessed two world wars and the genocide of the European Jews and Gypsies, assertions regarding a gradual progress toward eliminating violence cannot but appear paradoxical (Bauman, 1989: 107).[2] This paradox is as real as it is instructive.

It is quite clear that Western societies in their attitudes and words strongly disapprove of violence. However hypocritical this may be, its existence cannot be denied. Moreover, it has been exacerbated by the so-called "political correctness" that now prevails. At least in our pronouncements, the use of coercion, force and violence are social taboos. This is in the new ways in which the traditional agencies that hold a monopoly on the use of legitimate force — basically the police and the army — are claiming to view their action. Phrases such as "soft policing," "community policing," "problem-oriented policing" and "soldiering for peace," which might have been termed oxymorons one or two decades ago, are now part of the amended corporate image these organizations are trying to project.

In dire contrast, it is evident that no century has been more bloody than the 20th and what we thought were the unassailable lessons of World War II have been ignored by the unrepentant willingness of certain nations to carry out some forms of "ethnic cleansing" with total impunity. Hate groups that are claiming as their totem figures the Nazis or white supremacists, once thought to be the scourge of humanity, are apparently multiplying. The "new world order" that was supposed to spring from the end of the Cold War has given way to violent chaos where conflicts mushroom. Finally, the celebration of violence in sports, film, the media and even in our children's games is unparalleled in history. Hence, there is a wide gap between the values of peace and tolerance that we claim, on the one hand, to believe in and, on the other hand, what is happening historically. Not only is the widespread violence flaunting these values, but it is the very stuff of entertainment for our leisure time and is presented to our children as a pattern of behaviour.

It is unlikely that this divorce between our values, our current reality and our fantasies would fail to be reflected in the way we define and implement our projects, particularly when the definitions are programs for peace and the implementation provides the means for war. This divorce is at the heart of what always threatens to undermine peacekeeping operations, as I shall try to show. Unless the deep historical basis of this threat is recognized, there will be a tendency to underestimate the scope of the effort needed to answer it. More concretely, this split between program and application is a profound source of confusion and role ambiguity for those, such as soldiers and police officers, who have to apply the program in the field. Confusion and ambiguity are multiplied in strongly hierarchical organizations, such as the armed forces, where they affect the rank and file deployed in the field in very different ways from those at various levels of command. The higher the level, the more remote from the actual operations.

Peacekeeping in the Canadian Context: A Question of National Identity

In its report published in 1994, the Canada 21 Council[3] asserted that "Canada has much more at stake than most countries as the international community struggles to organize future peacekeeping operations" (p. 54). In line with this statement, the authors of the report propose that the Canadian Forces should be charged with two essential tasks, protecting territorial sovereignty and participating in common security missions, usually under the aegis of the United Nations (p. 63). This emphasis on common security echoes the position taken in a similar report, *Changing Our Ways: America and the New World*, published in 1993 in the United States by the Carnegie Endowment for International Peace, which stresses that "collective security has now come of age."[4]

Whether or not Canada has more at stake than most countries in the organization of future peacekeeping operations could be debated. What is clear is that no other country has been more closely associated with peacekeeping than Canada. Indeed, it is generally recognized that the concept of peacekeeping was originally developed in 1956 by then Secretary-General of the UN Dag Hammarskjold and Canadian diplomat Lester B. Pearson, who were the architects of United Nations Emergency Force (UNEF) 1, the first UN peacekeeping mission, as such missions came to be known (see e.g., Moskos, 1976: 26; Diehl, 1993: 31; and Fetherston, 1994: 13). Mr. Pearson was eventually awarded the Nobel prize for peace for his contribution.

It is difficult to arrive at the precise number of UN peacekeeping missions, as authors take different positions on whether the Korean War, operations that took place before 1956, and observation missions and similar activities should be included in the total. Furthermore, Canada has participated in non-UN peace missions such as the European Community Monitoring Mission in the former Yugoslavia (1991 to present). After consulting different sources (Gardam, 1992; Jockel, 1994, based on Morrison, forthcoming; and preliminary work by Allen G. Sens for this Commission), I estimate the total figure to be between 35 and 40 missions. In his foreword to Gardam (1992), General de Chastelain states that Canada took part in more than 30 peace missions, 23 of them having taken place under the UN flag. According to de Chastelain, Canada participated in all the UN peacekeeping missions from 1947 to 1992. Jockel (1994, based on Morrison) lists 29 UN peacekeeping missions and five non-UN missions from 1949 to 1994. What stands out is that Canada's commitment to peacekeeping, whether under the UN flag or otherwise, is unsurpassed by any other country. To sum up in a phrase occasionally heard at the UN, Canada comes close to being a model member state (Cox, 1993: 4; see also Cox, 1994).

David Cox, one of the leading Canadian scholars in the study of Canada's role in peacekeeping (Cox, 1968a, b and c, 1983), also asserts in his 1993 study for the Canada 21 Council that internationalism has become the central value of Canadian foreign policy.[5] He proceeds to identify five Canadian interests in participating in UN and other peacekeeping missions.

1. Because of increasing ethnic diversity in the origin of its population, conflicts in other countries are bound to impact directly on segments of public opinion in Canada, thus generating pressure on the Canadian government to help resolve these conflicts.
2. Even when a conflict does not resonate specifically on one component of the Canadian mosaic, the portrayal by the media of intense human suffering leads to public demands that the government take action.
3. Canadian participation in peacekeeping missions may further Canadian interests in fields other than foreign diplomacy, as when U.S. President Lyndon Johnson facilitated the negotiation of the automobile pact as a token of his appreciation of Canada's participation — which had originally been withheld — in the peacekeeping mission in Cyprus.[6]
4. In contrast with other middle powers, such as Australia, Canada defines its political and economic interests more globally than regionally and is a member of the G7, albeit the smallest. This global approach to foreign and trading policy is extensively reflected by the Canada 21 Council report, which is often cited by government officials in debates on Canadian defence and foreign policy.
5. Finally and most important, Canadian contributions to the peacekeeping record of the UN have become an element of our national identity.

I can conclude that peacekeeping is an important component of the way Canadians are trying to define themselves and their country (Band and Young, 1988). Thus, an erosion of Canada's credibility as a peacekeeping nation should not be seen as a marginal development with minor impact on the image we are trying to project.

Concepts of Peacekeeping

As I will show in detail in subsequent chapters, there was a great deal of confusion not only in the Battle Group but at different levels of the

Department of National Defence (DND) on the interpretation of the mandate of Operation Deliverance. What I aim to do in this section is to show, as briefly as possible, that this confusion stems from very deep-rooted ambiguities in the notion of peacekeeping itself and its related concepts. I shall review the UN Charter, the UN Charter doctrine and the literature on peacekeeping.

The UN Charter. Like all charter legislation, the UN Charter is deceptive in its apparent simplicity; its implication cannot be fully grasped without resorting to authoritative commentaries. I will use two such annotated editions of the Charter: the dated but still classic edition by Goodrich et al. (1969) and the more recent edition by Cot and Pellet (1991; see also Müller, 1992).

UN legal reality: With three notable exceptions, all UN peacekeeping missions have been authorized under Chapter VI, which is a corollary to Article 2 of the Charter. Both refer to the peaceful [Art. 2(3)] or pacific (Chapter VI) settlement of disputes. The actual Chapter VI articles invoked vary from one mission to the other and are not always specified by the Security Council, the UN body with authority for the preservation of world peace (Bailey, 1988: 3-6).

The three exceptions are the Korean and Gulf wars and the operations in Somalia (UNITAF and UNOSOM II), which came under Chapter VII of the Charter (Action with Respect to Threats to the Peace, Breaches of the Peace, and Acts of Aggression). It must be noted that none of these operations was authorized under Article 42 of Chapter VII, which is more permissive in the use of force. Article 42 has never yet been the blanket under which a UN operation was initiated (Goodrich et al., 1969: 314-315; Cot and Pellet, 1991: 705.). The wars in Korea and the Persian Gulf were conducted under Article 39 for the former and articles 39 and 40 for the latter; the operations in Somalia that followed UNOSOM I were conducted under Chapter VII of the UN Charter without specification of an article. That no UN-sponsored operation was conducted under Article 42 of the Chapter testifies to a contradiction. On the one hand, the UN has been unwilling to sanction a full-fledged war, while on the other hand such wars have been conducted in the field under the uncertain aegis of the UN. Canada 21 Council (1994: 78) reflects this contradiction when it states that "both the Korean and the Gulf wars can be considered Article 42 peace 'restoring' actions." Actually, none of these actions was performed under the umbrella of Article 42, although they certainly qualified as such actions.

UN legal ambiguities: These ambiguities are epitomized by the fact that the word "peacekeeping" does not appear anywhere in the UN Charter and that peacekeeping is often referred to as "Chapter 6 1/2" in literature on the UN. According to Goodrich et al. (1969: 256), Chapter VI is the most poorly drafted section of the UN Charter. This is clear from its differing interpretations. Goodrich et al. (1969: 263 and 277-278) assert that peacekeeping operations stem from the provisions of Chapter VI of the Charter, particularly articles 33, 36 and 37. In contradiction, Cot and Pellet (1991: 617) claim that the foundation for these operations lies in Article 25 of Chapter V. Furthermore, Goodrich et al. (1969: 278) state that "there have been many instances where members have disagreed as to whether the Council was acting, should be acting, or had acted under Chapter VI rather than Chapter VII."

These ambiguities were clearly demonstrated at the beginning of UN peacekeeping operations in 1956. It was then argued in the Security Council that there was no legal problem about applying the concept of "provisional measures" provided by Article 40 of Chapter VII to the matters referred to in Chapter VI (Goodrich et al., 1969: 278-279; and Cot and Pellet, 1991: 617-618). In the same way, it was argued within the Security Council that measures listed under Article 41 of Chapter VII could be called for under Chapter VI (Goodrich et al., 1969: 279).

UN Charter Doctrine. UN Charter doctrine refers to interpretations of its dispositions by legal scholars or to adjunctions to the Charter by the highest UN officials, such as its Secretary-General. Needless to say, these adjunctions are not new statutes but are more akin to what would be a common law tradition of the UN. Goodrich et al. (1969) and Cot and Pellet (1991) provide a good example of doctrinal interpretations of the Charter. I have seen that these interpretations are, to some extent, in conflict.

A pre-eminent illustration of a common law addition to the Charter is provided by a statement made by Secretary-General Boutros Boutros-Ghali. This statement, entitled *An Agenda for Peace* (Boutros-Ghali, 1992), is illustrative of the ambiguities of peacekeeping in two ways. First, it greatly adds to the terminology of peacekeeping by *officially* introducing notions that were unofficially used before in the legal and research literature (for their initial use, see Galtung, 1976, on peacekeeping, peacemaking and peacebuilding). Such terms are preventive diplomacy, peacemaking, peacekeeping, peacebuilding and actions to restore international peace and security.[7] In itself this proliferation of the terminology may generate confusion. Confusion is all the more likely when the meaning of these terms is, to a certain extent, left imprecise. *Peacemaking* is defined by Boutros-Ghali (1992: chapter IV) as an activity achieved through the peaceful means provided by Chapter VI of the UN. Yet in that same chapter, the Secretary-General mentions *peace enforcement units* that would be acting under Article 43 of Chapter VII of the UN Charter.

I can quote, as another example of conflicts of doctrine, the previously mentioned disagreement between Goodrich et al. (1969) and Cot and Pellet (1991) about which articles of the UN Charter actually provide authorization for peacekeeping operations (articles 33 to 37 as claimed by Goodrich et al. versus article 25 according to Cot and Pellet).

The Literature on Peacekeeping. Without being overwhelming, the literature on peacekeeping is multiplying. In his 1989 annotated bibliography, Peter Jones summarizes and comments on over 250 works produced from 1945 to the time of publication of his bibliography (most titles were written before 1986, at least with respect to the listed Canadian literature). There has been a surge in research literature on peacekeeping since 1990.

A review of the entire literature on peacekeeping far exceeds the limitations of this report. I shall briefly refer to it to the extent that it is pertinent for our purposes. In his analysis of the literature, Jones (1989: ix-xxxiii), divides its production into four periods (1945-1955, 1956-1966, 1967-1972 and 1973-1989). I find it more convenient to divide it into two periods, i.e., from 1956 to 1987 and from 1988 to the present. There was no peacekeeping literature in the strict sense of the word before 1956. I shall explain why I believe 1988 will serve as a new threshold.

1956-1987: The hallmark of this period, immediately following the successful UNEF 1 peacekeeping mission in Egypt (Suez), is that it begins in euphoria and ends with what is nearly a sense of failure. In 1987, for example, there was very little attention paid to peacekeeping in the Canadian Defence White Paper. Berdal (1993: 8) refers to a revival of peacekeeping since 1988.

The most general way to describe the literature on peacekeeping during this period is that it appears to comment and expand on the basic principles of peacekeeping as expressed by Hammarskjold and Pearson in a document outlining the features of UNEF 1.[8] These principles are the consent of the parties to a dispute to the establishment of a peacekeeping force to be sent into the theatre of conflict; the non-use of force except in self-defence; the establishment of a peacekeeping force of contingents contributed on a voluntary basis by countries neutral to the dispute; impartiality and non-intervention of the peacekeeping force as a participant in the conflict; and day-to-day monitoring of peacekeeping operations by the Secretary-General.

Of all these principles, the non-use of violence except in the most limited of circumstances has received the most attention, generally favourable. The most problematic is the principle of consent, although it was initially deemed the most basic. Landmark books and articles were Rikhye et al. (1974), Moskos (1975 and 1976), Galtung (1976) and Galtung and Hveem (1976), and the *Peacekeeper's Handbook* (International Peace Academy, 1984: 34), which claims that the enforcement model based on Chapter VII had been virtually abandoned.[9] In line with this emphasis on non-violence, Segal et al. (1984) concluded that paratroopers were not the right personnel for a peacekeeping operation.

There is one common feature underlying many of these writings. This is the reference to *policing*, which is clearly implied in the title of Rikhye et al. (1974) *The Thin Blue Line*. Moskos (1976) succeeded in granting this analogy a theoretical status. He (1975, and 1976: 93) described the emergence of a *constabulary ethic* characterized by behavioural adherence to the minimum-of-force concept. More significantly, he succeeded in constructing an empirical measure of the constabulary ethic from responses to two items in a questionnaire.

1. "Does an officer who is well trained in military skills and who has broad leadership in his home army still require additional skills for peacekeeping service?"
2. "Can a soldier be effective in the peacekeeping job if he/she cannot use force except in self-defence?"

This questionnaire was administered to peacekeepers in Cyprus (UNFICYP), and the military officers who gave affirmative answers to both questions were categorized as constabulary (Moskos, 1976: 105). Importantly, Moskos tells us that his sample broke evenly into officers adhering to the constabulary ethic (51 percent) and adhering to a non-constabulary ethic (49 percent).

There is a second point made in the first spate of literature that deserves to be mentioned, particularly in light of what has followed. Wiseman (1983) has coined the word "ad-hocracy" to refer to the kind of peacekeeping that took place in those days. This ad hoc character of peacekeeping operations is one of its standard features in the pre-1990 literature. However, Byers and Slack (1983) were early critics of the enthusiasm for peacekeeping, claiming that peacekeeping should be understood in its wider context, and that peacekeeping missions were conceivable outside the mantle of the UN.[10]

1988 to the present: Any categorizaton into periods is relatively arbitrary. However, beginning in 1988, peacekeeping enjoyed a revival. There are also several important works which are openly critical of the past (Haas, 1990; Urquhart, 1987; Urquhart and Childers, 1990; Diehl, 1993; Berdal, 1993; and Fetherston, 1994).

These works differ in one basic regard: they do not come to the same conclusions about what counts as the failure or success of a peacekeeping operation. Diehl (1993: 89) stresses external factors (opposition of third-party states and of subnational groups) that may thwart an operation, whereas Haas points out that, under certain conditions of agreement between the superpowers that sit permanently on the Security Council, the factors identified by Diehl may not be as decisive.

Whatever their differences, there is one point of agreement among the authors that have published since 1987. This is a common denunciation of ad-hocracy in peacekeeping and the realization of the need for an integrated theory of peacekeeping. Moving beyond ad-hocracy will require a major effort, because ad hoc pragmatism lies at the root of international peacekeeping as it was initially envisaged by Hammarskjold and Pearson. Urquhart (1987: 136) records Hammarskjold's reaction to the establishment of a standing UN peacekeeping force, as was advocated by John Foster Dulles and the U.S. Congress:

> Our approach to this problem is guided by the strictest respect for the rules of the Charter. It is entirely pragmatic in nature...It does not try to freeze a pattern of action, nor would it give rise to arrangements conducive to a premature or inappropriate use of similar means in the future. It does not presume to lay down legal rules binding in all circumstances. But it does, I hope, create a preparedness for such action as may later be found necessary, in so far as our previous experience of more general application can be utilized.

According to Haas (1990: 150-151), it is for want of an explicit and commonly shared theory that peacekeeping went awry in the 1980s. The need for a comprehensive and integrated theory of peacekeeping has been strongly advocated by Haas (1990), Diehl (1993), Jockel (1994) and Fetherston (1994), not to mention Boutros-Ghali (1992) in his *Agenda for Peace*. In addition to these calls for revision, strong voices are still heard in favour of the principle of the non-use of force (Liu, 1992; Morrison et al., 1993; MacKenzie, 1994) or are trying to find a middle ground between peacekeeping and peace enforcement (Goulding, 1993; and Cox, 1993 and 1994).

Problems in the Field. I believe I have shown that there are no clear guidelines with regard to the different versions of peacekeeping or the use of force during UN and UN-sponsored operations. These deficiencies ultimately reflect the ambiguities and omissions in the UN Charter itself and are, in turn, reflected in the UN Charter doctrine and in the literature on peacekeeping.

In the chapters of this report that will follow, I will document the problems generated in the field during the course of the operations in Somalia by the lack of clarity in the concept of peacekeeping. I will immediately provide a couple of general illustrations of these problems to indicate how real they were.

1. The UN commission of inquiry appointed by the Security Council tried to elucidate the differences between the mandate of the United Task Force — in which the Canadian Battle Group took part — and operation UNOSOM II which became a virtual war at the end of June 1993.[11] In its report (COI, REP), the UN commission of inquiry contrasts the UNITAF and UNOSOM II mandates:

   > Although operating under Chapter VII of the Charter of the United Nations, UNITAF did not interpret its mandate as requiring it to *enforce* disarmament of the Somali militias. (COI, REP: p. 13, par. 38, emphasis added)

   > In this context, the United Nations Security Council adopted its resolution 814 (1993) on 26 March 1993, expanding the size and mandate of UNOSOM to include not only the protection of humanitarian relief supplies and personnel but also to compel the Somali militias to disarm. *Thus unlike UNITAF, whose participation in the disarmament process was subsidiary and derived*

> *from the ceasefire and disarmament agreements of 8 and 15 January 1993, the New UNOSOM (UNOSOM II) was mandated to disarm Somali militias under Chapter VII of the United Nations Charter.* (COI, REP: p. 14, par. 44, emphasis added)

> Before UNOSOM II could accomplish these objectives, the events of 5 June occurred plunging it into *a peace enforcement role* which greatly compounded its problems. (COI, REP: p. 14, par. 48, emphasis added)

What comes out of these quotations from COI, REP is that although UNITAF was undertaken under Chapter VII of the UN Charter, it was *not viewed as a peace enforcement operation*, at least by the UN itself. It could be inferred from Berdal's (1993: 22) analysis that UNITAF was a paradigm case of an operation that fell in that no-rules-zone known as Chapter 6 1/2 by the commentators on the UN Charter.

> The US-led *Operation Restore Hope* appears to suggest that under certain circumstances member-states are prepared to accept that the delivery of humanitarian supplies to a civilian population can only be achieved through forceful action. This assumption was enshrined in the mandate for UNOSOM II, the largest operation in UN history, which in May 1992 assumed the task of providing security for humanitarian aid deliveries from the United Task Force. UNOSOM II is the first UN operation in which "peacekeepers" have been directed towards "coercive disarmament" and has highlighted the military and political risks alluded to above in a more extreme form.

According to Berdal's analysis, what "appeared to be suggested" in UNITAF's mandate was later incorporated in UNOSOM II's terms of reference. In other words, what was *assumed* implicitly by UNITAF was explicitly *asserted* by UNOSOM II.

However, it appears in retrospect that the Canadians interpreted UNITAF's ambiguous mandate as a straightforward and clear-cut operation. To all practical purposes, they viewed UNITAF as the blueprint that UNOSOM II was to reproduce. In its narrative, the de Faye Board of Inquiry (BOI) first describes the initial operation Cordon (UNOSOM I) that was to be a peacekeeping operation undertaken under Chapter VI of the UN Charter. Operation Cordon subsequently became Operation Deliverance (UNITAF). Here is how this change is described by the Canadian BOI:

> In response, the Security Council of the UN passed a further resolution on 4 December 1992, endorsing a US-led coalition whose mandate was to use all necessary means to establish a secure environment for humanitarian relief operations in Somalia. The UN-sponsored coalition, named the United Task Force (UNITAF), was authorized to operate under the provisions of UN Charter Chapter Seven, *making it a peace-enforcing mission.* The size and capabilities of the forces assigned to UNITAF were considerably greater than those for the UNOSOM operation.(BOI, Vol. XI: p. 1, par. 3 of the Narrative, emphasis added)

The UN COI Commissioners operated under the assumption that it is possible to conduct both a peacekeeping (UNITAF) and a peace-enforcing mission (UNOSOM II) under Chapter VII of the UN Charter, whereas the members of the Canadian BOI seem to assume that peacekeeping is conducted under Chapter VI of the Charter and that whatever is conducted under the provisions of Chapter VII automatically becomes peace enforcement.

2. The terminology used by the BOI fluctuates a great deal. In the important Annex G to its Narrative — bearing on the rules of engagement — the BOI translates the transition from Operation Cordon to Operation Deliverance in the following way:

> The process of promulgation and understanding of the ROE began during training prior to deployment. The Board noted that, due to the change of mission *from peacekeeping to peacemaking* (OPERATION CORDON to OPERATION DELIVERANCE), the issuance of the mission specific ROE occurred as forces were being deployed. (BOI, Vol. XI: Annex G, p. 1, par. 3, emphasis added)

> The Board noted that the Battle Group in Somalia experienced a profoundly different operational environment from that previously encountered by Canadians during other peacekeeping missions. *This was peacemaking, which might informally be described as falling somewhere between peacekeeping and low-intensity war.* (BOI, Vol. XI: Annex G, p. 2, par. 4)

This introduction of "peacemaking" in lieu of the former "peace enforcement" is troublesome. In his *An Agenda for Peace*, Boutros-Ghali (1992: Chapter II, Definitions, par. 20), defines peacemaking as an action to bring hostile parties to agreement essentially through the peaceful means foreseen in Chapter VI of the UN Charter. However, Boutros-Ghali (1992: Chapter IV, par. 44) devoted a section of his chapter on peacemaking to the creation of standby units for the *enforcement* of cease-fires. This wavering of the General-Secretary between a Chapter VI conception of peacemaking and a Chapter VII conception of peace enforcement is evidently reflected in the BOI report, which uncertainly switches between peaceful peacekeeping, confrontational peace enforcement and the grey area between peacekeeping and low-intensity war.

Perceptions in the Field. The last illustration that I will provide for now is taken from the courtmartial testimony of Captain Paul William Hope. Captain Hope was an intelligence officer with the CAR, and he testified as the first witness for the prosecution during the court-martial trial of Lieutenant-Colonel Mathieu. Asked about the work he performed in preparation of Operation Deliverance, Captain Hope made the following statements:

Q. So going to Operation Deliverance, did you, yourself, have a look at the Security Council's resolution authorizing Operation Deliverance? R. I didn't look at the actual U.N. Charters as they're written down, but I was very aware of the difference between the two chapters, and in my specific case, as an intelligence officer, there was a significant difference in my operations or my job between one and the other. So I was very aware of the difference between the two. (Mathieu, First Court Martial, Witness Hope, Folio Views, p. 1614)[12]

And what was, it might be asked, the difference between a Chapter VI and a Chapter VII operation, as it was *perceived* by this intelligence officer who never consulted the actual UN resolutions nor the UN Charter?

Q. And how was your job different for Operation Deliverance as compared to Operation Cordon? R. Under Operation Cordon, a straight United Nations operation they don't like even to talk about intelligence. They don't think that intelligence is the kind of activity that the U.N. should be involved in. This is the general attitude. Whereas a Chapter 7 operation, it's a straight military operation. I went as the intelligence officer. I had a full range of sources, et cetera, at my disposal, and I could operate and do my job, essentially as I was trained to do, *without restrictions.* (Mathieu, First Court Martial, Witness Hope, Folio Views, p. 1615, emphasis added)

Two things must be stressed about this testimony. First, the witness, an intelligence officer for the CAR, relied on his own perception of the UN Charter and resolution, without referring to any text. In his mind, a Chapter VI operation is a peacekeeping undertaking, whereas a Chapter VII operation is a "straight military operation." Chapter VII operations are associated with the Korean and Gulf wars; Chapter VI operations refer to regular peacekeeping operations. Second, this officer's misperception, however wrong it may appear to a UN legal expert, is not without foundation. Whether I refer to the UN Charter ("Chapter 6 1/2"), to the UN doctrine and the pronouncements by its Secretary-General, or to the contradictory literature on peacekeeping, I find no clear-cut dividing line between peacekeeping and peace enforcement. The line separating the two becomes completely blurred for operations conducted under the provisions of Chapter VII of the UN Charter.

Conclusions

I provided a brief analysis of one part of Captain Hope's testimony neither to show that he ought to have consulted the UN Charter and its commentary nor to argue that he was completely justified in seeing in Operation Deliverance a "straight military" operation because it was conducted under the provisions of Chapter VII of the UN Charter. I wanted to stress that the confusion that surrounded the interpretation of UNITAF's mandate cannot be reduced to the individual shortcomings of the Canadian Forces personnel who took part in this operation. This confusion, which will appear evident when I discuss the thorny issue of the ROE, stems from much deeper sources.

An examination of the UN Charter, its commentaries and the literature on peacekeeping yields the following conclusions. (1) Not only is there presently no agreement on the definition to be given of peacekeeping in its different versions but, more fundamentally, there is a growing suspicion that definitions are not what we need. For the practical purposes of this study, I shall use the provisory definitions of preventive diplomacy, peacemaking, peacekeeping, peacebuilding and actions to restore international peace and security as they are provided by the Canada 21 Council (1994: 77-78). They will be listed in Appendix 1 of my report. (2) What is needed is a theoretical framework that will permit us to go beyond adhocracy in thinking about peacekeeping and develop a comprehensive conceptual framework that allows us to fill the gaps in the seminal

Hammarsjkold/Pearson view, which are becoming increasingly obvious in the post-Cold War era. (3) Two threads running through the literature point to the two most pressing issues: the circumstances under which force can be legitimately used and the identification of the factors that impede or facilitate the success of a peacekeeping mission. The way in which these issues are approached presents striking analogies with developments in the field of domestic policing. (4) Finally, there is a consensus that these issues and the conceptual framework needed to resolve them are questions to be addressed collectively by teams of researchers working in a stimulating environment rather than by lone academics working inside libraries with few contacts with the outside world.

RACISM AND RACIAL-ETHNIC PREJUDICE

According to a media analysis performed for the CIDCFS, the issue of racism is one of the most salient in the press coverage of the events in Somalia.[13] More important, *a 1978 UN declaration against racism stated that it was of paramount importance that no force serving under a chapter of the UN Charter be guilty of racist acts* (United Nations, 1978). There are numerous problems with the concept of racism, not the least of which is that racism is exceedingly difficult to define. This difficulty is not only a serious hindrance to efforts to develop policies against racism, but it also serves as a screen to cover the fact that there is no political nor organizational will to go beyond the status quo.

The research literature on racism is very extensive. Reviewing it in its entirety would exceed the boundaries of any one report, if not a book. In line with the principle enunciated previously, I shall only address the questions raised in the literature that are of immediate pertinence for this report.

However, before addressing these questions, I shall briefly comment on some of the features of the literature on racism. I have already mentioned the first of these features: not only is this literature quite extensive, but it has proliferated significantly since the 1990s. However, despite its extensive nature — this is the point of these introductory remarks — this literature is of limited interest for the problems facing the CIDCFS.

First, racism is understood in this literature as a phenomenon occurring *within* a society. With the exception of historical works on colonialism (e.g., Delacampagne, 1983; and Todorov, 1984 and 1989), there are very few analyses dealing with racism as an international or cross-national phenomenon. Unfortunately, what is at issue with respect to the events in Somalia are incidents potentially involving racism in a cross-national context, specifically, Canadian peacekeepers acting in Somalia.

Second, the issue of racism within the military is generally not addressed.[14] This issue may be conceived in two ways. It may be raised with respect to racism *within* military organizations, in which case it mainly is an issue of racial integration and equal opportunity within the military. There are some studies of this issue, particularly with regard to the racial integration of the U.S. armed forces (e.g., Morris and MacGregor, 1981; and Myrdal, 1944). These studies are not numerous and most of them have little bearing on the terms of reference of the CIDCFS.

The issue of racism within the military may also be raised in regard to racial prejudice in the context of international conflict. Excepting the literature on war crimes, which I address in a separate report, there are very few studies on this subject. There are, of course, classical studies on well-known cases such as the Dreyfus Affair in France. But barring these studies, which have very little to do with the matter at hand, there is scant guidance in the literature. The paucity of the literature on racism and the military was one of the unexpected discoveries of my research.

Most of the literature on racism focusses on discrimination against two groups, American Blacks (Afro-Americans) and European Jews. In most books, discrimination against these two ethnic groups is treated as a single problem, the researcher attempting to find common features between racial prejudice against Afro-Americans and against Jews. Since these two phenomena are vastly different — discrimination against Afro-Americans is ultimately rooted in slavery whereas prejudice against Jews is much more ancient and originates at least, in part, in religion — attempts to fit them into a single mould are bound to result in discourse of a very high level of generality. Finally, the research on racism is intensely value-laden and mixes high-minded denunciation with factual analysis. Taguieff (1988 and 1992) has shown compellingly that this mix is generally not a happy one and that, perhaps not unexpectedly, the major weakness of this research was its failure to articulate a cogent and effective strategy to counter the spread of racial prejudice.

The questions that I will briefly address are the definition of racism and its different logics; the different kinds of racism; the various degrees of intensity of racism; its scope; the role of language in communicating racism; and racism and the military.

Issues of Definition

The issues that surround the definition of racism are very complex. The expressions of racism, such as anti-Semitism or racial discrimination against a particular ethnic group, are so diverse that it is doubtful that this diversity can be encapsulated in one neat and short definition. Even if it were, one would still be faced with the difficulty of applying this definition to particular cases. We have had numerous incidents of extreme police brutality where racism and racial prejudice against ethnic minorities seemed *prima facie* and in the eyes of public opinion to have been a factor in the behaviour of the police, if not its primary motivation.[15] It is a striking fact that the first statement made by persons possessing official authority (e.g., a chief of police, a mayor or a minister) after these incidents is usually to deny that racism played any part in it. Such statements are generally upheld in the proceedings that follow these incidents, such as a trial or an official inquiry. There are many commissions of inquiry who felt justified by their collection of evidence to assert that police abuse — or that any kind of official abuse — was in part or wholly rooted in racial prejudice. As I previously said, the problematic nature of the definitions of racism are often used as an alibi for the failure to develop significant policies to remedy racism.

The Words "Racism" and "Racist." Before saying anything about the fact and concept of racism, it is important to make a few remarks about the words "racism" and "racist," and how they are presently used. First, it is difficult to deny that the social trend often described as "political correctness" is having a real influence, not only in academic circles, but also on the media. Hence, in the present context, the accusation of "racism" is particularly serious; similarly, being labelled a "racist" is one of the most shameful things that can happen to an individual or an organization. In consequence, it should be expected that individuals and organizations will *deny,* with the greatest energy, being responsible for any act that would justify branding them as racist.

Second, ethnic tensions between the white majority in Western countries and the different racial-ethnic minorities which are also part of these countries have not really diminished since the great confrontation of the 1960s. Nor have they diminished between countries that belong to different ethnic communities. If anything, the level of verbal violence has actually increased. One of the consequences of this hardening of attitudes, is that the words "racism" and "racist" — and most particularly the epithet "racist" — are used in much the same way as disparaging racial epithets in the name-calling war. They are used as words of insult which are meant to hurt and cause anger, as racial slurs ordinarily do.

I do not mean to imply by my last remark that words such as "racism" and "racist" and their derivatives are always used as insults. What I do mean is that when these words are used as insults they effectively lose nearly all their descriptive content. Since, as I just noted, they are increasingly used as words of insult, this creates an additional difficulty in trying to describe their meaning and to use them rigorously in accordance with their meaning. Conceptual investigations into racism and its definition are rarely as dispassionate as inquiries into less controversial topics.

Racism and Xenophobia. Before dealing with racism itself, let us try to clear a preliminary hurdle. There may be a conceptual difference between xenophobia and racism. There certainly is one if we contrast the milder forms of xenophobia — for instance, the suspicion of strangers — and the most extreme form of racism — for instance, lynching and genocide. Whatever the need may be to distinguish between the two notions within the framework of a comprehensive theory of attitudes toward persons perceived as "others," I believe that it is neither necessary nor desirable to try to articulate a distinction between racism and xenophobia in the context of this report. Without prejudging the issue of the part played by racial prejudice in the extreme abuse against a Somali, for example, the beating death of Shidane Arone, I can exclude attributing such abuse to xenophobia. Extreme behaviour is usually motivated by extreme attitudes. Hence, if xenophobia is to play a role in the explanation of such excessive behaviour, it must be viewed as extreme. However, when xenophobia is conceived as extreme, it merges into ethnic prejudice and becomes indistinguishable from a form of racism. Thus, making a distinction between racism and xenophobia is not only unnecessary but useless in trying to explain extreme violence.

I believe that drawing such a distinction is undesirable for the analysis of less excessive forms of behaviour, abusive verbal behaviour or the use of racial slurs, for instance. Viewing these lesser forms as a token of xenophobia rather than as an expression of racial prejudice would smack of an attempt to belittle its seriousness, xenophobia not carrying the deplorable overtones associated with racism. Moreover, the Canadian public, the primary addressee of the report of a public inquiry, would possibly perceive the drawing out of a fine distinction between expressions of xenophobia and of racial prejudice as a prime example of the kind of semantic casuistry found in large bureaucracies. Such a distinction would, at best, cloud the issues and, at worse, bathe unacceptable conduct in the cleansing waters of euphemism.

Current Definitions of Racism. I shall now discuss current definitions of racism. The purpose is to suggest that it is better not to seek a concise definition of racism but to try to approach the meaning of this notion by making a series of points about it.

A current definition of racism is to be found in Armitrage and Kennedy (1989: 799). "The term racism as used in this paper, refers to actions, attitudes, or policies determined by beliefs about racial or ethnic characteristics."

This definition, as the authors make clear, is borrowed from *The Penguin Dictionary of Sociology* (Abercrombie, 1984: 173), hence, its wide currency. There is an obvious problem with this definition: it does not state that the beliefs referred to are negative and the actions, attitudes and policies based on these beliefs are harmful to the persons against whom they are directed. Taken at face value, this definition would imply that measures taken to remedy systemic discrimination (e.g., redefining the height needed to be recruited by a police force in order to give equality of access to Asians, who are generally not as tall as Caucasians) and even affirmative action programs could be termed "racist."[16] The character of this type of definition, of which many other examples can be found, is that it is basically *cognitive,* often resting on concepts such as (erroneous) *belief* or (unwarranted) *generalization.* One of the most concise definitions of racism is, in this regard, given by Cambridge and Feuchtwang (1990: ix): "In general we will argue for a conception of racism as a discursive fact, more precisely, we say that racism consists in categorisation of capacities for social life."

Another widely quoted definition would seem to remedy the previous difficulty. Wilbanks (1987: 27) first breaks down racism into the twin components of racial prejudice and racial discrimination. The key concept is racial prejudice, racial discrimination just being the acting out of the prejudice. Racial prejudice is then defined as the attribution of negative traits and motives to other ethnic or racial groups, viewed as "them" against the "us" of the racially prejudiced person or group. On first inspection, this proposal seems reasonable enough. Yet it allows a paradoxical inference, which is actually drawn by Wilbanks (1987: 27):

By this definition blacks are more racially prejudiced than whites. In recent years the negative attribution process (prejudice) has decreased among whites but increased among blacks. Finally, the tendency for blacks to assign negative traits and motives to whites (but not to blacks) is largely responsible for the black perception of criminal justice as racist.

Following this line of reasoning, both the Jews in Nazi-occupied Europe and the Blacks in South Africa (when it upheld apartheid) were or are as racially prejudiced as their oppressors. Wilbanks fails to notice that in racist contexts, there are aggressors and there are victims. It is ingenuous to blame victims for attributing negative traits and motivations to those who are harming them. This second kind of definition takes racism as a kind of attitude, an unfavourable attitude. In one of the great classics on racial prejudice, Allport (1979: 13)[17] tells us that any adequate definition of racial prejudice must contain two essential ingredients, namely an *attitude* of favour or disfavour and a relation to an overgeneralized *belief.* The previous definitions are incomplete in that they contain only one of these two ingredients. Allport (1979: 9) offers his own definition of ethnic prejudice, which remains in my opinion the most adequate, and which I shall use in the context of this report. "Ethnic prejudice is an antipathy based upon a faulty and inflexible generalization. It may be felt or expressed. It may be directed toward a group as a whole, or toward an individual because he is a member of that group." It is important to emphasize that the object of this definition is *not* racism per se but rather ethnic prejudice.

The last example comes from a UN declaration that resulted from a world conference against racism held in Geneva (14-25 August 1978). Racism was defined as a doctrine of racial superiority. It cannot be doubted that such a doctrine was once the core of racism and may still be for many racists. However, there is another form of racism, which asserts that the inferiority of certain ethnic groups is not rooted in biology but in culture, the cultural and social institutions of these groups (or the lack of such institutions) moulding their members into inferior beings. This kind of racism is no less obnoxious than its biological counterpart, but it is more pervasive and more in need of condemnation. Actually, there have been through history many conceptions of race, each giving birth to a particular form of racism. Banton (1987) gives an illuminating discussion of how race was successively conceived first as lineage, biological type, genetic subspecies and, afterwards, was linked with social status and economic class.

The point of my remarks is not to say that the attempts to define racism I have just reviewed are worthless — they all express some aspect of the truth, most particularly Allport's — but to stress that the complexity of racism does not permit a concise form of a definition.

Logics of Racism. Because of the strong backlash against immigrant workers in several European countries, the political far right has risen again and is enjoying unprecedented popularity among voters. This re-birth of the far right with its racist platform has generated new interest in the study of racism. Some of the most stimulating work on racism was conducted in France (see e.g., Delacampagne, 1983; Sibony, 1988; Taguieff, 1988; and Wieviorka, 1991, 1992, 1993, 1994). A main conclusion these researchers reached is that racism is governed by two quite distinct kinds of logic. I shall refer to the first as *the logic of exclusion* and to the second as *the logic of domination*. Racism is misguidedly represented as intolerance toward differences. This characterization misses the point that such intolerance can take two very contrasting forms. People different from us can be rejected, when they are seen as a threat to our own identity and even to our own being, or they can be forcibly cloned, as it were, when we feel confident enough to assimilate them. The first strategy is undergirt by a logic of exclusion, while the second rests on a logic of domination.

The difference between these two kinds of logic can be illustrated with reference to one of the very few books written on racism and the military. Its title is *War Without Mercy* (Dower, 1986), and it documents the war against the Japanese in the Pacific during World War II. Dower quotes some of the first articles written by Ernie Pyle, one of the most influential American war correspondents:

> What Pyle told this impressive audience, right away, was that the enemy in Asia was different. "*In Europe we felt that our enemies, horrible and deadly as they were, were still people,*" he explained in one of his first reports from the Pacific. "*But out here I soon gathered that the Japanese were looked upon as something subhuman and repulsive; the way some people feel about cockroaches and mice.*" (Dower, 1986: 78)

This perception of the Japanese as animals was widely shared in the Pacific theatre, where Admiral Halsey called the Japanese "bestial apes" (Fussell, 1989: 116). Fussell also quotes war correspondents to the same effect as Dower:

A marine on Guadalcanal could perceive the Japanese only as beasts of various species. He told John Hersey, "*They hide up in the trees like wildcats. Sometimes when they attack, they scream like a bunch of terrified cattle in a slaughter house.*" Another said, "*I wish we were fighting against Germans. They are human beings, like us...But the Japs are like animals...They take to the jungle as if they had been bred there, and like some beasts you never see them until they are dead.* (Fussell, 1989: 116-117)

This kind of racism rests on the perception of the other as an alien. Thus, ethnic groups are not merely seen as different or inferior but as being radically alien. This otherness is such that the racial group is excluded from the human race. This logic may eventually lead to a program to exterminate the non-human group, as it very nearly did when the Americans dropped nuclear bombs on two Japanese cities and as it actually did in the case of the European Jews. Dower (1986: 37, 54) makes clear that there was a strong link between the exclusion of the Japanese from the human race, "exterminationist rhetoric," and the dropping of nuclear bombs.[18]

The situation changed abruptly after the surrender of Japan and its occupation by the United States. The Western powers' racist treatment of the Japanese did not disappear overnight but was governed by a logic of *domination*, once the Japanese had surrendered. Both Dower (1986: 294-317) and Fussell (1989: 120) make it clear that this kind of racial prejudice does not lead to the same extremity as the former one, although it is no less repugnant. The ultimate result of exclusion is to exclude from life, whereas domination preserves the existence of its object in order to perpetuate itself.[19]

The racial logic of exclusion is grounded on a pathological fixation on one's own superior identity and the consequent obsession to preserve it from any external contamination. It unfolds into panic in the face of difference and its attendant behaviour, namely segregation, ethnic purification and, ultimately, extermination. The racial logic of domination stems from a belief in the inferiority of others and generates or reinforces inequality, discrimination, exploitation and oppression (Taguieff, 1988: 163; and Wieviorka, 1991: 141). Paradoxically, the incipient expressions of exclusionary racism appear to be less harmful than their counterparts in a racism of domination. For instance, segregation is often perceived to be less aggressive than outright discrimination, although the ultimate limit of the logic of exclusion — extermination — is overwhelmingly the bleakest. Needless to say, these two logics often intersect: South African apartheid was based on both exclusion and domination, although the main undercurrent was domination since the need for Black labour precluded an overt policy of extermination.

Realization of the differences between these two brands of racism has important practical consequences. Despite dramatic exceptions, the general purpose of war is conquest and domination, with the attendant reaction of self-defence. Consequently, the kind of racism that may be displayed by warriors would be governed by the logic of domination. Since soldiers on a peacekeeping mission are not supposed to be acting as warriors (and as a rule do not), it would follow that they are less at risk of falling into the same kind of racism that threatens warring soldiers. What threatens them more is the exclusionary kind of racism, of which the incipient expressions are the more insidious for not seeming the more overtly aggressive. Hence, a training or sensitization program for peacekeepers should be geared toward preventing the kind of racial prejudice that is based on a logic of exclusion.

In fact, most programs of cross-cultural sensitization fall dramatically short of this goal. They stress the differences between ethnic groups, while falling short of providing a meaningful reason for accepting these differences. In so doing, these programs actually run the risk of reinforcing exclusionary racism rather than curbing it. This kind of racism does not require more attention to the existence of ethnic differences. This is actually its starting point and its fuel. Once noted, these differences may be inflated to the point that no common ground will remain between the persons affected by these differences and those attributing the differences to them. What needs to be relentlessly emphasized in a cross-cultural sensitization training program is that, despite (and because of) these differences, the persons and groups involved are not aliens; they have much more *in common* with those viewing them as other than they have attributes that separate them.

Diversity should not be "tolerated" because it is part of accepting the "white man's burden," but because it is the substance of humanity and life. Only manufactured objects are, on their surface, alike; that they are not completely similar is owing to the fact that they are produced by human beings. Indeed, we should talk about cultural differences, but sensitization programs should present these differences as varying responses to recurring problems confronted by human beings who share common attributes. Although a balance between emphasis on what is different and what is common is difficult to preserve, it must be the goal of any training or sensitization program that aims to be fruitful.

I quickly make three further points with respect to the two logics of racism and racial prejudice. First, the logic of domination came, as far as I know, before the logic of exclusion. The racist logic of domination can be traced back to *The Politics* of Aristotle, where it is asserted that some persons are meant *by nature* to rule and others to serve (Aristotle, 1988, Book I, sect. 5: 6). However, this logic of domination really began to flourish during the period of the conquest of the Americas by the European powers. The basis for this dominating racism was the perception that the "Indians" lacked the kind of religious and state institutions that were being consolidated in Europe (Delacampagne, 1983: 22; Todorov, 1984: 35 and ff.; see also Todorov, 1989).

> The drama of the Indians, which puts them in a situation congenitally inferior to that of the Europeans, is that they have no culture, nor State, nor customs, nor laws, nor society, nor religion: (Christopher Columbus wrote) "...I believe that they (the 'Indians') can be easily christianized, since they do not belong to any sect..." (Delacampagne, 1983: 22, my translation)

When Columbus found out that the "Indians" could neither be culturally assimilated nor Christianized he wrote to a fellow Spaniard (Antonio de Torres) that "torn away from their inhumanity, they would make the best possible slaves." (quoted by Delacampagne, 1983: 55). The important point of these remarks is to stress how easy it is to fall into racism because of a false perception that an ethnic community has no institutions at all, when its institutions are simply different from our own. This process is a real threat to peacekeepers who operate in so-called "third world" countries, such as Somalia.

Second, these two kinds of racist logic are not upheld by the same people. Classic studies of the lynching of Blacks (Raper, 1933: 1; Cantrill, 1941) stress the difference between a Bourbon (lower middle class) and a proletariat (poor white) lynching (Cantrill, 1941: 94). This distinction was taken from Raper (1933) who wrote the report of the Southern Commission on the Study of Lynching created by the Commission on Interracial Cooperation. In his book, Raper (1933: 1) writes that:

> ...though there were a few notable exceptions, most of the lynchers, chiefly young men between their late teens and twenty-five, were from that unattached group

of people which exercised least public responsibility and was farthest removed from the institutions and agencies determining accepted standards of conduct.

Before the abolition of slavery, the logic of racist domination was primarily represented in the United States by the plantation owners, who were aspiring patricians using slave labour and protecting it as an essential commodity. After the American Civil War, the lynching mobs were, by contrast, generally poor Whites of precarious status and identity who were competing with Black labour. Their main concern was not to dominate Blacks, for which they no longer had institutional support, but to exclude them from breaking into their own sphere. This connection between exclusionary racism and "unattached groups of people exercising least public responsibility" and, to speak more plainly, the potential White underclass, is explicitly drawn by Wieviorka (1991: 59 and 180) and others.

I previously said that it was exclusionary racism more than domination that posed a threat for peacekeepers. It remains to be seen whether there is a connection between the alleged racism within the ranks of Canadian peacekeepers in Somalia and what would be the rank and file equivalent of poor Whites in the Canadian Forces.

Finally, if I did raise the question of the logic that governed xenophobia, I would find that it is essentially a logic of exclusion. It is this type of logic which begets the worst excesses of racism. Hence, it is questionable whether substituting xenophobia for racism to explain the events in Somalia would put the behaviour of some members of the Canadian Airborne Regiment (CAR) in a better light.

Symbolic Racism. There is one definitional issue that I will discuss, because it could find a potential application in the context of Operation Deliverance. A number of U.S. researchers have developed the concept of symbolic racism (see e.g., Sears and McConahay, 1973; McConahay and Hough, 1976; Weigel and Howes, 1985; Gabrenya, 1988; and St. Pierre, 1991). I will quote McConahay and Hough's (1976: 38) definition of symbolic racism and then describe the most salient features of this concept, which is now used in the sociology of the armed forces. "Our most concise definition of symbolic racism, then, is that it is the expression in terms of abstract ideological symbols and symbolic behaviors of the feeling that blacks are violating cherished values and making illegitimate demands for changes in the racial status quo."

This concept was coined to describe the growing resistance in White America to the civil liberties movement that aimed to improve the condition of life for Blacks. It may have a wider application. Here are the essential features of the concept.

1. Symbolic racism is expressed through abstract ideological symbols and through symbolic behaviours. With respect to attitudes, it is a set of moral assertions about Blacks' behaviour and situation (e.g., "Negroes are getting too demanding in their push for equal rights"; "Negroes have it better than they ever had it before"). In terms of behaviour, it consists in symbolic gestures such as voting against a Black candidate or opposing desegregation in housing and education. Such behaviour may be rationalized on a non-racial basis, but its effect is to maintain the racial status quo. The character of these symbolic expressions of racism is precisely that they are non-violent, as opposed to the more traditional redneck kind of racism.
2. Symbolic racism is also directed, as I just suggested, against behaviour, which has taken a symbolic dimension: Black persons running for election and soliciting Black support, desegregation, welfare, affirmative action programs and, generally speaking, action taken to ameliorate the social and economic conditions of minorities. The important feature of this symbolic behaviour is that it is *group action or collective behaviour*, as compared to the target of traditional racism which dissolved the individual into stereotypes but was nonetheless directed against individuals (e.g., in lynching, to use an extreme example). The focus of symbolic racism is the collective effort of minorities and of those who support them to better their situation, the most common condemnation being that they are too demanding. Indeed, one of the most frequent moral assertions made by persons inclined to symbolic racism is precisely that the civil liberties movement has given priority to collective rights over individual rights, the precedence of individual rights being what made the United States and similar Western countries "great" (Weigel and Howes, 1985: 131).
3. Symbolic racism is the expression of a conflict between values and perceptions of values. The personal welfare of the mostly affluent people who display this kind of racial prejudice is not threatened by the improvement of the living conditions of the minorities. What is threatened is their values, their way of life and their sense of propriety. Consequently, symbolic racism "tends to find its most vociferous expression in the political arena where such issues as unfair government assistance to Blacks are involved" (Kinder and Sears,

1981: 416). This is in contrast to the racism of poor Whites found in parts of a country where there is a high level of unemployment and where the personal welfare of individuals that belong to the majority is threatened by minorities and immigrant labour.

4. Finally, empirical studies conducted on symbolic racism tend to confirm that racial prejudice is but one symptom of a general tendency to disparage outgroups (Weigel and Howes, 1985: 131).

The concept of symbolic racism is not without its difficulties. It may, for instance, be debated whether being against affirmative action programs is an expression of racism. The answer depends on how one shows opposition to affirmative action programs. To say that such programs are pampering lazy people who are already overbenefiting from government assistance may be a manifestation of symbolic racism.

One interesting feature of this concept for our purposes is that this kind of racism finds its most acute expression in clamorous denunciation of government programs aimed at helping underprivileged minorities. It may follow that international humanitarian aid — and peacekeeping missions designed to support such humanitarian aid — will become the object of a condemnation so fierce that it becomes an expression of symbolic racism.

The Kinds of Racism

The literature (e.g., Friedman, 1975, who was widely followed) generally distinguishes between two very different kinds of racism — person racism and system racism (systemic discrimination). Person racism is the result of racial prejudice held by individuals and groups that can be identified with some degree of precision; system or systemic discrimination is the relatively anonymous product of institutional norms (e.g., the requirement that all applicants for a job be under a certain age, when first generation immigrants are usually over that limit by the time they receive Canadian citizenship).

This is the standard distinction. It can be made more precise by pairing it with the distinction between what is intentional or intended and what is not. Cross-tabulating these two pairs of distinctions, we get the following outcomes.

Intentional person racism: This category refers to well-known acts of racist behaviour such as the refusal to provide services to members of a particular ethnic community or the desecration of cemeteries.

Unintentional person racism: This is better explained through an example. It may be possible that during the first part of the 20th century many White communities in the U.S. South used the infamous word "nigger" to refer to Blacks without even suspecting that there were alternatives. This vocabulary was simply transmitted from generation to generation without being questioned and was used routinely rather than intentionally. Not infrequently, racial prejudices may be handed down from generation to generation as part of a racist heritage that is transmitted and received without premeditation. Needless to say, the poison of racial prejudice is frequently spread quite intentionally.

Unintentional system racism: The words "system racism" or "systemic racism" are now used instead of the term "institutional racism" originally coined by Robert Friedman and others (Friedman, 1975: 384-424). Following Friedman (1975: 386), systemic racism may be defined as any action, policy, ideology or structure of an organizational system or of several interdependent systems that work to the disadvantage of one ethnic group compared to another. According to Friedman, any feature of a system that works to the relative *advantage* of one ethnic community compared to others is also systemic racism. The difficulty with this adjunct to the definition is that it defines affirmative action programs as racist. The causal factors that influence a system are numerous. This leads Friedman (1975: 386) to write that "where causation is so diverse, it becomes futile to speak in terms of intent," intent being seen as a specific discriminatory decision. I agree with Friedman that in most cases not only is intent difficult to specify and even more so to attribute, but that a discriminatory intent was probably not even originally present when a system was implemented (e.g., before a particular wave of immigrants).

Intentional system racism: Nevertheless, I believe that there are historical examples of the implementation of a system of discrimination and even of extermination that were fully intentional and where the mark of the action of one (or several) will(s) is quite evident. The Nazi example comes to mind. So does the example of South African apartheid (but without an extermination component). Furthermore, what may have been at some point an unintentional system of racism may become fully intentional when it is recognized that a system is racially discriminatory but nothing is done to remedy the situation.

The Scope of Racism

A third sort of question with regard to racism concerns its degree of pervasiveness in a society, a community or a large organization. For the sake of brevity, I will limit the discussion to the scope of racism within a society considered as a whole, although my conclusions can be adapted and applied to other contexts.

Following Wieviorka (1991: 83 ff.) and to a lesser extent Allport (1979), I propose to distinguish between four levels in the social scope of racism. It is important to realize that these distinctions refer to the breadth and not to the intensity of racism, which can range from verbal behaviour to physical violence. In all the following categories, physical violence may occur.

Infra-social racism: Allport (1979: 27) believes that within all of us, there is a propensity to prejudice, in that we have a natural tendency to form generalizations which represent an oversimplification of our world of experience. This is particularly true of our relationships with others, where emotions play a prominent part. Infra-social racism represents this stage of racism, where it is socially more virtual than real. At the level of beliefs, such racism is not yet cogently articulated to form an ideology or a world view through which all our experience is filtered. This is why Wieviorka (1991: 83) refers to it rather as infra-racism than as infra-social racism. I prefer the latter designation, because this incipient stage allows for the expression of racial prejudice and also for the occurrence of overt acts of racial violence. Yet, these expressions of racism remain limited to the actions of marginal individuals or small gangs living on the fringes of society. The equivalent of such a stage of racism in a large organization would naturally be infra-organizational.

Fractional racism: At this level, racism is still very fragmented and has not contaminated the whole of society. However, it is in the process of coalescing into a recognizable ideology, which is afforded a certain amount of visibility by the media, and its spread can be measured, for instance, in public opinion polls. It is no longer held by marginal individuals and deviant gangs but by identifiable groups which are loosely organized, undisciplined and, consequently, lacking true stability, keep splintering. The defining feature of these groups is that they do not yet have mass appeal and are still actively recruiting among the fringe elements of society.

Political racism: This phase is crucial and can easily be described as the stage when racism is more or less overtly the platform of a stable political party with mass appeal. The activities of such a party (or parties) are regularly covered by the media, and it competes openly in the legal forms of political life such as public elections. The hallmarks of such racism are its gain of political legitimacy and its attendant public acceptance, and the fact that it is promoted by an efficient leadership. It becomes progressively embedded in the structure of social life and racist violence acts as a widespread tool of political intimidation.

Total racism: This level is reached when a racist political party captures the power of the state, either through legal or non-legal means (e.g., a coup d'état). It is then in a position of developing and enforcing systematic policies of racial discrimination and segregation, up to the point of perpetrating genocide against an ethnic community. Although total racism is systematic, there is an obvious difference between total and system racism, as I described above. As a deliberate process, total racism is always intentional, whereas system racism often is not.

As I just suggested, the crucial threshold in this enlargement of the scope of racism is when it is given overt, if not official, recognition by leaders who are part of the legal political structure of a society. In the context of an organization, this threshold is crossed when official authorities within its structure openly support racist positions (e.g., when the officers in a military organization openly condone racist behaviour and engage in it themselves).

The Intensity of Racism

As much as racism may vary in its breadth, it may also vary in its depth or in its intensity. Allport (1979: 14) has provided us with the most influential classifications. He distinguishes between antilocution, avoidance, discrimination, physical attack and extermination. Allport (1979: 49) further condensed this classification into verbal rejection, discrimination (including segregation)[20] and *physical attack* (of all degrees of intensity and breadth).

As most of these terms are transparent, I shall not define them all individually. Antilocution refers to negative verbal behaviour. Avoidance refers to behaviour resorted to in order not to be in the physical presence of members of the disliked group. Physical attack and extermination differ both in intensity and scope. They differ in intensity, in the sense that physical attack may not always result in death (e.g., physical ejection) and it may be directed against property and symbols (e.g., the desecration of cemeteries) as much as against persons. In contrast, extermination is always directed against persons and results in the death of various numbers of persons (lynching, pogroms, massacres and genocide). There is one definition given by Allport (1979: 51) that deserves to be quoted. It is his definition of discrimination: "Discrimination comes about only when we deny to individuals or groups of people equality of treatment which they may wish."

In his categorization of the different degrees of intensity that racism may take, Allport lists verbal discrimination or antilocution as the least serious form. However, in several parts of his book, he stresses the essential roles of rumour, verbal realism and linguistic tags (Allport, 1979: 63, 186, 304). In contrast to Wieviorka (1991: 89), Allport (1979: 57-58) develops a theory of the essential continuity between the different forms of racist behaviour, each expression of racial prejudice serving as a condition for the occurrence of a more intense one. This makes it seem as if there was some inevitable progression from verbal discrimination to the more extreme episodes of physical attack. I shall return to this issue, which deserves separate discussion, in the next section of this chapter.

The Role of Language in Communicating Racism

The question of the role of language in generating, communicating and perpetuating racism is both central and controversial. One high ranking officer of the Canadian Forces was quoted in the press to the effect that using words such as "niggers" and other racial slurs to refer to members of ethnic communities was benign and had no effect on dehumanizing them. Several witnesses testified before the BOI that they used the word "nigger" in a "descriptive" sense. Assuming that such statements are made in good faith, we may wonder whether it is not worse to believe that racial slurs can be used descriptively than to use them knowingly to stigmatize and hurt.

On the other hand, as I have already mentioned, Allport (1979) believes that language plays a major part in creating and maintaining racial prejudice. Actually, there is a consensus among authors who have researched racism on the role of rumour in generating the most extreme forms of physical attack (such as lynching; see Raper, 1933: 20; and Cantrill, 1941: Chapter 4); on the fact that "words are weapons" and that "naming power" is at the heart of racial prejudice (Hughes and Hughes, 1952:

139-140; Sibony, 1988: 14-15); on the fact that "racism was a discursive fact" at least in part (Cambridge and Feuchtwang, 1990: 2223). Such quotations could be endlessly multiplied.

There is one researcher who has developed these insights into a comprehensive theory that demonstrates that prejudice is socially reproduced through discourse (Van Dijk, 1987: 30; see also Van Dijk, 1991). I will summarize the gist of his argument, for its conclusion is highly relevant to this report. First, in a way that is analogous to Allport (1979), Van Dijk (1987: 27) distinguishes between *opinion* and *attitude*. Opinion is defined as single evaluative beliefs held by an individual person. In contrast, attitudes are forms of *social cognition* which "denote a complex schematic structure of general opinions stored in long-term memory" (Van Dijk, 1987: 27, and 391-392). In other words, whereas beliefs are simple and unstructured, attitudes are complex and structured; and whereas opinions are individual, attitudes are social.

Second, social attitudes take the form of a series of topics or tropes that differ marginally from one Western society to another one. These attitudes are strikingly summarized by Van Dijk (1987: 385-386) at the end of his book:

> What people tell us about the "foreigners" in the Netherlands or the "illegal aliens" or "immigrants" in California is essentially the same, with variations depending on context: Too many of them are (coming) here, immigration should become stricter, they make us feel unsafe on the streets, the neighbourhood is being run down by them, they are aggressive and involved in crime, some of them work hard but many of them are lazy and on welfare (for which we pay taxes), they take our houses and jobs and are unfairly favoured by the government, they do not adapt to our ways, do not speak our language or do have strange religion and other customs, they do not value education as we do, have too many children, do not respect their women, live in dirty places, and in general are different and have a different mentality...

On reading the characterization of these attitudes by Van Dijk, it is obvious that they are widely shared by the Canadian public (in varying degrees of intensity); it becomes equally obvious that they can be carried by Canadians travelling to foreign countries merely by substituting "natives" for "aliens."

Attitudes, by nature social, solidify into a consensus through which racism is reproduced, confirmed and accepted in society (Van Dijk, 1987: 336):

If the consensus is predominantly prejudiced, then the communications are, and if the communications are, the consensus is permanently confirmed. Few people reject negative communications, and still fewer do so explicitly and publicly. So, there is no communicative countermovement, no counterexamples, no counterarguments, and so no antiracist models and schemata.

The upshot of this argument is the assertion that *one of the most notable properties of a racist society is that it is not antiracist* (Van Dijk, 1987: 394). That is, a racist society is *passively* acquiescent to a form of "racial common sense" that is never questioned, precisely because it is so deeply embedded in its ways of talking about foreigners that it remains *unnoticed* by all who share in it. People are, in general, extremely reluctant to accept any responsibility for what can be described as action through omission, and any allegation of racial prejudice is met with angry denial. What I called the racial common sense of a community is, in consequence, constantly reinforced by this stern denial.

This analysis can admittedly be contradicted on several points. Yet it can also be confirmed in numerous ways. Raper (1933: 8) characterized a lyncher or pro-lyncher under the spell of the "mob mind" as a person who tended "to accept unqualifiedly any and all kinds of reports which fitted into certain preconceived notions about Negroes and the kinds of crimes which warrant lynching." This psycho-linguistic make-up is precisely what Van Dijk (1987) tried to describe as a complex social schemata used for putting a racial interpretation on reality. These social schemata could also be defined as organizational patterns of interpretation that filter reality through racial prejudice in certain circumstances.

Having stressed the deleterious influence of verbal aggression, there is one final question that we must address. As we previously saw, Allport (1979: 57-58) developed a theory of what might be referred to as the racist continuum. According to this theory, the outbreak of racist violence is preceded by a number of stages, each acting as a condition for a transition to the next level until the level of physical attacks of various degrees of intensity is reached. Allport identifies nine consecutive steps, the process beginning with the building of racial stereotypes and ending with physical violence, triggered by some precipitating incident. Allport's theory, although useful, is marred by ambiguity. Allport (1979: 57) writes that "in cases where violence breaks out we can be fairly certain that the following steps have prepared the way." Such an assertion can be made in *retrospect*, when we begin with the outbreak of violence and look backward to its previous conditions. The problem is to know whether we can look forward, undertake *prospective* research and, starting with the milder occurrence of racism (antilocution), whether we can predict the onset of physical violence. Although he is careful to avoid rigid determinism — "although most barking (antilocution) does not lead to biting" — Allport provides support for a deterministic perspective by insisting that "there is *never* a bite without previous barking" (Allport, 1979: 57, emphasis added; the two quotes are part of the same sentence). By stressing that violence is *always* an outgrowth of milder states of mind, Allport (1979: 57) imperceptibly switches from the retrospective to the prospective. When reflecting on Nazi anti-Semitism, for instance, he declares: " Here we see the not infrequent progression: antilocution > discrimination > physical violence." I am strongly inclined to side with Wieviorka (1991: 89 and 127-128) in this debate and assert that although the linguistic expression of racial prejudice and verbal aggression greatly facilitate and even prepare the way for the outbreak of violence, they cannot be seen as links in a causal chain that inevitably leads to violence. Furthermore, it must be stressed that, in contrast to the "not infrequent progression" from antilocution to violence, the occurrence of physical attacks also frequently implies a radical break with past events, even when these events are the expression of racial prejudice. In these cases, one must seek an alternative explanation for the outbreak of violence rather than accept the seemingly normal progression from words to deeds.

Racism and the Military

If I were to identify one study of American racism against Blacks as the classic study, I would probably cite Gunnar Myrdal's *An American Dilemma: The Negro Problem and Modern Democracy.*[21] It was published in 1944, at the end of World War II. One of the book's main theses is that American wars always have been crucial to the liberation of the Blacks. Hence, America was once more at the crossroads at the end of World War II. Whether or not it was perceived as controversial, Myrdal's thesis should have generated a spate of research on the military and racial minorities.

It did produce some research on the racial integration of the U.S. Armed Forces (e.g., Foner, 1974; Moskos et al., 1980; MacGregor, 1981; Binkin and Eitelberg, 1982; Young, 1982). A small proportion of this research addressed the problem of American Indians in the U.S. Armed Forces (Holm, 1986). Research was also conducted on the armed forces of other countries with a mixed population (Grundy, 1976; Guyot, 1976; Wheeler, 1976; Dominguez, 1976). Similar studies were undertaken in Canada

(Dreisziger, 1990). In this last book, the chapter by Armstrong (1990: 178-189) on the plight of a Black unit in the Canadian expeditionary force in Europe which was denied the right to serve as a combat unit and was used to harvest timber in Scotland, is of particular interest. Armstrong (1990: 179) quotes the explanation of the officer commanding the 104th Overseas Battalion for turning away a group of 14 Black men: "I have been fortunate to have secured a very fine class of recruits and I did not think it was fair to these men that they should have to mingle with (N)egroes (sic). I might state further that some of these (N)egroes arrived here very much the worse of liquor."

Despite these exceptions, few studies have investigated racial prejudice as such within national armed forces (see Lane, 1981 and Ball, 1994 for the Soviet Armed Forces and more particularly Doubnov, 1988, re-edition, for the life story of a Jewish soldier in the Russian Tsarist Army). Although we hear much about genocide and ethnic cleansing nowadays, still fewer studies are devoted to racism within the context of conflict between nations and, as the case may be, between ethnic communities (see Tinker, 1977 for a very general introduction). One of the reasons for this neglect of racism in inter-ethnic conflicts is that some of the more racist troops (the Russian Cossacks, the British Imperial Forces in South Africa, the French Foreign Legion, the German SS and *Waffen* SS, the French *Parachutistes* in Algeria and the U.S. Green Berets in Vietnam) have been shrouded in the so-called mystique of the "elite" corps, which fostered hagiographic writing rather than dispassionate study.[22]

Racism Against the Japanese in World War II. As I have previously mentioned, the work of Dower (1986) and, to a lesser extent, of Fussell (1989) are among the few systematic inquiries into racism in the context of a conflict between nations.[23] There are two important conclusions that stem from these books, both of which have already been discussed or alluded to.

Name calling and racism: The first conclusion bears on what Allport (1979) calls "antilocution" (verbal aggression). It is frequently asserted that the use of pejorative and demeaning language to refer to the enemy in the course of conflict is a common way to depersonalize him, thus making the killing job easier to perform. As anyone familiar with military literature can testify, this is largely true, the use of such words as heinies, huns, jerrys, krauts, fritz, fridolins to refer to the Germans during the two world conflicts being duly acknowledged. However, as Dower writes (1986: chapters 4 and 5) there is a vast difference between the names that were used to refer to the German or the Italians during World War II and those used to strip the Japanese of their humanity.[24] As we saw with the illustration of the different logics of racism, the words applied to the Japanese were quite often animal words; in a similar way they were depicted as bats, apes, monkeys and insects in war cartoons. This Anglo-American practice with respect to the Japanese was not followed when it came to applying names to their enemies in Europe; they were still seen as human beings.

It might actually be said that the Japanese were vilified because of their treachery and the humiliation they had inflicted on the Americans at Pearl Harbour. As Dower (1986: 151-152) makes clear, this explanation is not sufficient. In their conquest of the Philippines (1899-1901), the Americans were already using words such as "niggers," "gugus or googoos" (which became "gooks" during World War II and was widely popularized during the conflict in Vietnam) to refer to Filipinos. Dower reports that U.S. soldiers in the Philippines told journalists that "the country won't be pacified until the niggers are killed off like the Indians" and that "the only good Filipino is a dead one. Take no prisoners; lead is cheaper than rice." (Dower, 1986: 152). Not only is this strong language — genocidal, actually — but its recurrent feature is that it is only applied to non-Whites (American Aboriginals, Filipinos, Afro-Americans, Japanese and Vietnamese). That the use of such vocabulary was at the time indicative of racism seems to us undeniable. That it still would be at present also seems to us equally undeniable.

War trophies and extermination: The war in the Pacific was waged on a level of ferocity unparalleled on the Western front in Europe. Both Dower (1986) and Fussell (1989) refer to body mutilation as a way of taking war trophies (slicing the ears off an enemy corpse did not begin in Vietnam). I previously described (see endnote 18) the extent to which the annihilation of the Japanese people through the use of nuclear devices was approved by a significant proportion of the American public. In this case as in the previous one, the attitude and the behaviour toward non-Whites was markedly different from behaviour toward the foe in Europe, where it did not reach this level of savagery and did not imply an open willingness to exterminate whole populations. Genocide was however contemplated not only in the case of the Japanese but also for native Americans and Asians such as Filipinos, Communist Chinese during the war in Korea and North Vietnamese during the Vietnam War. The purpose of these remarks is not to condemn any particular Western nation but to stress that racial differences drive opponents to extremities not envisaged in conflicts where the belligerents perceive themselves as sharing the same racial background.[25]

The Gurkha Syndrome. The most immediate relationship between the military and ethnicity lies in what Enloe (1980) described as the "Gurkha syndrome." This theory immediately caught on among researchers and is now widely influential. I shall quote one of its concise formulations and provide comments. This formulation is provided by Enloe (1980: 26):

> Political institutions not only reflect ethnic conditions, but shape them. Military planners, far from bemoaning ethnic pluralism, may deliberately *foster new ethnic identities for the sake of achieving military goals*. State fostering of ethnicity is the first aspect of what is referred to here as "the Gurkha syndrome". The second aspect involves the inter-ethnic patterns derived from such military practices — *patterns that create a relationship of dependency between the peripheral group and the state military*. (emphasis added)

According to this formulation, the Gurkha syndrome has two main ingredients. Gurkha is actually the name of the ruling dynasty in Nepal, a small kingdom that lies to the north of India; it also refers to Nepalese soldiers recruited to serve in the British Indian army to fight the colonial war against the Indians on behalf of the Empire. In agreeing to side with the British in a colonial war against their Indian neighbours, the Gurkhas assumed a new ethnic identity as storm troopers for a foreign power both in the eyes of the British who used them and the Indians who feared and despised them. The Gurkhas had a reputation for their martial qualities, although the British, who called them "teddy bears," had no intention of allowing them to rise in their ranks by becoming officers. By accepting a new identity as storm troopers for a foreign power, the Gurkhas also became dependent on this British power for their livelihood and subsistence as a nation: they were rejected by the other nations that they were helping the British to conquer. For all practical purposes, a Gurkha was an institutional mercenary: by being institutionalized by a foreign power, he acquired a *new ethnic identity* and by being recruited as a mercenary he became a *dependant* of whomever he served.

The Gurkhas were not the only nation to be used in such a way. According to Kirk-Greene (1980), the British had ranked the different peoples of their Empire according to their hereditary martial qualities. Nor were the British the only colonial power to use the equivalent of Gurkhas. The French created the Black African *Tirailleurs* and used them in numerous conflicts from 1857 to 1958 (see Echenberg, 1990). Closer to the present time, Frisch (1993: 51) has shown that the Israeli have militarized the Druze minority and used it to police the Palestinians. Because of their growing dependency on the foreign power that employs them as storm troopers or riot police, the troops thus used have to outperform eventual competitors in order to keep their job. Frisch (1993: 60) has calculated that the percentage of Druze killed in the line of duty since 1948 is significantly above that of the Jewish population of Israel. In general, the peoples who have been used in the same capacity as the Gurkhas — Berbers, Bedouins, Cossacks, Kurds, Mongols, Montenegrins and Sikhs — all share a reputation for military valour and bellicosity.

The Gurkha syndrome is mentioned in the research literature with respect to North American Aboriginal members. For instance, in his introduction to *Ethnic Armies*, Dreisziger (1990: 11) writes that:

> The legacy of Indian warfare, and an exaggerated emphasis on the Indians' so-called "special martial qualities", have influenced the American Army's use of Native soldiers and their units. To some extent, Indians were victims of the Gurkha syndrome, much as were the Croatian-Slovenian *Grenzers* in the Habsburg Army, or the Bambara in the ATA [African Tirailleurs]. The results were predictable. In many battles, particularly in the two world wars, Indians were sent on lifethreatening assignments because of their real or supposed "special abilities," often with fatal consequences. In reality, the Native Indian had few special skills, though in many campaigns, good use was made of those he had.

Actually, Aboriginal members were used to make radio transmissions for there was little probability that the enemy would understand the language they were speaking if the communications were intercepted. But White (1990: 79) also believes that the Aboriginal soldiers who served in the U.S. Army during World War II were to some extent victims of the Gurkha syndrome.[26] However, he is quick to point out that the 25,000 Aboriginal men and 350 Aboriginal women who served during World War II were never segregated, in contrast with Black units. In fact, Aboriginal members found it desirable to disassociate themselves from Blacks and Mexican

Americans. On the east coast, mixed-blood Aboriginal people were numerous and they rebelled at being classified as Blacks by local boards, despite the official military policy that all recruits other than Blacks should be classified as white (White, 1990: 83).

Gurkhas and their equivalent generally serve in segregated units. The only true instance where Aboriginal members were used as the equivalent of Gurkha units in North America occurred at the beginning of the colonization, when Aboriginal tribes were pitted against each other and against colonial rivals.[27] The Iroquois were used by the British and the Hurons by the French at the beginning of the colonization of Canada. Whether Canadian Aboriginal peoples — or any other Canadian ethnic community — were, in later Canadian history, victims of the Gurkha syndrome is difficult to determine in the absence of any research on this topic in Canada.

Whether a concept such as the Gurkha syndrome can be used to explain the behaviour of Aboriginal soldiers serving in Somalia will be discussed in a later part of this report. Since the two members of the CAR most directly implicated in the beating death of Shidane Arone were Aboriginal or part Aboriginal — Corporal Matchee and Private Kyle Brown — there is a question as to whether or not they were under pressure to show how tough they were in order to win the acceptance of their peers. Enloe's theory about the Gurkha syndrome could be explored to see whether it sheds some light on their behaviour. Let us immediately formulate one reservation about the use of Enloe's theory. This theory bears on the behaviour of whole organizations and communities rather than on the behaviour of individuals. Naturally, individuals are members of communities and organizations, and to this extent we can draw an inference from group theory to individual behaviour. However, such an inference remains highly speculative, particularly when the group in question numbered less than 10 persons distributed in a whole regiment and when there is no indication that they actually behaved as a group.[28]

There is, however, another way of trying to answer this question, which takes us away from social science research on the military. We might explore research on another profession legally empowered to use lethal force. This profession is, of course, the police. There is considerable research on the pressures applied to members of ethno-racial minorities when they are recruited by a police force. Furthermore, this research often rests on interviews conducted with individuals and may be more germane to our problem. I postpone the review of this research to the chapter of this report that will deal with the behaviour of the CAR in-theatre. In addition to a review of the literature, I have myself conducted interviews with police recruited from ethno-racial minorities and from the White majority.

The Penetration of the Extreme Right in the Canadian Forces. Of all the issues related to racism and racial prejudice, this was one that attracted a very large share of media attention. Since it is more of a factual issue — what was the extent of this penetration — than a conceptual one, I shall address it in the later parts of this report.

ACCOUNTABILITY

In my discussion of accountability, I shall follow the same strategy as in my analysis of racism. I begin by raising the issue of the definition of accountability and make some theoretical points about this concept. The problem with the concept of accountability is not so much its complexity, as it was with racism, but the fact that it has an interface with the concepts of responsibility and control with which it is easily confused. The points that I will try to make in relation to accountability can be summarized with respect to the threefold question of *who* accounts *for what* and *to whom*. As I shall try to show, there is more than one answer. More important, its three elements are so intrinsically related that the kind of answer that is given to one part of the question determines the answers to the other parts, e.g., who should do the accounting may vary substantially depending on what is being accounted for and to whom the account is presented. Finally, I will try to see which government agency or agencies might be used to provide elements as a model for a structure that would enhance accountability within the Canadian Forces.

Issues of Definition

In her introduction to the bibliography that she compiled on *Accountability in the Administration of Criminal Justice*, Matthews (1993: vii) defines accountability as "the process by which government and its agencies are held accountable or responsible for their action." There are similar definitions in the literature, for instance, Wagner (1989: 8) who defines accountability as "being obligated or subject to giving account" or Larson (1978: xvii) who states that accountability is "the quality or state of being accountable, liable or responsible." What is unsatisfactory with these definitions is that they define accountability by a word that shares the same root (accountable), which only produces a tautology, or by closely related concepts such as responsibility and liability, while failing to spell out the difference between accountability and the concepts with which it shares an interface. (If accountability is just another synonym of responsibility, why has so much energy been devoted to exploring its meaning and practical implications?)

Etymology. It is, in some cases, rewarding to review the etymology of a word. "Accountability" and its French equivalent *reddition de comptes* both come from the Latin word *computare*, which not only means to count or to calculate but also to reckon with (*com*-putare) the value of a person or of something. The meaning of the verb "to account" actually evolved from the purely descriptive (to give a detailed account of a transaction) to the more normative (to produce a balance sheet matching gains against expenditures) and then to the explanatory or the justificatory.[29] In his analysis of accountability, Uglow (1988: 144) testifies to this evolution. There is yet a further implication of the normative component of the meaning of the verb "to account," which is perceptible in the "reckoning with" part of the meaning of computare: the person providing the account is under some form of obligation to produce it and its addressee has a right to request it.

This brief etymological analysis of the verbal root "to account," with its derivative, reveals that there are at least four semantic components to its meaning: providing information, meeting certain standards of information in providing it (e.g., a balance sheet), explaining and ultimately justifying the extent to which these standards have been met, and in doing so, satisfying a legitimate demand for accountability.

Accountability as a Cultural Trend. The fact that the search for more accountability has been a growing trend since the 1960s does not provide us per se with a definition of accountability. However, reviewing this trend may caution us against the mistake of confusing accountability with much older concepts. If there is any meaning to this rather recent trend, it cannot be merely reduced to an old demand under a new name.

There are many signs that the search for greater accountability is more than a passing fad. In 1993, the Privy Council of Canada (PCO) reissued a 1977 study on constitutional responsibility; its last chapter formulates "The Principles of Accountability." There is actually a large body of literature that bears on government accountability (for an insightful review of this literature, see Light, 1993 and 1995; and Doern, 1993).

Not only is there research on government accountability taken in general, but there are few fields of government that have escaped requests for more accountability. For example, there are numerous works on accountability in health care (Palmer, 1977; Malek, 1977; Greenspan, 1980; and Carbine, 1990). Others examine accountability in the context of services such as welfare and social work programs (Rehr, 1979; Butcher, 1995; and Martin, 1996), and education appears to have been the original field in which accountability was studied (Borich, 1974; Cronbach, 1982; Sallis, 1988; and Frymier, 1996). One of the earliest works on accountability as such looked at private foundations and charitable trusts (Taylor, 1953). Now the private sector, e.g., banking and the communications industry, are also scrutinized in relation to accountability (Roulier, 1995; Buttney, 1993; and Horton, 1992). The emphasis on accountability also seems to be an international one (Abdallah, 1984).

However, for reasons which are easily understandable, no government agencies have been submitted to more pressure to increase their accountability than those that operate in the field of criminal justice (Stenning, 1995a). As these agencies wield the most intrusive and coercive power, it is to be expected that attempts to limit unwarranted discretion in the exercise of these powers have been the most sustained. In their review of the research literature, Matthews (1993) and her assistants examined no less than 2,800 titles from 1965 to 1991. Of these, they retained 250, which they annotated. It is the police (for instance, Solomon (1985) reviews the large number of proposals made by the Law Reform Commission of Canada for the reform of police powers) that have been the object of the majority of inquiries with respect to accountability. Yet, the judiciary (Friedland, 1995; and Marshall, 1995) and corrections (Maguire et al., 1985; Rhine, 1993; and Vagg, 1994) have also received much attention.

The armed forces are the other government agency that can wield lethal power. However, since this power is rarely exercised by a democracy against its own citizens, there have been relatively few studies of military accountability outside the field of war crimes and the international laws and regulations on armed conflict (Kalshoven, 1973, 1989a, 1989b and 1991; Wakin, 1986; see also the references in Leman-Langlois and Brodeur, 1995, to the crimes perpetrated at My Lai by elements of the U.S. Army during the Vietnam war). There was at least one study on crime within the armed forces (duly entitled *Khaki-Collar Crime*, Bryant, 1979). Finally, there is an increasing amount of research conducted into the role of Parliament in defence policy making within a democratic framework (Laurance, 1980; Martin, 1994; Segal, 1994; and Stanley, 1994). Generally speaking, social science research on the military is limited to a few specialized journals such as *Armed Forces and Society,* and military accountability is still largely *terra incognita.*

There are two points that follow from this very brief résumé of the research trends on the concept of accountability. First, inquiries into the concept of accountability have now developed into an independent field of research, with its own problems and methods. Since this field now stands by itself, it should not be made to disappear under the guise of more traditional fields of inquiry such as legal studies of the concepts of criminal responsibility or civil liability. Second, although the inquiries into military accountability are still in an early stage, there are concepts that can be borrowed from the research on government and organizational accountability and applied with relevance to military organizations. One such concept is the thickening of government, i.e., the uncontrolled proliferation of intermediate levels at the top of an organizational hierarchy, with its attendant diffusion of accountability (Light, 1995). Since military organizations tend to be "top heavy," such a concept might be put to fruitful use in developing a better framework for accountability.

I shall now raise the specific issues that I referred to previously.

Accountability: Who Should Be Accountable?

The question of who should be accountable, which appears straightforward enough, has several formulations. It can be asked first in its nominal sense: who is the particular individual that should be held accountable under a given set of circumstances? This question can be answered specifically just by giving a name. This kind of answer can be given only if the particular circumstances under which it is asked are known. Since I am only concerned with a *general* theory of accountability in this part of my report, I will not address this question. In relation to the circumstances that defined the situation in Somalia, the full answer can only be given by the Commission after the completion of its inquiry.

The question of who should be held accountable in government or in one of its agencies can also be asked with respect to whole categories of individuals. For instance, is the need for accountability greater in the armed forces at the level of the rank and file, non-commissioned officers (NCOs), junior or senior officers? I shall postpone answering this question, because its resolution depends, in great part, on the answer given to a much more fundamental question: is it individual persons, groups or whole institutions that should primarily bear the burden of accountability?

There is, in Canada, a traditional answer to this question. It was officially formulated in a study published in 1977 by the PCO and reissued in 1993. This doctrine was also developed by Canadian researchers such as Kroeker (1978 and 1981).

Keeping in mind that our system of government derives from a time when the Crown provided the government, the PCO study (1993: 7) declares that:

> Personal accountability provides the foundation of our system of parliamentary and cabinet government. It derives from the individual responsibility of Ministers, which is in essence personal as opposed to institutional. It is shared by no one. It is the Minister, not the office, who is responsible, and it is this that vests in each minister the unique *constitutional responsibility* for the use of power. (emphasis in text)

The document (PCO, 1993: 4) had previously stated that: "*Collective ministerial responsibility,* a complex arrangement involving the personal responsibility of each minister and of ministers as a group, is of recent vintage in our constitution, dating back not much more than 100 years" (emphasis in text).[30]

What is remarkable about these texts is that they conceive of accountability wholly in terms of responsibility. If in addition to the constitutional history of Canada, which is fairly short, we also refer to the history of ideas, it becomes obvious that responsibility is as a rule predicated of individuals rather than of groups and that the meaning of institutional responsibility is not yet clear in present laws. The real question, however, is whether accountability is best thought of in terms of responsibility, although both notions are indisputably close. This is a crucial question for government agencies which resist being called to account for the way they exercise their considerable power (police forces, security services, armed forces). When such organizations, lacking in a tradition of accountability, come under pressure to be accountable, accountability tends at first to be confused with criminal responsibility or civil liability and with penal sanctions and disciplinary measures.[31]

The complete answer to the question of accountability cannot be given by reflecting only on who should be held accountable (responsible) in government and government agencies, but implies that we also inquire into what government authorities should account for and to whom. Nevertheless, I shall immediately proceed with my argument.

First, let us remember that according to my previous analysis, accountability is perceived to be a notion different from responsibility. Responsibility is one of the oldest ethical and legal concepts, whereas the field proper of accountability has only been recently developed.

Second, the analysis which is produced in PCO (1993) addresses the question of government accountability within the framework of a parliamentary democracy and concludes that the minister is personally responsible before Parliament. Individual responsibility ensures that for every act or neglect of his or her department, the minister must personally answer for what an unnamed officer does, this officer being protected from attack by the rule of anonymity (Kroeker, 1978: 12). In other words, the minister cannot hide behind the officers of his or her ministry. This analysis is correct in theory, but does not apply in many specific contexts. The events in Somalia illustrate this point. It is true that the Minister of Defence had to answer personally before the Canadian House of Commons for the behaviour of the CARBG in Somalia. Nevertheless, the persona of the Minister that was accountable to Parliament was *his ministerial and not his private self.* In certain contexts, the difference between the two can be easily drawn. For instance, when the parents of Shidane Arone received a financial compensation for the death of their son, this money did not come out of the pocket of the Minister but out of the budget of the ministry. In other contexts, it is much more difficult to distinguish which of the different personae of an elected official — private self, ministerial self, member-of-the-Cabinet self, this latter self being bound by rules of Cabinet solidarity on matters with which it may disagree — may be held accountable for a particular course of action.[32] This undermines the clarity of the conception of accountability as the equivalent of personal responsibility. What I am now saying is that this conception which was developed in relation to political accountability also applies when it is used in other domains of accountability (e.g., administrative or military).

There is yet a third reason to resist the identification of accountability with personal responsibility. In criminal matters, the intent to perpetrate an offence is deemed a necessary condition for conviction. The notion of intent is perhaps too narrow to be seen as a precondition of political responsibility. Indeed, government policies are known to produce unintended effects for which it would be difficult to make an official responsible, unless he or she had been repeatedly forewarned of that probability. In this latter case, it would be more difficult for the official to have any responsibility. Hence, even if intent does not act in all contexts as a precondition for the assertion of responsibility, it would at least appear that knowledge of the consequences of one's action is a very strong presumption in favour of the attribution of responsibility (or guilt). Conversely, *the absence of knowledge makes it very difficult to attribute responsibility*. However, in the context of large organizations, it is nearly always possible for someone in a position of authority to claim ignorance and to avoid responsibility. Not only does the size of the organization make it often truly impossible for its head to be informed of the behaviour of his or her subordinates, but there is, in certain organizations, especially those involved in the protection of national security, an implicit understanding that information should be kept from the higher authorities, and particularly from the ministers, in order to preserve the possibility of having recourse to "plausible deniability" when things go wrong.[33] Taken together these two facts — that responsibility can only be attributed where previous knowledge existed and that there is very nearly always a possibility for someone in authority to claim ignorance in large organizations — may in certain contexts operate to eliminate accountability, when it is identified with the personal responsibility of those who are in power, for what goes wrong in their organization. It is in this sense that Vandervort (1979: 129) spoke of "the myth of ministerial accountability" in her study on the political control of independent administrative agencies undertaken for the former Law Reform Commission of Canada.

> Widespread assertions that the minister is accountable for his actions and those of his subordinates do not become true simply through repetition. To the contrary, reliance on the principle to show the existence of accountability to Parliament for the performance of a myriad of government functions only underscores its mythical qualities by overburdening the principle beyond all credibility. (Vandervort, 1979: 129-130)

It must be emphasized that Vandervort's initial reservations about ministerial accountability have received confirmation in later research undertaken in the United Kingdom, which set the original constitutional pattern for our type of parliamentary democracy. Diana Woodhouse (1994: 296), for instance, concludes her recent book on ministers and Parliament thus:

> The starting-point for reform in Britain must be the recognition that, while the convention of ministerial responsibility cannot ensure accountability for all the

powers exercised by modern government, it should provide effective accountability for policy choice and overall administrative management. Its failure to do so does not necessarily mean that it is an unsuitable mechanism for accountability, but it does indicate that adjustments need to be made if the constitutional requirement of accountable government is to be a reality .

The proving ground for the theory of ministerial accountability is the clear determination of when a minister should resign for administrative mismanagement. This determination is particularly problematic when the minister is held accountable not only for his or her own actions but also for the actions of officials that fall under his or her ministerial responsibility. As Geoffrey Marshall (1989: 7; see also, Giddings, 1995: 223 and ff.) stressed in his introduction to a collection of papers on ministerial responsibility, this is "the major element of uncertainty in the doctrine of individual ministerial responsibility."

What Vandervort and later researchers say about ministerial accountability generally applies, I believe, to attempts to vest, in individual authorities, the personal responsibility of accounting for complex organizations or parts thereof.

At the end of my discussion of the etymology of the verb "to account," I came to the conclusion that the provision of information was actually the first component of the meaning of that verb. Notwithstanding the practical circumstances in which revealing what one knew may be self-incriminating and may thus clash with one's personal interest, it can still be asked whether an individual in a position of authority can be held personally responsible for the provision of information by a whole organization. There is no doubt that such a person can exercise a decisive influence, at least negatively. Indeed, should this person decide that the organization will provide as little information as possible, it is then highly probable that the organization will not officially provide information to the public (although there may be a fair amount of leakage to select individuals or organizations such as the media). There is again no doubt that this person could exert a positive influence in deciding that the organization will provide as much information as can be disseminated to the public. There are, however, powerful obstacles to the efficiency of personal, as opposed to collective action. First, as was stressed by Weber (1946), who developed the concept of bureaucracy, all bureaucracies are by nature secretive: it is precisely by keeping crucial information to themselves that bureaucrats rule. This is, of course, truer of organizations that collect and process information that is classified and legitimately kept confidential because its release may endanger national security. It is these organizations that have developed a culture of plausible deniability, which is extremely difficult to modify. It would seem unfair to blame a person holding authority for the lack of transparency of his or her subordinates because the chances are that, not being transparent to others, they will not be transparent to their superior. The latter may not suspect the extent to which information is being withheld from those who have a right to it, including himself or herself. Transparency is perhaps more a way of being than a way of doing. If persons can be compelled by regulation to act in a certain way, it is doubtful that any one person can pressure them to change the way they are. Because of the very nature of transparency and of the kind of professional culture found in large government bureaucracies, I would tend to believe that the quest for transparency in administration, with the attendant provision of information to all those who are entitled to it, is much more a matter of collective and institutional responsibility as opposed to personal responsibility. This argument reinforces the point made previously that accountability should not be confused with personal responsibility.

What then is the difference between responsibility and accountability? Fundamentally, when you are said to be responsible, you must answer for your personal (individual) behaviour, although you may *also* be made to answer for the behaviour of others and even of a whole organization. As the concept is presently used in the literature on government accountability, when you are said to be accountable, you primarily answer for an agency or an organization in which you hold a position of authority and you *also* answer by implication for your own actions. What I am in fact saying is that persons are properly said to be responsible but it is institutions which are, as such, accountable.

There is a second difference that points in the same direction as the previous one. As the words are now presently used, there is no contradiction in saying that although someone was fully responsible for something, that person was not made to account for it. Initially, the concept of responsibility signified both agency and answerability. Agency tends to be associated with responsibility, which is considered a matter of fact (you are either responsible or not for a particular result). It is accountability which has taken on the meaning of answerability, answerability being a matter of convention and legislation. This split between agency and answerability is less obvious with regard to personal behaviour than with regard to institutions. When we say that someone is responsible for his or her behaviour, we still mean that this person must answer for what he or she did. However, both authoritarian thinkers and politicians have held that the state is responsible for everything but accountable for nothing, particularly in relation to the use of force. With respect to state institutions, answerability can be completely divorced from responsibility, this divorce being proportional to the coercive nature of the institutions. Thus, it is not merely by reasserting the responsibility of persons that accountability will be promoted within organizations such as the police or the armed forces. It is the whole institution that must be made accountable and not just its figureheads.

Accountability: What Should Be Accounted For?

Traditionally, organizations have been made accountable for two things. The first is the way in which they spend and balance their budget. Financial accountability has been the cornerstone of accountability and the subject of the most systematic research (Kroeker, 1978). Since it lies outside the terms of reference of the Commission, I shall not address this issue.

The other usual object of accountability is the individual or collective behaviour of the members of a government agency. This is the part of government accountability that has received the most intensive coverage by the media, because it focusses not so much on performance as on misconduct. Not only have most of the public inquiries been triggered by allegations of misconduct, but one of the main questions they have had to investigate was whether the alleged misconduct was limited to a few deviant individuals or whether it was rooted in, or facilitated by, the structure of the organization of which they were a part. Historically, public inquiries tended at first to focus on individual deviance; the tendency is now to look at the broader organizational picture. In sum, it is the organizational behaviour of an agency which has now become the target of accountability rather than the misconduct of its members taken as individuals.

These preliminary remarks on the kinds of things for which government agencies have been held accountable can be developed in a more systematic form in relation to the research undertaken on accountability. One of the most interesting proposed frameworks was developed by Light (1993 and 1995). Regarding government accountability, Light (1993: 11-22) makes a distinction between three types of accountability.

1. *Rule-based compliance* consists in monitoring the compliance of individuals within an organization and the accounts given by the organization of its general performance with its stated rules and policies.

2. *Performance incentives* encompass a wide spectrum of measures that range from merit pay and promotions to occasional investments in civil service reform. In contrast to the previous kind of accountability, this kind of accountability stresses the positive rather than censuring the violation of rules.
3. *Capacity building,* defined as an attempt to improve basic government or agency capacity, builds on previous work by Fesler and Kettl (1991: 317) who make a distinction between accountability and ethical behaviour. Accountability is defined as faithful obedience to the law, to the directions of higher officials and to standards of efficiency and economy. This definition basically corresponds to Light's rule-based compliance. Ethical behaviour is defined as adherence to moral standards and avoidance even of the appearance of unethical behaviour. These authors reject the conventional compliance path to accountability and believe that ultimately what is important is to retain individuals that are "dedicated to public service, respectful of its call for bureaucratic accountability and ethical behaviour and both knowledgeable about and committed to the constitutional democratic system" (Fesler and Kettl, 1991: 335). This coupling of rule-based compliance with ethical behaviour is given by Light (1993: 13) as an example of a capacity-based view of accountability.

Light's table (see Table 1.1) summarizes the differences between these three perspectives on accountability. Needless to say, they are not mutually exclusive and Light refers to them both as definitions and as aspects of accountability (Light, 1993: 14).

This table is self-explanatory, except perhaps for the primary mechanism of capacity building, which is said to be technology.[34] Technology is broadly defined by Light (1993: 15) to include "people and the tools of management," the latter comprising state-of-the-art technology as it is currently understood (software and hardware). It is crucial to notice that Light makes a welcome distinction between management and oversight, thus stressing that the internal management of an organization has an important part to play in the achievement of accountability. In most of the literature, accountability is associated with oversight rather than with management.

The model of accountability that has by far been the prevailing one is the compliance model. It is also characterized by Light (1993: 12) as the "command and control" model. Accountability is then viewed as a systematic attempt to limit discretion within an organization, these limits

Table 1.1
Aspects of Accountability

Comparison of the methods:
Compliance, performance, capacity building
Aspects of accountability
Definition of accountability

	Compliance	*Performance*	*Capacity Building*
Characteristic			
Point of intervention	Post-activity	Mixed pre- and post-activity	Pre-activity
Primary target	Individuals and accounts	Individuals and programs	Agencies and government
Primary mechanism	Rules	Incentive	Technologies
Role of sanctions	Negative	Positive	Positive
Role of management	Supervision and discipline	Goal setting and reinforcement	Advocacy and stewardship
Role of oversight	Detection and enforcement	Evaluation and bench marking	Analysis and design
Complexity and strategy	Simple	More complex	Most complex
Durability of effects	Short term	Intermediate	Long term

Source: Reproduced from Light (1993), p. 14.

flowing from clear rules, which act as commands, and formal procedures of monitoring and enforcement that control compliance. The problem with this command and control model is that it creates an artificial trade-off between accountability and values such as creativity and innovation (Rourke, 1979: 103; Mosher, 1982: 72; and Light, 1993: 12). Accountability is not commonly associated with inventions or novelty but rather more so with carrying out assignments which are more or less specific, in a manner that is both effective and efficient. Therefore, relying exclusively on a definition of accountability as compliance may generate a conflict between the values associated with accountability and the values of originality, experimentation and risk taking.

Accountability: To Whom Should an Organization Be Accountable?

This question can be approached in two ways. We might try to identify the different bodies to which an organization should be accountable. To do that, we first need to know what the organization is to be made accountable for. The selection of bodies to which it will answer depends, in great part, on the nature of this organization. This is easy. All we need to know is the name of the organization. Second, we need to resolve the issue of whether these bodies should be part of the organization or external to it. In other words, we need to decide whether accountability is better conceived as an internal or an external process. As this question is more complex, I will now briefly address it.

In discussing the etymological meaning of accountability, I found that being accountable meant satisfying a legitimate demand to provide an explanation or a justification of one's performance. According to this analysis, we ought to conclude that accountability cannot be anything else than an external process. After all, demands come from the outside and there is no need to provide an explanation or a justification of one's behaviour to oneself, except in the psychological sense which does not concern us here. However, if by saying that accountability is an external process we basically mean that we can only account to *others*, then this conclusion, which is sound, can be reformulated.

1. If these others are members of the same organization as the person providing the account, we then have *internal* accountability, although the account is presented to others.
2. If these others are not part of the same organization as the person providing the account, we then have *external* accountability in the strict sense of the word.

Which one of these forms of accountability is preferable? Actually, both are necessary. Internal accountability, however, cannot be, by its own nature, a relationship *between* organizations or agencies. Hence, it is more of a relationship between different persons or subunits within this organization. According to my previous discussion of who should be accountable, such a relationship would qualify more as responsibility than as accountability as I have tried to define it. Such a network of internal responsibilities should properly be associated with the *management* of an organization, which must be conducted from within. Managing an organization from the outside is absurd; internal management is an organizational necessity. In Light's table of the different aspects of accountability (Table 1.1), management occupies one level. Internal accountability can thus still be conceived as a form of accountability, although in a weaker sense.

I previously came to the conclusion that it was organizations, considered as a whole, which were rightly said to be accountable. If we consider the organization as a whole — as a single unit — then it would have to reach outside of itself to be accountable, since we can only account to others. It would then follow that, in its true and stronger sense, accountability is an external process. This inference is in accord with the results of research on accountability in the police and similar organizations, where peer accountability consistently failed as a mechanism to control the abuse of power (Kerstetter, 1979 and 1985).

It is also in line with current practice. There is a growing consensus to the effect that serious abuse of discretion and of power should be investigated by an external body, to which an organization suspected of such abuse becomes accountable. According to McCubbins and Schwartz (1984), such investigations are an exercise in "fire alarm oversight." In his table on accountability, Light (1993: 14) also associates oversight with "detection and enforcement." Through such investigations, a relationship is established between the notions of oversight and of external accountability. Referring again to Light's table, we could say that internal accountability is an exercise in management, whereas external accountability is an exercise in oversight.

This matching between management and internal accountability on the one hand and oversight and external accountability on the other hand is not only consistent with current definitions found in the literature, but it is also helpful for avoiding serious confusion. Guyot (1985 and 1991: Chapter 10) defines political accountability as the power exercised by government officials outside an agency to formulate policy and obtain adherence to it. The agency Guyot is referring to is the police. There is a tendency in the literature on policing to refer to this power to formulate policy as a power to *control* the police (e.g., Cunningham, 1982; Goldstein, 1967; and Greenberg, 1990). Guyot argues correctly that the margin between political control and political interference is rather thin and that the directives given by government officials to public agencies should follow a set of criteria — legality, effectiveness, efficiency, equity and fiscal responsibility — to avoid politicizing unduly the action of the police. This argument is also valid for other government agencies. Although I do not formally object to the framing of issues of accountability as issues of control, I believe that the two notions should not be confused. In the same way that external accountability is closer to the true meaning of accountability than internal accountability, I would argue that internal control is more germane to the real sense of control than external control.

Models of Accountability

With the exception of financial accountability and war crimes, there has not yet been any sustained discussion of military accountability in the sense in which I am using the word.[35] It is normal to look for structures of accountability which have been developed by other government agencies in order to see if some of their elements could provide a model for military accountability. Police forces first come to mind, because they are, in certain respects, similar to armed forces. Their structures are closely patterned after the military. Most important, armed forces and police forces are the two main agencies of the state which have the legitimate power to use lethal force. Finally, armed forces are increasingly used in policing missions, peacekeeping often being referred to as international policing.

However, despite these similarities, there are very obvious differences, the following being relevant for our purposes (the list is far from exhaustive).

1. Despite all claims to the contrary, the police are not trained for combat but for problem solving. Their main weapon consists of a simple handgun, and they are not skilled in resolving large-scale armed confrontations, such as the one that occurred at Oka in 1990.[36]
2. The physical space of armed combat is very different from the physical space of policing. In armed combat, there is usually a front where the lines are drawn and where the combatants are facing each other and fighting over territory. Furthermore, the combatants are easy to identify since they are, for the most part, in uniform. Finally, the civilian population is normally evacuated to the rear and does not witness the combat, although there may be large civilian casualties in military operations.

   This description is admittedly basic and does not apply to many situations such as guerilla warfare. Its only purpose is to contrast policing and soldiering. Despite all the talk about the "war on crime," the expression is only a metaphor. The police may believe they are waging war on criminals, but criminals know better than to wage war on the police, despite the fact that police officers are occasionally killed in the course of duty. In normal circumstances, policing occurs within a physical space where no lines are drawn and where the fight is never over territory. In fighting crime, the basic task of the police is to identify their potential adversaries who try as much as they can to avoid detection.

In order to succeed in their task of identifying, finding and arresting criminals, the police are largely dependent on information provided by citizens. Hence, policing is an activity which is fully *integrated* in the physical and social space, where it often takes place in plain view of citizens and the press. This integration is such that the key principle of policing, as it was formulated by Sir Robert Peel, is that the police are the public and the public are the police (Reith, 1975: 163), from which the idea of "community policing" is largely derived.[37] By contrast, warfare is entirely structured by *dichotomies* (us and them, our territory, their territory and no man's land, combatants and non-combatants, soldiers and civilians, and so forth).

3. Finally and most important of all, policing is a *service,* afforded to the general population, that is very highly constrained by the law. Despite current metaphors, offenders are not treated as foreign enemies, and they enjoy the same legal protection as other citizens. The legal constraints on warfare are, in contrast, much less exacting.

This contrast between the armed forces and the police is not only oversimplified, but it is also artificial. It compares soldiering and policing in their most dramatic forms, namely warfare and crime fighting. However, soldiers are now engaged in peacekeeping missions which are specifically designed to be a countermeasure to warfare. On the other hand, the most commonplace assertion in policing research is that the police spend most of their time doing something else than fighting crime (e.g., Bayley, 1994: 17). Yet I still believe the contrast I have tried to develop serves a useful purpose.

First, whatever might be said about the actual time which is devoted by soldiers to combat and by police to fighting crime, it cannot be denied that they respectively represent the core of the military and of the police ethos, this word being understood here in its widest sense (ideals, mind set, professional culture and so forth). This fact is reflected in their respective training, although it must be recognized that police forces are now devoting huge efforts to direct training toward the real needs of the job.

More basically, this contrast has important implications for developing frameworks of accountability. I tried to show that whereas policing was integrated into society, soldiering occurred in a context where national and international dividing lines were no longer delineated and the existing order had disintegrated. The fact that policing is embedded in the community has decisive consequences for police accountability. Police are now believed to be, first and foremost, accountable to the community itself, rather than political authorities. The main mechanisms being developed to promote accountability are civilian review boards, which are independent both of the police and of political officials, and complaint committees to which citizens may address a grievance against the police.[38]

Since policing takes place in close contact with the citizens who are witnesses to the performance of the police, complaint committees seem a sound mechanism to ensure police accountability. It is doubtful that the general context of soldiering is amenable to such an accountability mechanism. But, would such a mechanism be more viable in the context of peacekeeping missions?

This last question can be given a general formulation: how adequate for the armed forces is the accountability framework developed in relation to policing? The preceding remarks suggest that this framework could not be exported to the armed forces — if it ever could be — without major adjustments. There is, however, a kind of policing which may be more germane to the armed forces in the sense that its explicit mandate is the protection of Canadian national security. The agency responsible for such policing is the Canadian Security Intelligence Agency. It was created following the recommendations of the second report of the McDonald Commission (Canada, 1981a). The McDonald Commission has developed an elaborate structure for accountability in the context of intelligence gathering and political policing. This structure is based on the four key elements of political accountability both to the responsible minister (the Solicitor General) and to a group of government officials who have a stake in security intelligence; parliamentary oversight; an inspector general who reports to the minister; and an independent civilian review committee with strong investigative powers (the Security Intelligence Review Committee). I shall examine this structure in the part of my report devoted to the post-deployment phase of the operation in Somalia.

CHAPTER TWO

Data

In this chapter, I present data concerning the Canadian Forces (CF). These data are in part quantitative and in part qualitative, qualitative data here encompassing historical data. Some data were collected in relation to the Canadian Airborne Regiment (CAR) and some in relation to the whole Canadian Forces. When specific data on the CAR were missing, I tried to infer from general data what might have been the situation in the CAR. The data presented were selected strictly on the basis of their relevance for my purposes, and I do not claim to present a complete picture of the Canadian Forces.

THE CANADIAN AIRBORNE REGIMENT: HISTORICAL DATA

There is a question about the CAR on which opinion seems to be divided: is the CAR an "elite" military unit or not? This is an important issue, because it has an impact on how the CAR's behaviour in Somalia might be interpreted.

After its investigation, the Board of Inquiry (BOI) came to the following conclusion. There was a widely shared belief by junior officers and junior ranks that the CAR was an elite unit (BOI, Volume XI: Annex K, par. 4). This view was not shared by the CAR's higher ranking staff or those with more years of service: "For them, being elite means being significantly more skilled than the average well-trained infantry soldier. They felt that the Airborne soldiers were not that much better than a good infantry soldier" (BOI, Chapter XI, Annex K, page 3, par. 7).

This analysis agrees with the previous conclusions of the Hewson Report.[1]

> The glamorous, heroic life style of "the para" is appealing even to the present generation of young men. This is how airborne forces are perceived; in Canada, the reality is that they are no more than highly-spirited dismounted infantry.

Hence, although the Cdn AB Regt is simply a battalion-sized infantry capable of being inserted by parachute, soldiers on the outside perceive it as being an elite unit. Indeed, most of the members of the Regiment perceive themselves as being part of an elite unit. (Hewson Report, p. 24, par. 61-62)

b. this perceived elitism attracts young men who associate the "airborne mystique" with the essentially fictitious content of military/para-military television programs, movies and magazines;

c. although the Cdn AB Regt is probably not an elite unit, it is perceived by soldiers on the outside as being elite. Furthermore, the soldiers in the Regiment believe it is an elite unit; (Hewson Report, p. 26, par. 66)

I believe that a brief look at the CAR's history may shed light on this issue of whether it is or is not an elite unit.[2] The Canadian airborne military history began during World War II, when the First Canadian Parachute Battalion was created in July 1942. The Second Canadian Parachute Battalion was also created in July 1942; it was redesignated First Canadian Special Force Battalion to become part of the First Special Service Force, a joint Canadian-U.S. formation. Both units served with great distinction in combat during World War II. They were both disbanded shortly after the end of the war.

In 1948, an airborne brigade group called the Mobile Striking Force (CASF) was established, consisting of a battalion from each of The Royal Canadian Regiment, Princess Patricia's Canadian Light Infantry and the Royal 22nd Regiment — the VanDoos (the CASF also comprised an artillery battery and signal squadron as well as service corps and medical elements). The CASF was reduced to one infantry company group from each of the regiments in 1958.

It was only 10 years later that the airborne role was revitalized with the creation of the Canadian Airborne Regiment in Edmonton, on 8 April 1968. The principal roles assigned to the Regiment merit quotation. They were:

Defence of Canada Operations against small scale enemy incursions in the North, primarily where road and rail facilities are non-existent or in support of a larger operation elsewhere.

Provision of short notice response to United Nations requests for peace keeping or peace restoring forces.

Operations in limited or general war within the context of a larger allied force, particularly a variety of "special service" missions including pathfinders, deep patrolling and winter operations.

Domestic operations in assistance to civil authorities. (Grimshaw, 1981: 33)

The structure of the regiment underwent numerous changes, but its basic elements became similar to the previous CASF, consisting of one commando from each of the Royal Canadian Regiment, Princess Patricia's Canadian Light Infantry and, later in 1979, the Royal 22nd Regiment — the Vandoos (with a headquarters and signals squadron, an airborne service company, an artillery battery and an engineer field squadron; both the artillery battery and the engineer squadron were removed from the Order of Battle in 1977). The CAR trained extensively in terrain varying from Alaska to the Northwest Territories, from Eastern to Western Canada, from the California desert to Jamaica. It served in the following operations:

1. It was called out in Quebec, in aid of the civil authority to counter the terrorist threat posed by the Front de libération du Québec (FLQ) in October 1970.
2. In 1974, it received its baptism of fire when it served in Cyprus, on United Nations (UN) duty. The Turkish Army invaded Cyprus during this tour causing the regiment to become involved in combat as it interposed itself between the belligerents and was also given the mandate to protect Canadian nationals. Two soldiers were killed and 28 were injured. The CAR earned two Stars of Courage and six Medals of Bravery.[3]
3. In 1976, it took part in Operation Gamescan 76 as an immediate reaction force in the case of a terrorist incident during the Montréal Olympic games.
4. From 19 March to 31 September 1981, the CAR was deployed during another peacekeeping mission in Cyprus.
5. It was once again deployed in Cyprus as the 47th Canadian Contingent to the United Nations Force of Intervention in Cyprus (UNFICYP) from September 1986 to 9 March 1987. It suffered one fatal casualty during this tour.

6. In July 1982, 2 Commando was engaged to aid the civil authority in providing security at the Toronto International Airport during the Canadian-hosted Economic Summit.
7. It was part of the security apparatus deployed for the G7 summit held in Toronto on 19-21 June 1988.
8. It also participated in the policing action in Québec during the Oka Crisis (July-November 1990).

However brief, this survey of the history of the CAR yields two conclusions. The first is that peacekeeping and policing operations are not foreign to the CAR's military experience and tradition. In the BOI and the media, there are questions about whether, as an "elite" combat unit, the CAR was the appropriate unit to send to Somalia. Not only did the CAR previously participate in numerous operations that were similar, but the only time it saw combat was during events that occurred during a peacekeeping operation in Cyprus, in 1974, where it served with distinction.

This last remark leads me to suggest an answer to my original question: was the CAR an elite unit? As we previously saw, the junior and senior ranks are of differing opinions on this question. In all the literature on elite units, there is one fact that stands out: the only proof of the elite character of a unit is in combat.[4] With regard to this test, it can only be concluded that claims to the effect that the CAR is an elite unit are to the present day untested. The fact is that the CAR was never in combat, with the exception of its interposition between Turks and Greeks in 1974, when it was posted in Cyprus in the context of a peacekeeping operation. None of its more challenging skills — pathfinding, deep patrol, winter operations — were ever tested in combat, although they were in training. According to the test-in-combat criteria, the CAR would not qualify as an elite unit but only as a would-have-been elite unit. Indeed, a main problem of the CAR may have been the complete divorce between self-perception and performance that is experienced by would-be elite combat units during protracted periods of peace. To have the right stuff is different from having the right look.

In the recruitment and attrition section of this chapter, I address the issue of the recruitment of the CAR's members from parent infantry regiments. As we shall see, it is far from obvious that this recruitment and screening is in any way up to the standards that one would expect in selecting applicants for an elite unit. Because of the operational need to maintain the strength of the CAR at a 90 percent manning level, any person applying from the parent regiments would be accepted as long as the physical fitness criteria was met.

There is one final point to be made on this issue of whether the CAR is an elite unit. An elite unit should be in an utmost state of readiness. This would imply that its membership comprise only regulars who are in a position to devote all their time to being in this state of complete readiness. However, due to the imperative of downsizing, the CF developed the concept of the Total Force, i.e., the integration of Regular Force and Primary Reserve units. The Total Force concept also applied to the CAR and a number of Primary Reserve members were integrated into it and went on to serve in Somalia. The integration of reservists in the CAR was the subject of a tense exchange between Lieutenant Colonel P.R. Morneault, who was its Commander before it departed for Somalia and General Beno (in his testimony before the BOI, Colonel Morneault describes the day on which this exchange took place as "the worst day of his life"; see BOI, Colonel Morneault, 14572). Again, the issue of whether the integration of reservists — whose state of combat readiness is by definition inferior to that of the regulars — conforms to the nature of an elite combat unit can be raised.

The gist of all these remarks is to suggest that any explanation of what occurred in Somalia which is solely based on such notions as "airborne mystique" and manly feelings about being part of a so-called elite unit is at best incomplete and at worst seriously deficient.

THE CANADIAN FORCES AND THE CANADIAN AIRBORNE REGIMENT: QUANTITATIVE DATA

Before 1995, the official position of the CF was in line with the position of the other Canadian agencies, which was that you do not collect racial-ethnic sensitive data. This position was clearly expressed in the First Report of the National Defence Consultative Committee on Social Change (1992: 8-9; henceforth NDCC-SC).

When the Committee first asked for information on the representation of visible minorities in the Canadian Forces, it was told:

Where specific responses are not available, it is tq be noted that the absence of such information is the result of Canadian Forces policies which have been designated to ensure that enrolment and career decisions are absolutely free from non-qualitative factors.

This position is still largely held by the CF, to the extent that I could verify it when I tried to obtain data sensitive to racial-ethnic diversity.

The CF has not yet been subject to employment equity, legislation. However, Bill C-64, an act respecting employment equity has been introduced in the House of Commons and will include the CF. The CF has developed the Diversity Survey Questionnaire over a two-year period and Chouinard and Chiasson (1995) have provided an Analysis of the Canadian Forces Diversity Survey.[5] The information thus collected makes up the CF Employment Equity (EE) data base. Chouinard and Chiasson (1995) have systematically compared data collected through the Diversity Survey to data contained in the Military Personnel Information System (MPIS). The NDCC-SC was provided with a briefing book entitled Demographic Profile of Department of National Defence Personnel (Canada, DND:1991, henceforth DPIS, 1991), which I shall also use for comparison purposes.

The Canadian Forces Diversity Study (CFDS): Method

The CFDS is the result of a questionnaire that was sent to all Regular Force and Primary Reserve members of the CF in January 1995 and was to be returned by March of the same year. The questionnaire consisted of two brief sections, the first section asking for personnel identification and the second part for self-identification as a member of a racial-ethnic group (either majority, Aboriginal or visible minority). The net response rate of 75.4 percent was reduced to 67.1 percent because some respondents did not complete the self-identification in the racial-ethnic group part. The net response was 43.8 percent for the Primary Reserve and 76.9 percent for the Regular Force. The CFDS is based on 66,089 responses (of a total of 98,483).

Table 2.1
Summary of Fully Completed Forms Received

	Regular Force		*Reserve*		*Total*	
	Number	*%*	*Number*	*%*	*Number*	*%*
Majority	(50,938)	95.6	(11,739)	91.6	(62,677)	94.8
Aboriginal	(1,123)	2.1	(272)	2.1	(1,395)	2.1
Visible Minority	(1,219)	2.2	(798)	6.3	(2,017)	3.1
Total	(53,280)	100.0	(12,809)	100.0	(66,089)	100.0

Source: Chouinard and Chiasson (1995) Table VIII.

Chouinard and Chiasson (1995: 6) take pains to warn us that the assumption that the proportion of Aboriginal people (or members of a visible minority) is the same within the non-respondents as among the respondents underlies most calculations made in their analysis. Compared, for instance, to public opinion polls, which operate with a sample of the population that represents less than 0.03 percent of a given population, a response rate of 67.1 percent is remarkably high and warrants extrapolating, to the entire CF personnel, the results of the analysis of the CFDS. As a measure of cautiousness, the results of the CFDS will be compared to the figures given to the NDCC-SC in 1991, whenever they are comparable.

Ethnic Diversity in the CF

As I just said, I compared the results of the CFDS (1995) to the data presented to the NDCC-SC (1991).

Table 2.2
Ethnic Diversity in 1986, 1990 and 1995

	1986	*1990*	*1995*	
	Regular %	*Reserve* %	*Regular* %	*Reserve* %
Aboriginal Soldiers	1.4	2.17	2.1	2.1
Visible Minority Soldiers	2.7	2.26	2.2	6.3

Source: Chouinard and Chiasson: Table VIII. DPIS (1991), parts I and II.

For the Regular Force, the ethnic profile presented in NDCC-SC (1991) is based on CF 1986 Regular Force Applicant Ethnic Groupings, (the CF did not have at that time an EE data base). The applicants numbered 2,933 persons, which is much lower than the 66,089 surveyed by the CFDS. For Aboriginal members, the comparison shows that their proportion in the Regular Force grew from 1.4 percent to 2.1 percent between 1986 and 1995 (the sample for 1986 being much smaller). The percentage of CF members from visible minorities was nearly constant from 1986 to 1995, falling from 2.7 percent in 1986 to 2.2 percent in 1995.

The proportion of Aboriginal members in the Primary Reserve remained essentially the same from the 1990 Primary Reserve ethnic groupings

(2.17 percent) to the data from the CFDS (2.1 percent). The percentage of members from ethnic minorities in the Primary Reserve nearly trebled from 1990 (2.26 percent) to 1995 (6.3 percent). All in all, the figures from 1986 for the regulars, 1990 for the reservists and 1995 for both the regulars and reservists do not show significant differences, with the exception of a very notable increase of the visible minorities in the Primary Reserve. Nevertheless, this comparison between 1986, 1990 and 1995 does not undermine the 1995 figures.

Chouinard and Chiasson (1995: 27-29) ask whether the level of representation in the CF of Aboriginal and visible minority members corresponds to their proportion in the general population. Generally speaking, it does not. The proportion of the *recruitable* Aboriginal population in Canada is estimated by Smith (1995) at 3.2 percent.[6] Their proportion in the CF is at most 2.1 percent. According to Chouinard and Chiasson (1995: 27), the discrepancy may be due to the fact that at least 21.7 percent of the Aboriginal population between the ages of 15 and 54 have disabilities (versus 10.6 percent of the overall Canadian population between the same ages Statistics Canada, 1993). The proportion of members of visible minorities in the recruitable population is 9.4 percent, which is much higher than the 3.1 percent acknowledged by Chouinard and Chiasson (1995). Taking into account factors relating to Canadian citizenship (e.g., whether you were born in Canada, the average age when an immigrant is given citizenship), Chouinard and Chiasson (1995: 29) manage to reduce the pool of recruitable members from visible minorities to 2.82 percent of the Canadian population, which is their present proportion in the CF.

Whatever the difficulties of evaluating the match between Aboriginal and racial-ethnic minority members, on the one hand, and their representation within the CF, on the other, it remains that there is a relatively close match with respect to Aboriginal and a significant discrepancy with respect to ethnic minorities.

Ethnic Diversity: The CF and the CAR

I now compare data from Chouinard and Chiasson (1995) and data obtained from the Somalia Inquiry Liaison Team (SILT) on the ethnic make-up of the CAR. The results of this comparison should be interpreted with utmost caution. (1) The number of Aboriginals and members from visible minorities in the CAR are so small (seven Aboriginals and 17 from visible minorities) that they make it virtually impossible to draw meaningful statistical inferences; (2) the data provided by SILT refers to the

CAR as of 20 January 1995; data on the members of the CAR that went to Somalia in 1992-1993 was not collected at that time. My assumption is that the numbers may have changed from 1992-1993 to the beginning of 1995, but not in a significant way. In other words, the situation in January 1995 substantially reflects the situation in 1992-1993.[7] (3) *An important point* is the numbers given by Chouinard and Chiasson (1995). They refer to the numbers of *respondents* to their questionnaire and not to the actual total of CF personnel, whereas the numbers provided to me by SILT do refer to the actual numbers in the personnel of the CAR; nevertheless, as I have already said, I accepted Chouinard and Chiasson's hypotheses that the percentages among non-respondents are the same as those among respondents. Hence I always put the actual numbers given by Chouinard and Chiasson (1995) in brackets, in order to remind the reader that these numbers refer to the respondents to a questionnaire and not to all CF personnel (we followed this rule throughout all the tables). (4) When it was deployed in Somalia, the CAR's personnel strength was reduced from 754 to 601 and it has fluctuated ever since. I have calculated my percentages with regard to the CAR on the basis of strengths of 600 and 700. The difference between the percentages is not significant for my inquiry.

Taking all these reservations into account, what we are looking for in the comparison between CAR and CF survey data are conspicuous anomalies rather than fine differences.

Table 2.3
Ethnic Diversity: CF and CAR

Minorities	*Canadian Forces*					*Canadian Airb. Regt*		
	Regular		*Reserve*		*Total*			
	Number	*%*	*Number*	*%*	*%*	*Numbers*	*% 600*	*% 700*
Aboriginal Members	(1,123)	2.1	(272)	2.1	2.1	7	1.16	1
Visible Minority (All) Members	(1,219)	2.2	(798)	6.3	3.1	17	2.8	2.4
African Canadian (Black) Members	(289)	0.75	(181)*	0.75	0.75	11	1.83	1.57

Note:
* The figure 0.75 percent is the ratio between the percentage of visible minority members in the CF (3.1) and the proportion of the percentage made up of Blacks.
Source: Chouinard and Chiasson (1995): tables VIII and XXIV.

To the extent that it can be undertaken, the comparison between the CF survey and the CAR data gives the following results.

- The percentage of Aboriginal members in the CAR is twice as low as in the CF survey.
- The percentage of members from visible minorities is slightly higher in the CAR than in the Regular CF (when the CF percentages for the Regular Force and Primary Reserve are averaged, then the CAR ratio becomes slightly lower).
- The percentage of African Canadians in the CAR is at least twice as high as in the CF survey (this is the most significant result of the comparison).

Because of all the reservations that I made, this last result is difficult to interpret. However, *it would contradict the assertion that the recruitment policy of the CAR was tainted by discrimination against African-Canadians.*

Table 2.4 identifies the distribution by component of the CF minority members.

This table shows that the percentage of visible minority members concentrated in the Primary Reserve is twice as high as the percentage of Aboriginal members.

Table 2.4
Minorities by Component

	Native Aboriginal Members		*Visible Minority Members*	
Minorities	*Number*	*%*	*Number*	*%*
Regular Force	(1,123)	80.5	(1,219)	60.4
Primary Reserve	(272)	19.5	(798)	39.6
Total	(1,395)	100.0	(2,017)	100.0

Source: Chouinard and Chiasson (1995) tables IX and XVIII.

Table 2.5 provides a breakdown of visible minorities.

The interesting feature of Table 2.5 is the diversity it shows in visible minorities. African-Canadians (Blacks in the CF survey) account for 23.6 percent of the presence of visible minorities in the CF. Asians accounts for 21.1 percent of the percentage, a figure which tends to be overlooked.

Table 2.5
Visible Minorities by Subgroup and Component

	Regular		*Reserve*		*Total*	
Subgroup	*Number*	*%*	*Number*	*%*	*Number*	*%*
Unspecified	(313)	25.7	(70)	8.8	(383)	19.0
Black	(289)	23.6	(181)	22.6	(470)	23.3
Chinese	(77)	6.3	(109)	13.7	(186)	9.2
Filipino	(42)	3.4	(59)	7.4	(101)	5.0
Japanese	(27)	2.2	(11)	1.4	(38)	1.9
Korean	(12)	1.0	(16)	2.0	(28)	1.4
South Asian/ East Indian	(75)	6.2	(70)	8.8	(145)	7.2
Southeast Asian	(24)	2.0	(25)	3.1	(49)	2.4
Latin American	(52)	4.3	(57)	7.1	(109)	5.4
W Asian, N African/Arab	(47)	3.9	(46)	5.8	(93)	4.6
Mixed Race	(239)	19.6	(125)	15.7	(364)	18.1
Other	(22)	1.8	(29)	3.6	(51)	2.5
Total	(1,219)	100.0	(798)	100.0	(2,017)	100.0

Source: Chouinard and Chiasson (1995): Table XXIV.

Officers vs. NCMs

Table 2.6 shows the distribution of officers and non-commissioned members (NCMs) for Aboriginal and visible minority members.

On the surface, Table 2.6 would show that the proportion of officers from visible minorities is twice as high as Aboriginal officers. The proportion of officers from visible minorities is actually comparable to, and even higher than the general ratio between officers and NCMs in the CF survey. This impression has to be put in context, as Table 2.7 will show.

Table 2.7 shows that there are very significant differences between the respective distribution of visible minority CF members among officers and NCMs, depending on the different ethnic groups to which they belong.

Table 2.6
Officers and NCMs: MPIS Data

	Regular		*Reserve*		*Total*	
	Number	*%*	*Number*	*%*	*Number*	*%*
Officers	(15,582)	22.5	(4,798)	16.1	(20,290)	20.6
NCMS	(53,687)	77.5	(24,506)	83.9	(78,193)	79.4
Total	(69,269)	100.0	(29,214)	100.0	(98,483)	100.0

Aboriginal Officers and NCMs Diversity Survey Data:

	Regular		*Reserve*		*Total*	
	Number	*%*	*Number*	*%*	*Number*	*%*
Officers	(112)	10.0	(20)	7.4	(132)	9.5
NCMS	(1,011)	90.0	(252)	92.6	(1,263)	90.5
Total	(1,123)	100.0	(272)	100.0	(1,395)	100.0

Visible Minority Officers and NCMs Diversity Survey Data:

	Regular		*Reserve*		*Total*	
	Number	*%*	*Number*	*%*	*Number*	*%*
Officers	(309)	25.3	(95)	11.9	(404)	20.0
NCMS	(910)	74.7	(703)	88.1	(1,613)	80.0
Total	(1,219)	100.0	(798)	100.0	(2,017)	100.0

Source: Chouinard and Chiasson (1995), tables III, X and XIX.

The performance of ethnic minorities such as Koreans and South Asians is very high (respectively 42.9 percent and 44.1 percent of their members among officers, which is twice as high as the general average of CF members) as compared to Blacks and Latin Americans (8.7 percent and 9.2 percent, respectively). Aboriginal members, with a score of 9.5 percent (see Table 2.6 middle section), would rank as third from the bottom group with the least representation among officers (slightly above Blacks and Latin Americans). In the absence of other data (e.g., the educational level of minority members in the CF), it is impossible to explain these discrepancies.

Table 2.7
Summary of Members of Visible Minorities by Subgroup and Rank

Subgroup	*Highest Rank*	*Officers*	*Total Members*	*% Officers*
Unspecified	LCol	46	383	12
Black	Major	41	470	8.7
Chinese	LCol	65	186	34.9
Filipino	LCol	20	101	19.8
Japanese	Major	12	38	31.6
Korean	Major	12	28	42.9
South Asian/ East Indian	LCol	64	145	44.1
Southeast Asian	Major	16	49	32.7
Latin American	Col	10	109	9.2
W Asian, N. African/Arab	Major	29	93	31.2
Mixed Race	Major	79	364	21.7
Other	Major	10	51	10.6
Overall		404	2017	20

Source : Chouinard and Chiasson (1995), Table XXV.

There is one more table (Table 2.8) on the subject of the CF hierarchy which deserves mention.

According to Table 2.8, which refers to all members of the CF, only 6.9 percent of all Regular CF members are beneath the lowest rank of corporal. This figure is similar to that for Aboriginal and visible minority CF members (it stands at 7 percent; see Chouinard and Chiasson, 1995: tables XIb and XXb). This percentage is tabulated as a ratio between privates and other NCMs (from corporal to chief warrant officer). It gets even smaller when calculated as a ratio between privates and all other ranks, including commissioned officers; it then is 5.3 percent.

Table 2.8
NCMs from MPIS Data by Rank and Component

Rank	*Regular* *Number*	*%*	*Reserve* *Number*	*%*	*Total* *Number*	*%*
Chief Warrant Officer	(756)	1.4	(199)	0.8	(955)	1.2
Master Warrant Officer	(2,123)	4.0	(481)	2.0	(2,604)	3.3
Warrant Officer	(4,438)	8.3	(792)	3.2	(5,230)	6.7
Sergeant	(8,664)	16.1	(2,057)	8.4	(10,721)	13.7
Master Corporal	(10,024)	18.7	(2,301)	9.4	(12,325)	15.8
Corporal	(24,000)	44.7	(6,517)	26.6	(30,517)	39.0
Private	(3,682)	6.9	(12,159)	49.6	(12,841)	20.3
NCM Total	(53,687)	100.0	(24,506)	100.0	(78,193)	100.0

Source: Chouinard and Chiasson (1995): Table IV B.

Two members of the CAR were primarily responsible for the beating death of Shidane Arone, which was the worst incident involving members of the CAR in Somalia. Master Corporal Matchee was one of these two persons. As he suffered brain damage following an unsuccessful suicide attempt, he was unfit to stand trial and we will never know his version of events. The other principal culprit for the beating death of Shidane Arone was Private Kyle Brown. It must be emphasized that as a private, Brown was part of a group that had 95 percent of the CF rank structure above him. Add to this the fact that he was relatively new to the CAR and therefore part of the "fucking new guys" (FNGs) within the CAR culture. The collective weight of the hierarchy, official and non-official and commissioned and non-commissioned, on a private, has to be a difficult burden to bear before one is moved up the next rung to corporal.

It is very difficult to draw conclusions from these data. As I previously said, this is not because the data is unreliable but because it is not collected on a regular basis. The data I am presenting were collected during a single survey. Because they were not collected on a regular basis, the data cannot be compared to similar figures in other countries. For example, Dorn (1990: 101) tells us on the basis of official statistics that Blacks are overrepresented in the United States Army, comprising 20 percent of the U.S. Armed Forces in 1981 (this phenomenon of Black overrepresentation is well acknowledged in the literature). Dorn (1989: 106) also tells us that in 1984 Blacks made up 6.2 percent of active duty commissioned officers, which was a dramatic increase from the 2.2 percent level in 1972. In 1981, Blacks accounted for one fifth of the U.S. Army and a little more than one twentieth of its officers. According to the figures given in Chouinard and Chiasson (1995), Aboriginal and members of visible minorities make up 2.1 percent and 2.2 percent respectively of the personnel of the regular CF. The ratio between officers and NCMs is tabulated as a proportion between Aboriginal and visible minority members, on the one hand, and their respective numbers in the CF, on the other hand (see Table 7). When the percentage of officers is tabulated as a ratio between CF officers as a whole and members of minorities, it is 0.21 percent (compared to 2.1 percent overall membership) for Aboriginal members and 0.57 percent (compared to 2.2 percent) for visible minorities.[8] The ratio between enlisted men and members of visible minorities is lower in Canada than in the United States (for Blacks), but it is still comparable (one in four in Canada as compared to one in three in the United States); for Aboriginal members, the ratio of 1:10 is very low. These comparisons are risky, because the data are not similarly categorized in the United States and Canada.

This lack of reliable data is, in my experience, a general feature of many, if not all, Canadian ministries and agencies. Although such a recommendation has been made in almost every context, it deserves to be repeated:

RECOMMENDATION I: **I recommend that the CF collect data on its racial and ethnic composition in a much more systematic way.**

Recruitment and Attrition

One question I wanted to answer was whether the attrition rate of the CF personnel or a decline in CF recruitment was such that the CF could not afford to be selective in recruiting new personnel.

Table 2.9
MPIS Data by Years of Service and Component

Years of Service (YOS)	*Regular*		*Reserve*		*Total*	
	Number	*%*	*Number*	*%*	*Number*	*%*
0 - 4	(8,927)	12.9	(18,068)	61.8	(26,995)	27.4
5 - 9	(18,409)	26.6	(6,956)	23.9	(25,365)	25.8
10 - 14	(16,161)	23.3	(2,254)	7.7	(18,415)	18.7
15 - 19	(13,397)	19.3	(1,033)	3.5	(14,430)	14.7
20 - 24	(6,284)	9.1	(524)	1.8	(6,808)	6.9
25 - 29	(3,976)	5.7	(242)	0.8	(4,218)	4.3
30 and up	(2,115)	3.1	(137)	0.5	(2,252)	2.3
Total	(69,269)	100.0	(29,214)	100.0	(98,483)	100.0

Source: Chouinard and Chiasson (1995): Table V.

The corresponding tables for Aboriginal and visible minority members are Chouinard and Chiassson (1995), tables XII and XXI. They depict a situation that is not dissimilar to the general CF. These tables suggest the following conclusions.

1. The attrition level is very high in the Primary Reserve, going from a membership of 61.8 percent in the 0-4 years of service (YOS) group, to 23.9 percent in the 5-9 YOS group, and then to 7.7 percent in the 10-14 YOS group. The situation is quite similar with the Aboriginal members (64.0, 21.7 and 8.0 for the same YOS groups) and the visible minorities (70.5, 19.5 and 6.3 for the corresponding YOS groups).
2. The level of attrition is much lower in the Regular Forces than in the Primary Reserve. The main point of attrition in the CF is between the 15-19 (19.3 percent) and the 20-24 (9.1 percent) YOS groups. Significant points of attrition with respect to Aboriginal and visible minority members occur at the lower transition points between the 5-9 and the 10-14 YOS groups (from 31.2 percent to 23.2 percent for the Aboriginal members and from 32.2 percent to 23.1 percent for the visible minorities.
3. The significant differences between the YOS 0-4 (range: 12.4 - 16.7 percent) and 5-9 (range: 26.6 - 32.2 percent) groups in the CF (general, Aboriginal and visible minority) show that recruitment is declining, either because of a deliberate policy to reduce CF strength or

because a CF career is becoming less attractive. This question deserves attention.

According to DPIS statistics presented to the NDCC-SC, the effective strength of the CF Regular Force was 84,243 (as of September 1991). In a 1992 paper, General de Chastelain assessed the effective strength of the Regular Force to be 86,000 (deChastelain, 1992: 14). The MPIS data estimated the strength of the Regular Force to be 69,269 in 1995. According to the latest DPIS statistics, this number had fallen to 65,387 (as of January 1996). According to information obtained in my interviews, it is expected to further decrease in the next years. These numbers indicate, without any doubt, that the CF is downsizing. The decline in recruitment should then be seen as the result of a deliberate CF policy rather than the loss of the appeal of a career in the CF.

To test this conclusion further, I compared the salary levels in Canadian police forces to that of CF members. As of 1992, the last year for which national figures are available, the average yearly salary earned by Canadian police was $44,768 (Boivin et al., 1995: Annex IV, p. 3). There is no such average figure for the CF. However, according to the latest 1996 DPIS statistics, the largest group in the CF is by far the corporals (22,288 of a total strength of 65,387). According to the 1994 Pay Rates Tables for the Regular Forces, an army corporal in the Military Occupation Group of Specialist 1 will earn $36,576 after two years. Although this is not an average, it does reflect the pay rates of the largest group in the CF. Furthermore, the number of CF members *above* the rank of corporal is much higher than the number below this rank (in January 1996, only 3,781 of the 65,387 members of the CF were not at least corporals, according to the DPIS statistics). Furthermore, the level of education required for recruitment by a police force is generally higher than for recruitment by the CF (according to my estimation based on 1996 DPIS data, only 16 percent of NCMs in the CF have more than a secondary (grade 12-13) education, whereas in Quebec, for instance, a post-secondary education is a prerequisite for applying to join a police force). Despite the difference between police and CF salaries, being a member of the CF remains a financially attractive and competitive proposition; it is even more so, when the CF pay rates are compared to the average wages paid to Canadians.

The answer to my initial question is that the CF can indeed afford to be selective in its recruitment policy. The decrease in recruitment is the result of a deliberate policy of downsizing and not of the lack of attraction of a career in the CF, which remains competitive, at least with regard to the important factor of pay rates. The attrition rate in the Regular Force is not such that it warrants the conclusion that the CF cannot keep its personnel and that it would have to relax its standards of recruiting in order to maintain its effective strength. However, this cannot be said with respect to the Primary Reserve, where there is a real problem of attrition. There was no statistical indication that this overall situation is different with regard to Aboriginal and visible minority members.

Recruitment and Attrition: the CAR

It is important to examine whether the analysis developed above applied to the CAR. Generally speaking, there is no particular reason why it would not apply, the pay rates, levels of education and other factors not being different in the CAR. However, when we come down to specifics, it does appear that the situation was rather special with regard to the CAR. Candidates for service in the infantry commandos of the CAR are normally selected from among personnel already serving in the Royal Canadian Regiment, Princess Patricia's Canadian Light Infantry and the Royal 22ième Régiment.

In response to highly visible criminal acts involving members of the CAR, Major-General Hewson chaired a team that was to review disciplinary infractions and antisocial behaviour within Force Mobile Command (FMC) with particular reference to the Special Service Force and the CAR. In September 1985, this team issued its report, precisely entitled *A Report on Disciplinary Infractions and Antisocial Behaviour within FMC with Particular Reference to the Special Service Force and the Canadian Airborne Regiment* (known as the Hewson Report). In his covering letter to Lieutenant-General Belzile, the FMC Commander, Major-General Hewson wrote: "It may also be necessary to remind the three Regular Force Infantry regiments of their continuing responsibilities to provide 'good' soldiers to the Canadian Airborne Regiment" (Hewson to Belzile, 26 September 1985, ref.: 5200-1 TD 5237 CIS).

On 25 November 1985, Lieutenant-General Belzile, the FMC Commander, reacted to the Hewson Report. In Annex A of his letter addressed to the Chief of the Defence Staff, he acknowledges that the number and quality of potential candidates for the CAR is strongly influenced by the effective strengths of the parent infantry regiment. He then adds:

> When parent regiments are fully manned with strong performers it is not difficult to find the required number of high quality volunteers for the Cdn AB Regt.

But the operational requirement to maintain the AB Regt at a 90 percent manning level does lead periodically to a rather lenient application of the selection standards if the parent regiments are deficient in either numbers or quality of candidates. To insist upon rigid selection criteria that are applicable in all cases could, on occasion, leave us unable to man the AB Regt without ignoring some of the selection criteria. (par. 5)

This line of argument is to the effect that the selection of applicants to the CAR from parent regiments is far from rigid, due to the operational need to maintain the strength of the CAR at a 90 percent manning level.

On the other hand, the 1993 BOI Report tells us that a thorough screening of all the personnel assigned to the Canadian Airborne Regiment Battle Group (CARBG) for deployment into Somalia could be performed, because the CAR was overmanned with respect to the personnel ceiling that had been determined for Operation Deliverance (BOI, Vol. XI, Annex F, pp. 2-3, par. 4-6). The report describes this screening process as a "cautious pruning" of the three commandos' personnel. The evidence as to whether this cautious pruning of the personnel to be sent to Somalia ever took place is, as we shall see, very contradictory.

Although it can be said that the CF could generally afford to implement an exacting policy of screening in the 1990s, it would seem that the screening process of applicants for membership in the CAR, and in the CARBG that was to be sent to Somalia was not consistent.

Gender Distribution

Table 2.10 shows the distribution of males and females in the Regular Forces and in the Primary Reserve.

Table 2.10
MPIS Data by Gender and Component

	Regular		*Reserve*		*Total*	
Gender	*Number*	*%*	*Number*	*%*	*Number*	*%*
Female	(7,452)	10.7	(6,432)	22.0	(13,884)	14.1
Male	(61.817)	89.2	(22,782)	78.0	(84,599)	85.9
Total	(69,269)	100.0	(29,214)	100.0	(98,483)	100.0

Source: Chouinard and Chiasson (1995): Table VI.

The general proportion for the CF Regular Forces (89 percent males versus 11 percent females) is similar with respect to Aboriginal (86:14) and visible minority CF members (89:19). The ratios are substantially the same in relation to the Primary Reserve, although the distribution between males and females differs slightly for visible minorities (81:19).

THE EVENTS RELATED TO THE CAR'S MISSION IN SOMALIA: QUALITATIVE DATA

By qualitative data, I mean the discussion of features of the events that took place during the three phases of the operation in Somalia, namely pre-deployment, in-theatre and post-deployment. The features to be discussed are not exhaustive, but were selected on the basis of their significance for the topics of this report (peacekeeping, racial prejudice and accountability).

The Incidents that Occurred in Somalia: Their Roots in Time

One of the distinctive features of the disciplinary problems that occurred in Somalia in 1993 is that they were, to a certain extent, foreshadowed by similar breaches of discipline that occurred previously in Canada. That 2 Commando of the CAR was in serious trouble with regard to discipline in the months before its departure for Somalia is now well known through the testimonies of Lieutenant-Colonel Morneault, Lieutenant-Colonel Mathieu and General Beno before several bodies, such as the BOI or the Somalia Commission. However, these incidents have much deeper roots in time.

Major-General Hewson, then Chief of Intelligence and Security, was commissioned as early as 1985 to investigate disciplinary infractions and antisocial behaviour that occurred in Petawawa, where the CAR was stationed. The Hewson Report issued a dire warning about what was to be expected within the CAR if vigorous countermeasures were not taken.

> Unless the matter of taskings, turbulence and absentee junior leadership is addressed now, there is a very real danger that present cases of antisocial behaviour will escalate both in frequency and severity. (Hewson Report, p. 31, par. 78 h.)

> It should once again be emphasized that the prolonged absence of junior leaders eventually leads to a lapse and ultimate breakdown in discipline. (Hewson Report, p. 32, par. 81 c.)

This warning was apparently not heard. A thorough comparison between the Hewson Report and Volume XI (Phase I) of the BOI report reveals that they describe a situation that is essentially the same.[9] Because of the limits on the length of this report, I shall give only one example of overlapping statements:

> Lacking effective leadership from junior officers and NCO's, the majority of the soldiers must turn to something else. Many look to the informal leaders amongst themselves; unfortunately, some of these informal leaders cannot cope with the challenge in a responsible manner and problems can occur; their priorities and goals do not correspond to those of the CF. (Hewson Report, p. 29, par. 72)

> ...certain events occurring recently and over the last several years provided the Board with sufficient evidence to conclude that there was and *may still be*...an informal group at the junior rank level which poses a direct challenge to authority...The Board considers that the leadership has been unable, *to date*, to resolve this problem.

> It is quite natural that...peer groups with their informal leaders form at the junior rank level...The problem arises when (the) interests and values (of the peer groups with their informal leaders) run counter to accepted norms; ... At this stage, such individuals present a serious challenge to formal authority and the leadership must excise the inappropriate interests and values, the individuals involved, or, preferably both. (BOI, Vol. XI, Phase I, Annex D, p. 1, par. 2-3, emphasis added)

There are many other parallel paragraphs in both reports. This would be an indication that there was a relative failure to remedy a situation before it got out of control. It might then be asked why was there such a relative failure to take the necessary steps to remedy a threatening situation. I cannot provide a simple answer to this complex question. I will, nevertheless, raise the hypotheses that this failure is related to a feature of internal inquiries within the CF (this may well be a general feature of internal inquiries). This feature is the perceptible strain that exists between the findings of an inquiry or of an investigation *and* the significance granted these findings, given the general conclusions and recommendations in the final report. This strain is, on occasion, such that it begets true inconsistencies; the reader has to wonder whether the same inquiry generated the findings *and* the recommendations.

In the excerpt above from the BOI Report, it is stated that deviant informal leadership and values present a serious challenge to formal authority and must be "excised" (a surgical term) by the leadership. The whole of Annex D is devoted to a thorough discussion of the disciplinary problems within the CAR (particularly in 2 Commando) and the leadership of 2 Commando is chastised for having "failed to take sufficient action to rid themselves of a known challenge to formal authority" (Annex D, p. 6, par. 19,). In quoting this excerpt, I have emphasized the words used to indicate that the disciplinary problem was not yet resolved at the time of writing the BOI Report. Yet the first recommendation of Annex D is to the effect that the CF formally recognize the exceptional contribution by members of the Canadian Joint Forces in Somalia (CJFS) and the Battle Group to the operation in Somalia and that a contingent from the Battle Group be highly profiled at the Remembrance Day ceremonies. It is also recommended that the problems of leadership and challenge to authority existing in the CAR be resolved, these problems being, however, specifically limited to 2 Commando.

It does not appear that these problems were limited to 2 Commando. Neither were they solved after the problems of discipline of 2 Commando surfaced. Parts of videotapes made by members of the CAR to put on film what happened during hazing rituals were shown on Canadian television. These hazing rituals were carried on in 1 Commando. The Commission is in possession of two such tapes, one that recorded a hazing ritual that took place in July 1992, before the deployment to Somalia, and another that recorded a hazing ritual that occurred in 1994, after the CAR had returned from Somalia. The content of these tapes was so shocking that it created a major scandal in Canada; projection of scenes from them on national television was shortly followed by the Minister's decision to disband the CAR. On 23 January 1995, i.e., some 18 months after the publication of the BOI's phase 1 report limiting the regiment's disciplinary problems to 2 Commando, the whole regiment was disbanded. On 20 March 1995, some two months after the disbandment of the CAR, the CIDCFS was created to conduct an inquiry into Operation Deliverance.

The Hewson Report focussed on the Special Service Force and on the CAR, which were both part of the FMC. As we have seen, the Hewson Report diagnosed quite correctly that if the CAR's problem with the vacuum in junior leadership was not resolved, disciplinary incidents and antisocial behaviour would escalate both in severity and frequency and might ultimately lead to a breakdown of discipline, which it did. In his covering letter to Lieutenant-General C.H. Belzile, Major-General Hewson wrote:

You will find no particularly dramatic conclusions nor any radical recommendations; however, there are some suggestions you may wish to have your staff examine in further detail. Some of the recommendations, should you agree with them, will require staffing in conjunction with NDHQ and *others would require further study by officers with special expertise.* (Major-General Hewson to Lieutenant-General Belzile, 26 September, 1985, par. 2)

The Hewson Report begins with an executive summary, which ends with overall recommendations. The first of these overall recommendations begins thus: "There is no requirement for immediate action. (Hewson Report, 1985, Executive Summary, par. 26)

As is the case with the BOI Report, this initially reassuring conclusion is followed by a list of actions that might be taken (Hewson Report, 1985, Executive Summary, par. 27). It might be said that these introductory claims to the effect that the problem investigated was limited and required no immediate remedy are part of a perfunctory CF ritual which should not be taken at face value. There is some truth in this observation but it diminishes their impact far too much. The real effect is to remove any sense of *urgency* about taking significant action to remedy a situation. Ultimately, little or no action was taken, as is shown by the fact that the BOI Report describes, in 1993, the situation prevailing in the CAR in much the same terms that were used eight years before in the Hewson Report.

Allegations of Racism and Infiltration by the Extreme Right

Another feature of these events is the serious number of public allegations to the effect that members of the CAR may have been racially prejudiced against Somalis and Blacks in general, and the CF and the CAR, in particular, may have been infiltrated by elements from the extreme right. These allegations were not only made by the media. They can also be found in books and research literature (Barrett, 1987; Kinsella, 1992 and 1994). In its report on the Heritage Front affair, the Security Intelligence Review Committee (SIRC, 1994) also refers to these allegations (leaders of the extreme right have a tendency to boast that they have a large following in the armed forces, infiltrating the army playing an important part in their strategy for "world domination").[10]

A crucial feature of these allegations is that they are mutually supportive and have a maximum impact when they are made together. Before going to Somalia a member of the CAR — Corporal Matt McKay — had his photograph published in the *Winnipeg Sun* (13 September 1992). He was dressed in neo-Nazi regalia, giving the fascist salute. This news story, which was to haunt the CF, was entitled "Neo-Nazis Open Doors to Soldiers"; it came to be interpreted by many as meaning that it was the CF that was opening its doors to neo-Nazis.

The CF was at first very reluctant to view, as a real threat, the possibility that an undetermined number of its members might be part of extreme right groups and organizations. However, a July 1993 internal review of CF personnel policy entitled *Racism and the Canadian Forces* (Canada, Department of National defence, 1993) declared:

> Certainly, the revelation that one or more members of the regiment involved (note: this is the CAR) have been associated with a racist organization has shattered any complacency that might previously have existed with respect to racism in the Canadian Forces. (p. 14)

In the wake of the Somalia crisis, the CF produced a series of conflicting evaluations of the extent of right-wing penetration into its ranks. I shall not comment at this stage on all these documents. However, there is one analysis that I will briefly note, because it reveals, in exemplary fashion, the CF's ambivalence in acknowledging the existence of problems within the organization.

Within the context of project SIROS, a CF investigator produced an analysis of allegations made in Warren Kinsella's *Web of Hate* (1994) about links between members of the CF and extreme right organizations (particularly in chapter 12 entitled "Matt McKay's Youthful Folly"). After scrutinizing some 16 allegations made by Kinsella, the investigator comes to the following conclusions: six allegations are true, two are possibly true, two are possibly false, two are false,[11] and in four cases he could not determine whether the allegations were true or not. It would appear that Mr. Kinsella withstood the test of accuracy rather well, for of 16 allegations, the investigator only found two that were false (and he is wrong in one of these cases).

Relying on a very critical review of Kinsella's book by Bill Dunphy (*Ottawa Sun*, 10 April 1994; this article is a reprint from a paper that was first published in the *Toronto Sun*, Sunday, 3 April 1994, p. M 21) that alleged there were factual errors in one of Kinsella's chapters, the SIROS project investigator wraps up his analysis thus:

Ref D (Dunphy) indicates a degree of inaccuracy on the part of the author of ref A (Kinsella), in fact, ref D identifies factual errors in one particular chapter of ref A. As a result, and keeping in mind the inaccuracies identified above, caution is strongly urged in accepting the facts contained in the *novel* "Web of Hate" as entirely accurate. (Memorandum, 2112-2 (D Secur Ops 2), 14 June 94)

Actually, Bill Dunphy mentions having spotted 19 factual errors in the chapter dealing with the Ottawa and Toronto scene and does not talk about chapter 12 on the penetration of right wingers into the CF. Dismissing Kinsella's carefully researched study as a "novel" is, to say the least, cavalier. This dismissal is another indication of the strain that I identified previously between the analyses produced within the CF, which are generally thorough, and the watering down of the conclusions that are drawn from these analyses.

The Events in Somalia: An International Perspective

A third feature of the CARBG's mission in Somalia is that it was accomplished within the context of an international operation led by the United States and which involved several countries (Belgium, Botswana, Canada, Italy, Pakistan and the United States, among others). The operation in Somalia seemed to have occurred in a particularly difficult context, for there is not one contingent which did not experience serious trouble and was not involved in violent confrontations with the numerous factions that were warring in Somalia as well as with Somali civilians. To all practical purposes, operation UNOSOM II, the Second United Nations Operation in Somalia, which succeeded UNITAF, the United Task Force, degenerated from a peace-enforcement mission into what is now called a low intensity conflict.

I cannot review, at this point, all that happened to every national contingent present. However, I will briefly draw a parallel between the events that surrounded the participation of the Belgian contingent in operation UNITAF and what happened to the Canadian Forces.

Belgium sent a contingent of 850 soldiers to Somalia to serve in the sector of Kismayo, reputed to be a particularly difficult one. Despite the fact that these troops are referred to as "paracommandos" by the Belgian press, they were not part of such an elite unit and actually lacked combat experience.[12] The Belgians suffered 14 casualties (four killed and 10 wounded). They were accused by the organization African Rights of many human rights abuses, including 26 specific cases of alleged unwarranted killing (African Rights, 1993: 17-28, see p. 19). African Rights (1993: 29-30) has also a rather short section documenting abuses by Canadian troops, which I shall quote in part at a later stage of this report.

The Belgian government has investigated the allegations made by African Rights (Belgique, Auditorat Général près la Cour Militaire, 1993) and concluded that most of them were not founded on fact.

A Belgian military force partaking in an operation is usually accompanied by a legal and policing unit [une unité judiciaire], which monitors the troops' respect for Belgian and international law and for human rights. Despite legal and structural differences, there are analogies between the operation of Belgian and military justice with respect to these events.

1. Belgian military justice actually found that Belgian troops were involved in a number of incidents. Two hundred and sixty-eight were investigated. Fifty-nine of these incidents involved acts of brutality against Somali civilians (these acts ranged from homicide to simple assault). On the basis of these investigations, eight members of the Belgian contingent were charged with various counts of homicide, manslaughter and assault. No officers were among this first group of accused (there were two sergeants). A second group of 16 were also charged with assault. There were two captains and one lieutenant, in this second group, all the rest being NCOs.
2. The officers, NCOs and soldiers were all court-martialled. The first round of verdicts were perceived as lenient by public opinion. Of the 24 persons charged, 13 were acquitted, one of the persons acquitted having been charged with "attempted murder in order to cover up a theft." The two captains that were charged were acquitted. The only severe sentence — five years of imprisonment — was imposed on the soldier found guilty of murder in order to cover up a theft. The most scandalous sentence was imposed on a lieutenant and a corporal who had forced two Somali children caught stealing to dig their own grave, while one of the military held his weapon to the temple of one of the children and the other fired a shot in the air. Both were given suspended sentences, with the lieutenant having to spend eight days in jail (the corporal was not incarcerated). Fines and disciplinary sanctions were imposed on the two sergeants who had beaten a Somali and tortured him with electricity.

3. As I said, these dispositions were perceived as unduly lenient. The military justice authority came to believe that sterner measures had to be applied in order to dispel the troops' belief that they did not have to account for their misconduct abroad. The Auditeur Général has appealed the initial court-martial decisions in six cases. (In particular, the acquittal of the two captains, who were the highest ranking officers involved, is being appealed; the lenient sentence imposed on the lieutenant who staged the fake execution of two children is also being appealed.)

What happened in relation to the contingent from Belgium is similar to what happened in relation to the CAR in two respects. First, the kind of misconduct that occurred in the course of the mission is analogous (brutality against Somalis suspected of theft, unnecessary force used in crowd control, careless use of firearms and so forth). Some of the similarities are in fact fairly exact. The publication of photographs showing the bruised and blindfolded face of Shidane Arone was extremely damaging to the CF's image in Canada. The same phenomenon occurred in Belgium, where military guilty of serious misconduct began to show pictures of their abused victims to their friends as trophies. Some of these photographs eventually fell into the hands of the press and of military authorities, who were deeply shocked.

A second point of similarity between these two trains of events lies in the change of course of military justice. It first tended to be lenient, particularly toward officers, and then reversed itself.

There are more similarities, which I do not have the space to develop in these pages. They all lead to the same point: at least a significant part of the explanation for what happened to the CARBG during its mission in Somalia lies outside the Canadian contingent and must be sought in the nature of its mission, in the social, political and physical context of this mission in Somalia and in other external and general factors. If the explanation for the conduct of certain members of the CAR was wholly to be found within this regiment, it could not account in any way for very similar behaviour in other contingents of UNITAF. This issue is of crucial importance and shall be revisited in more details in my chapter dealing with the in-theatre phase of Operation Deliverance.

Needless to say, if there are analogies between the behaviour of the Canadian and Belgian contingents, there are also important differences, as we shall see. I believe that the acknowledgement by African Rights of the differences between the Canadian attitude with respect to accountability in the case of serious violations of human rights deserves to be emphasized.

There is one further similarity between the Belgian and Canadian contingents that I shall only briefly mention. It is the difficulty experienced by all military in interpreting and applying the ROE. This is one of the main reasons for the perceived leniency of the court-martial in Belgium. I have discussed this issue previously, and I will leave it for now.

CONCLUDING COMMENTS

The conclusions that can be drawn from the analysis of the quantitative and qualitative data were formulated as I proceeded with this analysis. There is no need to repeat them again in these concluding comments. I shall only underline what appear to be the main trends revealed by the data.

The main conclusion to be derived from the historical data is the disproportion between the time spent by the CAR in training and in proving itself in combat. They believed themselves to be Olympians in a time when there were no games. This disproportion between training and action was likely to generate frustration within the CAR and increase the aggressiveness of its members when sent on a mission. Due to the craving of these soldiers to prove themselves in combat, there was a significant probability that any mission in a country torn by internal conflict would have been interpreted as a straightforward military intervention implying the use of force. Segal et al. (1984) made a specific inquiry into the fitness of paratroopers for peacekeeping missions. They concluded that such troops were too single-mindedly trained for combat and too aggressive to be suited for this kind of duty. The issue is, however, much broader than the suitableness of Airborne troops for peacekeeping missions. Considered in all its scope, this issue requires a re-examination of the changing nature of the military and of their role in the constantly evolving political context we have experienced since the end of World War II and, most particularly, since the collapse of what used to be known as the Eastern block. Starting with Janowitz' classic work on the professional soldier (Janowitz, 1971), research on the military has paid growing attention to the accelerating pace of change affecting the military (Janowitz, 1977 and 1982; Janowitz and Moskos, 1975; Stahl et al., 1980; Segal, 1986; Caforia, 1988; Shields, 1993; and Dandeker, 1994). This list is far from being exhaustive; it is merely an indication of the interest of researchers in the need for reformulating the nature of the military profession in Western democracies.

As I said, it is very difficult to interpret with any degree of reliability the quantitative data, because they are not collected on a regular basis. Nevertheless, main trends do show up in the data. Generally, the ethnic composition of the CF is not representative of the ethnic make-up of the general population, the members from minorities being underrepresented. This last statement is particularly true with regard to minorities other than Aboriginal, such as African Canadians. The reverse situation seemed to be true in the CAR, where African Canadians enjoyed a greater representation than Aboriginal Canadians. However, because of the very small number of people involved, it is impossible to draw any statistically meaningful conclusions. On the whole, minorities were as much underrepresented in the CAR as elsewhere in the CF.

A key question with respect to the recruitment of persons from minorities is whether they get their fair share of promotions. It would seem they do when aggregate statistics on recruits from minorities are considered. However, when we differentiate between the various minorities we obtain the familiar picture of Asians outperforming all other minority groups. The data base that I analysed is much too narrow and incomplete to provide an explanation for this situation. The level of attrition was rather high in the Primary Reserve and relatively low in the Regular Force. Furthermore, it appeared that the decrease in personnel was the result of a deliberate policy of downsizing rather than a result of the inability of the CF to retain its personnel. This situation suggests that the CF is in a position to screen in those it wants to be part of the Forces and to weed out unsuitable applicants. However, this general conclusion did not apply fully to the CAR. Because of the requirement to maintain its manning level at 90 percent, the screening policies for being admitted into the CAR were not rigid and they were determined on an ad hoc basis. They offered no guarantee that only the best elements from the parent regiments would be sent into the CAR.

The examination of qualitative and documentary sources yielded two main conclusions. The first one is that the misbehaviour of members of the CARBG was not specific to the Canadian Forces and that many other contingents seemed to have experienced similar problems. The second conclusion, based in particular on a review of the Hewson Report and of an internal assessment of the seriousness of allegations of White supremacist penetration into the CF, is that the CF has a tendency to play down and even deny problems rather than taking energetic measures to remedy them. Both of these conclusions are to be considered as hypotheses that will be tested in the next chapters, which bear on Operation Deliverance as such.

CHAPTER THREE

Pre-Deployment

INTRODUCTION

Although I have used government and Canadian Forces (CF) documents in the two previous chapters of this report, this chapter and the two following are devoted to a greater extent to a discussion of factual issues directly related to Operation Deliverance. This discussion cannot wholly rely on documentary sources, and I must now refer to the transcripts of testimony presented before the Board of Inquiry (BOI) and before this Commission.[1]

The use of witness testimony raises several problems. First, there is the problem of the sheer weight of information which, for one person working alone, can be daunting. The BOI, courts-martial and the Commission files have respectively 26,038, 78,125 and 50,125 records. And it has to be noted in addition that the BOI computer file is not complete for, as of February 1996, the 50,125 Commission records only represent a part of its hearings on the pre-deployment phase. (More is to come on pre-deployment; the in-theatre and post-deployment phases as such have not yet been subjects of the Commission's hearings.) Obviously, my discussion of the issues will have to be greatly compressed. I will, for example, not be able to explore in detail in my written report the inconsistencies and contradictions between the different testimonies.

This last point brings me to the main problem in using testimonies, namely, the differences between answers given to the same question. These differences vary from contrasts in tone and emphasis to inconsistency and outright contradictions in the content of statements. Using a broad notion of disagreement, I can identify no less than six basic kinds of disagreement.

1. Disagreement with respect to the different *forums* where questions are asked and answered. The same witness may have a markedly

different attitude depending on whether he is testifying before a court-martial or the military BOI or before the Commission which is staffed by civilians and has a different mandate than either a court-martial or the BOI. This difference of attitude does not necessarily result in contradictions but determines various patterns or styles of answers (e.g., different degrees of evasiveness or straightforwardness).

2. Disagreement between what witnesses have actually said and the conclusions which are drawn from their answers. For example, Volume XI, Phase I of the BOI presents the findings and conclusions of the BOI's investigations. One may question whether the BOI's conclusions adequately reflect what it was told. In the same way, the verdict of a court-martial may appear to run counter to the evidence that was presented during its hearings.
3. Disagreement between the content of verbal testimony and the content of written documents which may or may not be official exhibits in a proceeding.
4. Disagreement between the statements made by different categories of witnesses, e.g., non-commissioned members of the Canadian Forces (NCMs), officers, civilian officials of the Department of National Defence (DND).
5. Disagreement between the content of individual testimonies on the same topics. This is the current meaning of inconsistency between witnesses.
6. Disagreement between the answers given by the same witness to an identical question asked at different times or by different bodies of inquiry (i.e., self-contradiction).

I have found examples of all these types of disagreement in the course of my work. In some cases, I was able to resolve the disagreement in favour of a particular version of events. In some cases, this resolution can only be achieved by reaching beyond a particular topic (e.g., racism) and considering all the evidence that may be directly or indirectly relevant to an issue. Such resolution can not be contemplated within the confines of this report.[2]

It is of crucial importance to bear in mind that most of the training of the Canadian Airborne Regiment (CAR) for Somalia took place in the months of September and October 1992. At that time Resolution 794 of the United Nations, which transformed the original Chapter VI peacekeeping mission in Somalia to a Chapter VII peacemaking under U.S. command, had not yet been approved (this Resolution is dated 4 December

1992). Hence the CAR trained for a Chapter VI UN mission (Operation Cordon), according to which it was to be deployed to the northern port of Bossasso. Later its mandate was changed to a Chapter VII mission, which is more aggressive in nature (Operation Deliverance). The place where it was to be deployed also shifted, from Bossasso to Belet Huen, which was an entirely different terrain much further to the south and far away from the sea.

The same circumstances apply to the selection and screening of the unit and the personnel to be sent to Somalia. This process was carried out entirely on the basis of an operation — Operation Cordon under Chapter VI of the UN Charter — that in the end was not the operation for which the CAR and the Canadian Airborne Regiment Battle Group (CARBG) were deployed (Operation Deliverance, under Chapter VII of the UN Charter and under the command of the U.S. Armed Forces). So, from the very inception of Operation Deliverance, there was a serious discrepancy between the selection and screening process and the training, on the one hand, and the operation that was actually carried out, on the other hand.

One general observation applies both to the issue of selection and screening and to the issue of training. Although demands that the armed forces receive more cross-cultural training in an era in which the number of peacekeeping missions is constantly growing are becoming more forceful (e.g., Fetherston, 1994: Chapter 7-8), not that much has been written on this subject. When issues such as race are discussed in the literature on the armed forces, the focus tends to be on the issue of equality of access to a military career (Knouse, 1991; and St. Pierre, 1991). Hence one must look elsewhere for information and models in cross-cultural screening and training. One of the most obvious places to look for such information is the literature of policing, where there is abundant material on the issue of ethnicity with regard to the screening of applicants and cross-cultural training (for a review of the Canadian literature, see Cryderman and Fleras, 1992). The military occasionally refer to the literature on policing (e.g., Last, 1992).

There is, however, another font of information which is still largely untapped. It is the literature on international development, where the issues of the screening and cross-cultural training of international advisors who will be sent to foreign countries is extensively canvassed.[3] Canada is not only involved in international peacekeeping, it is also heavily involved in international development. D. J. Kealey, a researcher within the Canadian International Development Agency (CIDA), has systematically developed instruments for the screening of personnel to be sent as advisors to foreign countries (Kealey, 1988 and 1990; see also Kealey and Ruben, 1983; Ruben and Kealey, 1979; Hawes and Kealey, 1980, 1981; and Torbiorn, 1982). Kealey's work is actually cited in the literature on the training of peacekeepers (Fetherston, 1994: 172-175). I shall refer to the work undertaken on screening and cross-cultural training in the field of international development. It is at least as relevant to international peacekeeping as the research on policing.

As is implied by the previous discussion, this chapter will address the issues of selection and screening and of training; I will also discuss hazing rituals and the alleged penetration of the CAR by white supremacists.

SELECTION AND SCREENING

In an article that summarizes his Ph.D. thesis on cross-cultural effectiveness, Kealey (1989: 423) writes:

> In summary, the results of this study should serve to re-establish an interest in personality traits as important in predicting behaviour overseas. This is not to say that the social skills training approach...is irrelevant to enhancing success overseas. It is to say, however, that unless an individual first possess certain personal traits and interpersonal attitudes, no amount of training will ever enable that person to effectively transfer skills and knowledge to people of another culture. (Kealey, 1989: 423)

Two points follow from his summary. The first is obvious: the selection and training of personnel are different processes, the first acting as a precondition for the success of the second. The second point is less evident: Kealey makes a distinction between personality traits and interpersonal attitudes. Personality traits are clearly distinct from social skills, which can be learned. Interpersonal attitudes, however, are less distinct from social skills, and there is at least a possibility that they might be developed by training. In other words, there is a certain amount of overlapping between selection and training in respect to the possession or acquisition of interpersonal attitudes and social skills.

Two words are used in respect to the pre-training phase, namely selection and screening. As I shall use these words, selection refers to a process of integration (a person is "selected in") and screening to a process of exclusion (a person is "screened out").

This section on selection and screening is divided into four parts: factual issues, questions of principle, comparisons and recommendations.

Factual Issues

Two main factual questions must be considered in regard to Operation Deliverance (previously Cordon). The first is whether the CAR was the right unit for this mission. The second is whether all the individual members of the CAR who were, on various grounds, judged to be unfit for this mission were in fact screened out of the operation.

I do not view it as my primary task to deal with these questions, which can only be answered through systematic inquiry by a team of investigators. I shall nevertheless raise a few issues which are connected to my own terms of reference.

The Choice of the CAR. The issue that I want to raise is not whether the CAR was too aggressive as a unit to be used in the context of Operation Cordon under Chapter VI of the UN Charter, which was the original operation for which it was chosen. The CAR was successfully involved in at least one other UN peacekeeping mission — in Cyprus. Most witnesses have testified before the BOI or the Commission that they did not believe that the CAR was too "gung ho" to be sent to Somalia; finally, the transformation of Operation Cordon into a Chapter VII operation under U.S. command — Operation Deliverance — could be interpreted as validation *ex post facto* of the choice of the CAR.[4]

Lieutenant-Colonel (L.Col.) Turner, who was at the time of Operation Deliverance a major responsible to the Commander for all operational and training matters pertaining to the Special Service Force, testified before the BOI that the selection of the CAR was based on reasons of unit morale: "While it is understood the decision to deploy the Canadian Airborne Regiment was a national one based on reasons of unit morale, the choice necessitated the reroleing from para to mechanized infantry." (BOI, Major Turner, 588; also see his CIDCFS testimony, 20: 3634).

In 1991 the CAR had trained hard for deployment in the Western Sahara in the context of Operation Python, and it was felt that serious problems of morale would arise if it were passed over for deployment in Somalia because it was not a mechanized infantry unit.[5] Like Major Turner, L.Col. Macdonald and Major Kampman also testified that training the CAR for its new role as mechanized infantry was a significant challenge (BOI, L.Col. Macdonald, 972; CIDCFS, Major Kampman, 28: 5306). All training objectives, as we shall see, could not be fully achieved, and the partial reroling of the CAR as mechanized infantry was achieved at the expense of training for other skills relevant to the mission because the time was too short.

The reason given by L.Col. Turner for the selection of the CAR instead of a mechanized infantry unit — the preservation of morale — is at least a plausible one. It is not immediately obvious why the CF would select a unit which lacked the basic technical skills — being able to operate armoured vehicles for the land transportation of troops — for its mission if it had not been seen as necessary for the preservation of the CAR's morale or for some analogous reason. To put it in more general terms, the question that is raised by the selection of the CAR for Operation Cordon, and afterwards for Operation Deliverance, is the balance to be achieved between motives internal to the CF and the competence suitable for the external requirements of a mission. My hypothesis is that internal factors took precedence over external factors in selecting the CAR to go to Somalia.[6]

The Screening of Individual Members of the CAR. There are two issues to be raised in relation to the selection and screening of the members of the CAR, who were all volunteers. The first concerns individual members who were identified by name. With regard to facts, this issue is very complex and only the Commission has the resources to clarify the situation. In particular, according to the testimony of Master Warrant Officer (MWO) Mills, who was the chief disciplinarian of the CAR's 2 Commando, several lists of individuals were drawn up either officially or unofficially, with the names of members of the CAR whose behaviour was problematic and who, in the mind of some, should not be sent to Somalia. Among the officers of 2 Commando who raised hesitations, the names of Major Seward and Captain Rainville are the most frequently mentioned in the testimonies and in documents of different nature. Both eventually went to Somalia and both were later accused and convicted of misconduct.

Several lists of NCMs have also been scrutinized either by the BOI or by the Commission: the Chairman of the latter publicly quoted such a list that contained the names of Brocklebank, Cooper, Thompson, Mercier, Matchee, Sinclair, Bibby, Ingram, Kraftscik, Smith and Leach (CIDCFS, Chairman, 23: 4326). Powers and McKay are also recurrent names. All went to Somalia, where some of them were involved in the most serious abuses of power, e.g., Matchee and Brocklebank. MWO Mills was asked whether he had any concern about the following persons: Matchee, Powers, McKay, Brocklebank, Smith, Sinclair, Leach, Murphy, Cooper, Thompson, Rainville and Seward. He had none about any of them (CIDCFS, Vol. 23: 4333). MWO Mills had his own mental list of potential troublemakers, namely, Corporal Ford, Corporal Powers, Private Gatzky and Private Johnstone. Of these four, only Corporal Powers made it to Somalia;

the three others were left behind (CIDCFS, MWO Mills, 23: 4349). LCol Mathieu eventually screened six members of 2 Commando out of the personnel to be sent to Somalia.

The more general issue raised by these lists and by the fate of the individuals mentioned in them is *the exercise of discretionary powers by commanding officers at different levels of the hierarchy.*

A striking example is provided by the action of Brigadier-General (B.Gen.) Beno. B.Gen. Beno was impatient with L.Col. Morneault's difficulties in re-establishing discipline in 2 Commando and finally replaced him with L.Col. Mathieu. B.Gen. Beno had very serious reservations about Major Seward, Major McKay and Captain Rainville (BOI, L.Col. Mathieu, 1186). He also expressed concerns about Major Seward, Captain Rainville and Sergeant-Major (S.M.) Vienneau of 3 Commando, as late as the second week of February 1993 (BOI, B.Gen. Beno, 1023). He finally wanted 25 NCMs of 2 Commando left behind in Petawawa or shifted to other posts within the CAR before deployment in Somalia. Yet, L.Col. Mathieu never took any action in connection with these officers and only left six NCMs of 2 Commando behind, before deploying it to Somalia. B.Gen. Beno did not insist on any more screening out.

It is difficult to find any method or consistency in this screening procedure: target lists are drawn several times by different persons and followed by little or no action at all.

The second issue is connected to judgments about the CAR's personnel considered as a group or in its components. Two kinds of judgments are at stake, namely a judgment of the CAR's NCMs and a judgment of its officers. Generally speaking, it must be said that these judgments are positive to the point of being self-congratulatory. I shall single out dissident voices, not merely because they are dissident but because what they point out is more precise than a blanket endorsement that everything went according to plan.

NCMs: When asked whether the Airborne was an elite unit, Captain Yuzichuk, who was its adjutant during Operation Deliverance, answered that a lot of people would say that it was (a lot actually did), but he believed that a more rigorous selection process would be needed to put the CAR on a par with special forces such as the British SAS or the U.S. Special Forces. As a special force, the Airborne could not be said to be an elite unit. However, with regard to general purpose combat capability, the CAR could be compared to the U.S. Marines, for instance (BOI, Captain Yuzichuk, 674 and 675). Captain Yuzichuk's basic criticism of the

CAR was that its members were too young and that they lacked self-discipline and maturity. He faulted the selection process for this situation:

In the Canadian Airborne Regiment, in my opinion, there is a lack of maturity in a sense that there's no real selection process to come to the Airborne Regiment. (BOI, Captain Yuzichuk, 672-673)

...I would take them from the other regiments after a certain number of years and then put them through a certain selection process as I said. Age is one. Another factor that I observed with the Airborne Regiment is people have the — there's a big esprit de corps in the Airborne Regiment; however, sometimes people get a big ego out of it. And they're what we call "bar commandos." They're not too good in the field but when they hit the local scene they're the rough, tough Airborne soldiers. (BOI, Captain Yuzichuk, 673-674)

These words fit well with the picture that is provided by the Hewson Report on the problems of discipline in the Airborne where it is urged that the parent regiments send "good soldiers" to the Airborne. They ring true with the record of bar brawls in which members of the CAR were involved in Petawawa and its vicinity. They are completely consonant with all the videotapes that I have viewed, either in connection with the beer and vomit-drenched hazing rituals to which new recruits had to submit or the videotapes shot in Somalia, in which one of the most familiar sights is an empty beer can. Finally, the absence of any strict selection process for admission into the CAR is an acknowledged fact.

Officers: One of the key witnesses in this respect is Colonel J.D. Joly, who testified both before the BOI and the Commission. Colonel Joly was Acting Deputy Chief of Staff Administration and Directorate of Infantry in Land Force Headquarters. Colonel Joly testified before the Commission that in his opinion, the CAR was not getting the best officers from its parent regiments, and particularly from the Princess Patricia's Canadian Light Infantry (PPCLI) (CIDCFS, Colonel Joly, 18: 3346). He had previously testified in the same way before the BOI:

Well, when I looked at the list of majors, I did an analysis. I'm not sure when it was. I think it was November or December [note: 1992], sir, that I looked at each of the majors in a comparative way. How long he'd been in the Regiment, how old he was, where he was on the overall infantry merit list, and what his potential was, and it became clear to me in this overview that we had, not in all cases, but in about 60 percent of the cases, I think there were about seven — seven majors involved. We had manned the Regiment with second and third string majors, and the third string ones clearly had no potential. (BOI, Colonel Joly, 1320)

The most interesting part of Colonel Joly's testimony is how he addresses why the situation was such as it was. The CAR's Commanding Officer had always been a full colonel, previous to the downsizing that occurred during the summer of 1992 (from a strength of 750 to a strength of 600). Following this downsizing, the Commanding Officer of the CAR could only be a lieutenant-colonel instead of the full colonel it had been, with dire consequences for its leadership. In contrast to full colonels, lieutenant-colonels do not participate in the Infantry Council process, which also had unfortunate consequences for the manning of the CAR.

Questions of Principle

By questions of principle, I do not mean normative issues (what ought to be done and what not) but the considerations and standards of the CF that were effective in selecting and screening personnel. There are many of these considerations and standards, and the testimonies in connection to them are often relatively contradictory.

Is the Selection Process Formalized? L.Col. Nordick was a career manager at the time of Operation Deliverance and was Director of Senior Appointments III, when he testified before the Commission. He declared that selection for UN and overseas duty was becoming a very formalized process. According to this witness, the steps in this selection process, listed in Appendix 1 of (Canadian Forces Administrative Order) CFAO-50, are as follows: a member's supervisor must say that he is trained and capable of doing the job; a medical officer must state that he is medically fit to do the job; the dental officer must guarantee that he is "dentally" fit to do the job; and the social worker must certify that the family and home situation meet the requirements (CIDCFS, L.Col. Nordick, 2: 287-288). It must be noted that only one of the four steps actually implies a judgment about the member's professional competence to do the job. We have previously seen that a very wide margin of discretion existed in making this judgment.

Furthermore, L.Col. Nordick's answer must be interpreted in the context of the testimony of two further witnesses. Ret. L.Gen. Foster who was the Commander of the CAR when it was first established in the late 1970s was asked if at any time there were specific guidelines or orders given as to the screening of officers and men for the regiment. He answered that he was not aware of any specific instructions that would say these are the screening steps (CIDCFS, L.Col. Foster, 13: 461). Major Priestman, an Infantry Career Manager, was also asked if there was not a policy in the regulations regarding what the screening process is or would be. He answered that to his knowledge there are no written regulations relating to the process of screening an individual, and particularly not in the case of selecting a major.

Judging from the screening process that took place before sending individual members of the CAR to Somalia and judging from the lack of written policy and directives for selecting an individual for membership in the CAR, it would seem that the selection and screening process was much less than formalized and was in fact very lax.

One Significant Criterion. There was a fairly high percentage of turnover of men in the CAR. Captain Walsh, who was the CAR's training officer at the regimental level, estimated that this figure was as high as 25 percent in one year (CIDCFS, Captain Walsh, 13: 2288). Similarly, Colonel Joly assessed this turnover figure as about 25-30 percent. (CIDCFS, Colonel Joly, 18: 3288). This problem was aggravated by the fact that for some of the sergeants and the junior officers, the CAR was used as a training ground: they would be recalled to their parent regiment after having reached a certain level of training (CIDCFS, Chief Warrant Officer (CWO) Cooke, 26: 4916-4917). This turnover generated a problem in view of the requirement to keep the CAR at a 90 percent manning level. In consequence, one of the significant criteria for selection into the CAR was the willingness of the person recruited to stay for at least two years, this requirement being particularly applied to officers (CIDCFS, Major Priestman, 15: 2831).

The Basis for Selection and Screening. To perform the selection or the screening of an individual, his or her superiors must rely on information. There are three sources for such information. A first source is the regimental merit list; a second is the courses reports of an individual (CIDCFS, CWO Cooke, 26: 4913-4914). The problem with these sources is that since they only provide a rating or a rank, they give minimal information about an individual. A third source of information is the written personal files of a CF member (see CIDCFS, Colonel Arp, Director of Personnel Careers, Colonel III, 11: 2156-2160; see also CIDCFS, CWO Cooke, 26: 4880-4911). The problem with these personal files was clearly emphasized by both the Chairman of the Commission and Commissioner Desbarats: these files can be very misleading, and there is a tendency to downplay the shortcomings of the members under examination.

In my experience, this is not peculiar to the CF. In my different investigations of police forces in the course of research undertaken for various commissions of inquiry, I have found that supervisors are markedly reluctant to enter damaging information in the personal files of their subordinates (see Malouf and Brodeur, 1993; and Brodeur, 1994).

Failures of Screening. Given the limitations of the bases of information from which selection and screening are conducted, it is not surprising that there are failures to screen out CAR applicants who are generally undesirable or who are unfit for peacekeeping missions.

Individuals with criminal records: L.Gen. Hewson, who conducted an inquiry into the CAR's discipline in 1985, found out that it was possible for a clever, level-headed criminal to enrol in the CF and lie about his criminal record. L.Gen. Hewson recommended that there should be an enrolment check of criminal records (CIDCFS, L.Gen. Hewson, 2: 340). Whether this recommendation was systematically applied is highly doubtful in view of the answers given by the Commission's witnesses (see in particular Colonel Houghton's testimony). More important however, it does not seem that actually having a criminal record or having behaved in such a way as to deserve a criminal record was an impediment for being selected to partake in a peacekeeping mission such as Operation Deliverance. There is one exception to this indulgence: two infantry reservists were returned to their units because they had been summarily convicted of fighting.[7]

Individuals involved in extreme right-wing activities and in hate groups: There was no general screening of individuals related to potential racism or right-wing extremist activities when they moved into the CAR (CIDCFS, Commander Jenkins, 6: 1227). Nor was there screening of far-right extremists before people were sent to Somalia (CIDCFS, Commander Jenkins, 14: 2662). Actually, Annex C of a Briefing Note for the Minister (MND), before a 1993 meeting with the Canadian Jewish Congress, identifies five members of the CAR suspected of racism and extreme right-wing activities who went to Somalia (not counting Corporal McKay).[8] On the subject of alleged racism, L.Col. Mathieu has claimed that there was representation from a wide selection of ethnic groups across Canada and that the CAR represented Canadian society very accurately with all its strengths and weaknesses (BOI, L.Col. Mathieu, 504). It is true that there were individuals from many ethnic groups within the CAR. However, according to the figures quoted in chapter 2 on the ethnic make-up of the CAR, their representation in the CAR was in no instance proportional to their presence in the general Canadian population.

Overly aggressive soldiers: There was no screening to keep overly aggressive soldiers out of the CAR (CIDCFS, Sergeant Major Jardine, 25: 1995).

Comparisons

The comparisons will be made in connection with the sources of knowledge previously identified, that is, governmental or non-governmental agencies for international development and police forces.

International Development. The research literature on international development is quite extensive, particularly here in Canada. There is a fairly long tradition, dating back to the early 1970s, of interest in selection and screening in this literature (Tucker, 1974; and Torbiorn, 1982). Kealey (1989: 416-417) developed a profile of the effective technical advisor and rated as "winners" the people rated highly effective by peers and researchers. This profile ranges over seven dimensions, but I shall only mention what seems to me the most relevant one: the "winner" demonstrates "out-of-self" orientation as indicated by control, flexibility, openness to new ideas and perseverance when assigned tasks.

Hannigan (1990: 107) reviewed the literature on traits, attitudes and skills that are related to intercultural effectiveness and their implications for cross-cultural training. The summary of his research reads thus:

> A number of traits play an important role in intercultural effectiveness. These include patience, tolerance, courtesy, persistence with flexibility, energy, self-confident maturity and self-esteem.
>
> Traits that correlate negatively with intercultural effectiveness include perfectionism, rigidity, dogmatism, ethnocentricism, dependent anxiety, task-oriented behaviour, narrow mindedness and self-centered behaviour.

Strikingly, CAR personnel lacked to a very significant degree the traits playing an important role in intercultural effectiveness; it would only be a slight exaggeration to declare that the CAR's members displayed to a significant degree all the traits that correlate negatively with intercultural effectiveness. It is unnecessary to review the traits one by one. There is however one trait — self-centred behaviour, with its various derivatives — that defines in an essential manner the CAR, which prided itself in being different from anything non-CAR and took steps, like hazing rituals displaying excessive brutality, to construct this difference. This extreme self-involvement was one of the worst attitudes that a peacekeeping unit could manifest.

In view of Kealey's and Hannigan's research, it is doubtful that an elite military unit should be sent on a peacekeeping mission, for such units are, by definition, self-centred and task-oriented (to the extent that the task's definition fits how they view themselves). Segal et al. (1984) made a convincing case that paratroopers, who train for heavy action and crave it, were not the best peacekeepers. His demonstration is all the more convincing for troops who, by reason of circumstance, could not prove themselves outside a training ground and so indulged in barroom violence to bolster their self-esteem.

Police Forces. There is not much to be said by way of comparison with the police. The police do not generally have the equivalent of the Airborne (a unit, which is theoretically "elite"), although they have specialized units (riot squads and technical units). The selection and screening process is no more stringent in police forces than in the CF; in the former, physical fitness plays a unique role.

There is however a screening process which is performed in police departments to a greater extent than in the CF. Most police acts, which fall within provincial jurisdiction with the exception of the Royal Canadian Mounted Police (RCMP) and Canadian Security Intelligence Service (CSIS), stipulate that no person can become a peace officer if that person has a criminal record. The requirements are similar in the RCMP and CSIS.

During my three years of experience with the Sûreté du Québec (SQ) external complaint review committee, it became clear that the regulation that a police officer should not have a criminal record was flexibly applied. The crucial determinant is whether, despite the fact that a police officer may have been convicted of an offence, the relationship of confidence between that police officer and the organization has been broken.[9]

This determinant only applies to police officers who have served for a certain number of years on a police force.

However, there is a mostly uncompromised screening for new applicants to a police force: no applicants are hired if they have a police record. Furthermore, in most major police forces, an applicant must undergo a battery of psychological screening tests before even being considered for admission.

Recommendations. It is difficult to make recommendations that would apply to a unit which has been disbanded. Yet there are perennial concerns about sending the right units and the right persons on a peacekeeping mission.

SUGGESTION: I have stated the hypothesis that the CAR was selected for Operation Deliverance with a view to preserving the morale of a unit, which had trained for other operations, such as Python, that were abandoned. Opinions vary as to the exact weight that was given to this factor. I suggest *that in selecting a unit for a peacekeeping mission — for any kind of mission, for that matter — more weight should be given to meeting the external needs of the mission than the internal needs of the organization.* Herman Goldstein, who is a prominent figure in the movement to reform the police, has coined the phrase "the means over the end syndrome" for a police strategy that gives more importance to solving the organization's problems than to providing the public with an external product, that is, security (Goldstein, 1979, 1987 and 1990). I believe that the means over the end syndrome has played a part, although difficult to precisely assess, in the selection of the CAR, and that this syndrome should, to all possible extent, be avoided in the future. Suggesting steps to curtail or eliminate the application of this syndrome would require an examination in depth of the decision-making process in the CF, which was not part of my terms of reference. Thus what might become a recommendation — create a partly civilian review board for decisions made in connection to peacekeeping — is only presented here as a consideration.

RECOMMENDATION II: **I recommend that the CF follow the example of some police forces and government agencies and use socio-psychological tests for screening purposes before personnel embark on a peacekeeping mission.**

In his testimony before the Commission, L.Col. Nordick expressed reservations about the use of psychological testing.

Mr. Chairman, in paragraph 9, we considered strengthening the determination of suitability of recruits. And in this the recruits tend to look at eligibility and suitability. And we felt that the suitability question could possibly be increased by the — be improved by increased psychological testing. But we determined that this was not practicable on the grounds of human rights considerations. The high potential for false-positives, that would be the unnecessarily screening out of otherwise suitable applicants due to the necessarily conservative test scores and also on grounds of cost-effectiveness. (CIDCFS, L.Col. Nordick, 2: 340)

The end of the paragraph is difficult to understand: the witness probably meant that the high potential for false-positives (e.g., persons who test as too aggressive, while they are not in reality aggressive) would result in the unnecessary screening out of suitable applicants.

While agreeing with the L.Col. that there is potential for making mistakes in using tests, I also believe that sending persons on a peacekeeping mission who are completely unsuited for this kind of mission can result in disaster — as it did in Somalia. This is why I believe that the CF could learn from other agencies using socio-psychological screening tests (such as the RCMP, CSIS, CIDA and various police forces) how to do so with the necessary flexibility.

RECOMMENDATION III: **I recommend that the disciplinary record of persons selected by their unit for participation in a peacekeeping mission should be thoroughly reviewed, particularly in connection with convictions under the *National Defence Act* (NDA) and the Criminal Code.**

As I previously said, this recommendation may be applied discriminately. Not all persons having been convicted of NDA or Criminal Code charges should necessarily be screened out of a contingent going on a peacekeeping mission. However, there are cases — when, for instance, a person has been convicted of a hate crime — where there can be a reasonable measure of confidence that proper behaviour and professional conduct cannot be counted on. In these cases, the potential troublemaker should be screened out.

RECOMMENDATION IV: **I recommend that the CF take the necessary steps to grant equality of access to all Canadian ethnic groups to the units that might be sent on peacekeeping missions.**

L.Col. Mathieu recognized in his testimony the need for the different ethno-racial groups that make up the Canadian mosaic to be represented in a peacekeeping force. There was an ethnic presence in the CAR, but it was far from representative of the true weight of ethno-racial minorities in the Canadian population. However, in making this recommendation, I would stay shy of promoting the use, on the short term, of affirmative action programs, as these are presently highly controversial and may seriously backfire.

RECOMMENDATION V: **I recommend that the CF conduct a study on the presence and role of women in the military in the context of peacekeeping missions.**

I remember vividly having predicted some 10 years ago in a police training course that the number of women in police forces would grow very significantly over the next years. This prediction generated a certain amount of incredulity. It is now standard wisdom.

I see many advantages in the presence of women, including the de-escalation of potential conflicts, in peacekeeping missions. Needless to say, there are contexts where the presence of women might not be well received. I believe that researchers should now begin to explore this issue.

TRAINING

The issue of training is generally complex. It is even more so in the context of the armed forces. First of all, the meaning of the word "training" is not as obvious as it may seem at first glance. It generally signifies the structured learning of perceptible skills. Actually, as I shall use the word, training refers to a continuum of activity that ranges all the way from physical drill to the teaching of attitudes and to the communication of knowledge, the latter activity being more often currently referred to as "education." Although there is much to be said about the potential difference between training and education, I shall use training to encompass both of these notions. This use of the word "training" is problematic in one respect that deserves to be mentioned. When the object of the training is neither the teaching of skills nor the communication of information, but the reinforcement of certain specific attitudes (e.g., tolerance, openness to diversity, an anti-racist commitment), the development of an adequate teaching program is a problem which becomes particularly difficult to solve.[10] In this chapter, I shall address the general issue of non-combat training and leave for the next chapter, which will raise the issue of racism, the specifics of the development of training programs targetting racial and ethnic prejudice and devoted to the promotion of non-racist attitudes.

It must also be kept in mind that training is accomplished through a relationship between trainers and trainees, both being responsible for the performance of their respective activities.

There are at least three different levels that must be distinguished with regard to training.

1. *Who is being trained?* This is a question which takes on a particular meaning within military organizations. Not only does formal hierarchy play a crucial role in structuring these organizations, but their personnel is divided between NCMs and officers. This fundamental divide running through military organizations determines almost everything which occurs within their context, especially training. Admittedly, officers often train with their men, but there are parts of training which are addressed only to them, with the view that they will transmit what is relevant of what they have learned to their subordinates.
2. *What is being transmitted?* In view of what I just said, this question has a double significance. It first bears on the content of the training. However, because training is in part a top down process in the CF, the question also admits of a more complex interpretation: to what extent is all necessary training transmitted to all the trainees?
3. *How successful was the training?* This is a standard question, which can be further subdivided:
   - Were the skills, attitudes and knowledge to be taught actually acquired by the trainees? It is crucial that this question be asked on the basis of a reasonable appraisal of the capability of the trainees to receive and integrate the training.
   - How relevant (useful) was the training deemed to be by the trainees?

The distinctions previously made and these questions provide the framework in which I will address the issue of training.

Training for Operation Deliverance: Divergent Perspectives

Instead of beginning with a description of what training took place before the CAR was sent to Somalia, I will start by quoting different assessments of the training received by the CAR before going. The reason for this approach is the following. As we shall see, there is a wide discrepancy in the assessment of the usefulness of the training. This discrepancy is only one illustration, albeit a dramatic one, of the systematic differences of perception with regard to the issue of training. These differences are such that they substantially affect the description, given in CF documents (such as Volume XI of the BOI), of what occurred during that training by witnesses who are officers and by witnesses who are NCMs. In other words, there is no general agreement about the content of the CAR's training for Somalia. Hence, there is no consensual answer to the question of what that training actually was. It is to show this lack of consensus — which applies to nearly all aspects of the training — that I begin with its most striking instance, its assessment by the different parties involved in it.

The View from the BOI. Annex E of the BOI's report (Volume XI, Phase 1) examines the training objectives and standards used to prepare for deployment in Somalia. The BOI's overall conclusion is formulated thus: "The Board found the Battle Group was very well-trained for its eventual task in Belet Huen because of its general purpose combat capability, the previous training for the contingency task in the Western Sahara, and the comprehensive training for Bossasso it underwent in Petawawa in the fall" (BOI, Volume XI, Phase 1, Annex E, p. 6, par. 25).

In line with this overall assessment, the BOI essentially recommends that past training practices be continued.

However, one member of the BOI, who later had his status changed to that of "special advisor," thought otherwise about the training and, particularly, about the cultural training:

Cultural differences briefing

At first glance, the briefing seems adequate. Briefers "delivered the goods" they had in "stock". But on second thought, it strikes this writer that there is not much in "stock". (Notes on the Proceedings of the Board of Inquiry on the Actions of the Battle Group of the Canadian Airborne Regiment in Somalia, by Jacques Bellemare, Solicitor, 28 July 1993, in English translation)

The comment is even clearer in the original French: what the writer says is that the persons who were responsible for the briefings gave all the information that they had (in "stock"), but that this information was thin enough [assez mince].

The Officers' View. Generally speaking, the BOI's conclusion reflects the view of the officers. For instance, the view of Major Moffat, who was G3 of the Canadian Joint Force Command (CJFS) in Somalia, when he was questioned by Major-General (M.Gen.) de Faye on the adequacy of the CAR's training was: "I'll be perfectly frank with you. I was pleasantly surprised to see how well-prepared that particular unit was for this kind of operation...They were very, in my opinion, very well prepared for the overall operation." (BOI, Major Moffat, 391).

More particularly, the officers testified to the effect that the training which they had received regarding the particular context in which they were deployed was adequate. These answers are interesting because they bear on that part of their training — cultural training — which received the least attention and might have been thought the weakest component.

For example, M.Gen. de Faye questioned Captain Nash, who was second-in-command of 3 Commando in Somalia, about the kind of training he had received pertaining to Somalian geography, politics, population, economy, etc. He answered: "I think there was a very high standard of training. Unfortunately whenever you go on a mission you can't train for everything that's going to happen because you don't know what's going to happen."(BOI, Captain Nash, 449).

Major Kyle, who was the CAR's Operation Officer during the pre-deployment phase and eventually replaced Major Seward as the officer commanding 2 Commando in Somalia, was explicitly asked by Commissioner Desbarats whether he did not think with the benefit of hindsight that more cultural training would have been helpful for Operation Deliverance:

> I will try to choose my words carefully here. Yes, in hindsight I think it would have been helpful. I don't think it was necessary, but I would like to maybe just put it in terms of recent experience and prior experience, in that I think the efforts we made and the type of training in that respect that was conducted is probably greater, if not equal, of any of the training conducted by units going overseas for training for operations in Croatia and in Bosnia in the recent years.

So if you want to compare it to something, I think we did something better than the average and, yes, it would certainly be very helpful if we had more information, but I still believe that we had sufficient to go and be effective in that respect. (CIDCFS, Major Kyle, 22: 4098)

Even if they recognize that training cannot anticipate everything that happens during deployment, and that there was scope for small — but unnecessary — improvements, both officers gave high marks to the training for Operation Deliverance. Their attitude reflects the general opinion voiced by officers, even if some individuals expressed some reservations about the training.

The NCMs' View. The view from the lower ranks is markedly different, as these quotations from the BOI demonstrate. M.Gen. de Faye is doing all the interrogation.

1. Q. - Tell me about the training that you took in Canada to get ready for the tour of duty over here?

   A. - It was useless. (BOI, Private Grant, 865)

2. Q. - Some people have suggested that Airborne training was too gung ho for the kind of work that was done over there on OP DELIVERANCE. What is your opinion of that statement?

   A. - I believe the training we did for OP DELIVERANCE was ...wasn't wrong. We did a lot of — it was not wrong, just not adequate. (BOI, Trooper Hodgson, 1092)

3. Q. - Would you characterize this training as being different from the normal regimental training that the Regiment would conduct?

   A. - No. not really...

   Q. - All right. Then bearing in mind that you were coming into this theatre here in the Horn of Africa, specifically Somalia, what specific training did you receive on Somalia as a country, the people of Somalia, the customs, things of that nature?

   A.- They brought us into a classroom and one of the supervisors came in. They'd visited Somalia and they came back with different studies saying that this is their culture, and this is what they eat, and what they drink, this is how they live, this is what they believe in, and that's about it. (BOI, Trooper Jacobs, 640)

4. Q. - Tell us about the training you took before coming to Somalia.

   A. - Before we left we learned the escort convoy, distribution of food and practiced the roadblocks. That's about all, sir.

   Q. - Did this differ from the normal regimental training?

   A. - We worked with the vehicles. It's different that what we were doing with foot training with rucksack but for me personally I arrived from 3rd Battalion, I was already used to work with the vehicles. That's about all.

   Q. - Did you receive training on Somalia, the country, the geography, and the people?

   A. - We received briefings on the people, how they live, the order of shootings... (BOI, Private Langlois, 482)

In his last answer, the witness confused his cultural training with his training on the rules of engagement (ROE), as M.Gen. de Faye noticed. This hardly speaks for the impression that his training on both topics made on him.

5. Q. - Sergeant Martin, what particular training did you receive, before you left Canada, on Somalia, the country, the geography, and the people?

   A.- There was an intense training that was given to us regarding all kinds of bugs that we could find here. There was no period, as such, given regarding racism. Because we were so in a hurry to do a mission... (BOI, Sergeant Martin, 771)

After M.Gen. de Faye explained to the witness what he meant by his question, the witness answered that there were briefings by the Commandant of the Commando on the politics of Somalia, its geography, the kind of operation they were supposed to undertake in Somalia. Again, this blurred memory does not indicate any sustained training in these matters.

6. Q.- Were there other sorts of training activities that took place that would be — I suppose you could put them into the miscellaneous category. I'm thinking of, and I'm just giving you these as an example. Oh, for example, maybe, combat first aid, tropical heat injury prevention, the geography and so on of Somalia, Geneva Conventions, any sort of miscellaneous items like that covered that you can recall?

A.- We did have some first aid stands. We do that all the time. We usually do that at least — at least once a month we do our combat first aid, where we went as far as administering IVs to each other.

Q.- Anything else come to your mind?

A.- Not right off, sir. (BOI, MWO Mills, 827)

These quotations are not in all cases strictly related to the assessment of the training received. The answers quoted also refer to the extent of the training received on Somalia, its geography, culture, politics and related topics. The picture that they present is markedly different from the BOI's overall assessment of training. First, the adequacy of this training is questioned. But more significantly perhaps, the NCMs do not appear to assess this training as being tailored to Operation Deliverance. They see little difference between it and their customary regimental training and, when queried on specialized training, they either confuse the issues (e.g., Somalian culture, first aid, bugs and ROE) or fail to identify with any degree of precision the special features of this training, with the exception of learning how to use motorized vehicles.

I have quoted these NCMs for several reasons. Despite the fact that they reflect what happened in the terrain during a tour of duty, their voices are generally drowned when official inquiries are held (for instance, the BOI interviewed six troopers, 17 non-commissioned officers and 57 officers from the CF). The officers, by contrast, have a tendency to adapt reality to the flow of papers and briefings that they received, seeking a match between the former and the latter.

This brief comparison between the conclusions of the BOI, the views of officers and of NCOs has shown that, as I originally claimed, their perspectives were different. It also shows that there was a significant difference between briefing the troops and communicating a lasting message through these briefings. Clearly, recalling the actual content of the training and assessing its usefulness are intricately linked. The less enthusiastic one is about the training, the less one recalls its content; the more one believes or claims to believe that it was in all respects adequate, the more one is able to describe it in detail. There is one last question which is raised by the NCMs' testimonies: were they receptive to the non-combat training (cultural sensitization, the laws of war, ROE and so forth) which they allegedly received and, granting the fact that this training was limited, would they have been more receptive to it if it had been more systematic? I shall address this question in some detail in a later part of this chapter.

The Content of the Training

In my description of the training I shall focus on non-combat training, which is more specifically part of my terms of reference. By non-combat training, I mean sensitization to the Somalian context (its geography, institutions, culture and politics) and training concerning the laws of war, the Geneva Convention and the handling of detainees.

As I said previously, there is some difficulty in inferring from the witnesses' testimonies what was precisely the content of this training. There are three additional difficulties in describing the content of this training, stemming from its structural traits.

First, the training was not, from one Commando to the other, horizontally uniform. As L.Col. Macdonald, who was in charge of running the main training exercise STALWART PROVIDENCE, testified, "the three Commandos are totally distinct organizations." This was confirmed before the Commission by L.Col. Turner (CIDCFS, L.Col. Turner, 20: 3623). It then followed that there was a basic training menu from which the officer in command of each Commando selected items for the training of his own men.

> The Regiment laid on an overall list of items that the Commander wanted covered, and then each Commando broke it down and the training was up to them to go and prepare their different stands, and it was up to them to do how they wanted to train for this. (BOI, MWO Mills, 823)

> The freedom really arrives in two ways. One is that the training plan is set, but there is not clear written direction to them as to exactly what training they have to carry out and, secondly, that they are given a lot of free reign because no one is observing the commandos, aside from their OCs, on a regular basis to determine how the training is going and evaluate it. (CIDCFS, Major Kyle, 21: 3866)

I presume that a commander's discretion in selecting items from the menu was not absolute in order to ensure a certain unity of training. Nevertheless, the flexibility allowed in the training of each Commando was considerable. Consequently, a fully accurate description of the content of the training would have to account for the training of each Commando.

On the basis of the sources presently available, such a description is not possible.

Second, the training was not vertically uniform. Particularly, all briefings on non-combat training were to be given to officers, who would in turn brief their men (see the CIDCFS Media Analysis memorandum by Ouafaa Douab, p. 3; the conclusion reached by Mr. Douab is fully supported by the testimonial evidence, which is to the effect that all briefings on non-combat training given to the NCMs were given to them by their officers). The officers had a wide margin of choice about what they would communicate to their men. This posed an acute problem when officers decided to keep information at their own level. According to L.Col. Macdonald, this was precisely the problem with the Commanding Officer of 2 Commando (BOI, L.Col. Macdonald, 972). The problem was serious enough for L.Col. Macdonald to recommend that the Commander of 2 Commando not follow his men to Somalia (BOI, L.Col. Macdonald, 975). This top down structure of training makes it difficult to assess precisely how much non-combat training each commando received.

Finally, not all training was performed during the period of pre-deployment. Some of it was performed in-theatre, particularly first aid exercises, crowd control (the use of pepper spray), driver training and weapons firing (BOI, Major Moffat, 2116).

The difficulties any assessment of the training poses are serious and should be borne in mind. Yet they are not so overwhelming as to make it impossible to reach conclusions on the content of the training.

General Description (Combat Training). The sources for the content of the training are documentary and testimonial. A document produced as an exhibit for the BOI is entitled *Canadian Airborne Regiment, Operation Cordon, Training Plan* (Canadian Airborne Regiment, 1992, henceforth CAR-TP); appended to this document as part of the same exhibit is a 22 September 1992 letter from B.Gen. Beno, Commander of the Special Service Force, to (L.Col.) Morneault, who was then the Commanding Officer of the CAR. This letter from B.Gen. Beno gives the CO of the CAR training direction for Operation Cordon. The Commission is, of course, in possession of many other documents.

The testimonial sources concerning training are abundant. A great number of the witnesses appearing before the Commission during its investigations into the pre-deployment phase have testified in varying degrees about the content of the training. Many of the BOI's witnesses also did. Key testimonies with respect to training are those of B.Gen. Beno, L.Col. Morneault, L.Col. Turner, L.Col. Macdonald, Major (then Captain) Kyle and Captain Walsh. All these officers were directly involved in the training of the CAR, and I have previously described their function. I base my description of the training of the CAR on all the previously mentioned sources.

As I am focussing on non-combat training, it is unnecessary to provide an extensive description of all the training that took place. Such general descriptions can be found in the testimony of L.Col. Turner and of L.Col. Morneault before the BOI and the Commission. They can also be found in the CAR-TP and other documents. Instead of providing a list of all the items covered in training, I shall present the conclusions which I reached after examining all the material on training that I found available.

1. The inescapable conclusion is that, whatever some witnesses may have claimed, non-combat training was negligible. The only kinds of non-combat training, as I use the word, mentioned by L.Col. Turner in his briefing to the BOI pertained to local customs/intelligence briefings (given only to officers), arrest and detainment procedures, incident resolution, public affairs procedures and briefings on the ROE that were yet to come. The only mention by the CAR-TP of information given about Somalia and its political situation is related to a "Family Orientation Briefing" (Summary of Regimental Level Training Activities, Item 14). The other non-combat training mentioned is the same as in L.Col. Turner's briefing. In his 22 September 1992 letter giving directives on training, B.Gen. Beno requires that the following be completed:

   > Individual training: Combat first aid, mine awareness, communications, map using, sentry duties, physical fitness, weapons training, surveillance equipment operation, gas mask drills, personal hygiene.
   >
   > Collective training: security of the base camp, security of port and airfield facilities, convoy escort, establishment and security of distribution centres, incident resolution, crowd control, rules of engagement, contingency operations, arrest and detainment procedures and public affairs procedures.

The only non-combat training mentioned in this letter is incident resolution, crowd control, rules of engagement, arrest and detainment procedures and public affairs procedures. Since B.Gen. Beno testified that there was no training in the lawful conduct of an operation and the laws of war, it must be assumed that the arrest and detainment procedures addressed other concerns than the respect of human rights (BOI, B.Gen. Beno, 269).

2. According to a majority of witnesses — particularly NCMs — the training for the Somalia operation was not markedly different from their usual regimental training (where there is no non-combat training as I have defined these words). There were actually two differences with regard to the CAR's usual training.
   - The first difference is an important one. Not being a mechanized unit, the CAR had to be "re-roled" from para-infantry to a mechanized unit and retrained in the use of armoured vehicles (BOI, L.Col. Turner, 222 and CIDCFS, L.Col. Turner, 20: 364). L.Col. Macdonald who supervised this transformation of the CAR into a mechanized unit testified that it was "a very considerable challenge" and that more time would have been needed to train the CAR's drivers (BOI, L.Col. Macdonald, 974).
   - The second difference is that Operation Cordon being a UN operation under Chapter VI of the UN Charter, the training followed UN peacekeeping guidelines in part. This is reflected both at the top and the bottom of the ranks in the testimony given by L.Col. Turner (BOI, 225 and also before the CIDCFS) and by Trooper Hodgson (BOI, 11, 1092).

With L.Col. Macdonald, I believe that the transformation of the CAR into a mechanized unit was a serious challenge. This transformation had to accommodate all the other "dismounted" training requirements of the CAR. L.Col. Macdonald believed these individual soldier skills were developed at the expense of mounted skills (BOI, Macdonald, 974). Meeting the competing needs of developing mounted skills and maintaining the individual and collective skills of the CAR left very little time for teaching non-combat skills. The part of the training that was the most sensitive to the Somalian context was health preservation and first aid training.

Non-Combat Training. There is consensus among the witnesses about some of the non-combat training activities that took place during pre-deployment. All these activities, with perhaps two exceptions, took the form of briefings given to officers, which I shall describe.

- L.Col. Morneault testified to the effect that an initial briefing was given to all ranks, where 11 to 12 pictures of Bossasso were shown. L.Col.

Morneault "tried to give the guys a sense, an idea of what they were going into" (BOI, L.Col. Morneault, 1446-1448).

- In his testimony before the BOI, Major Kyle mentioned that a critical stress team from Ottawa briefed the entire Battle Group (Major Kyle, BOI, 524). The existence of those briefings is dubious, given Major Kyle's subsequent testimony before the Commission. He stated unambiguously that a briefing by the RCMP concerning negotiating skills in crisis situations was given to a select group of officers and NCOs and not to the entire regiment (CIDCFS, Major Kyle, 21: 3896).
- There was a briefing given to all the officers of the CAR and the company sergeant-majors by a Somali national, Mr. Mohamud Hassan, who had recently immigrated to Canada.[11] This briefing was held on 18 September according to L.Col. Morneault (BOI, 1401) and portrayed different aspects of the Somali culture. Major Kyle, among many others, also refers to this briefing (BOI, Major Kyle, 524 and CIDCFS, Major Kyle, 2: 3897; see also Captain Koch, CIDCFS, 23: 4231).
- L.Col. Morneault testified before the BOI that the CAR held a family day on 26 September, which was also the day of the meeting of the Regimental Executive Committee. On this day, a lieutenant-commander from the militia, who had spent several years in Somalia, gave a briefing to the CAR's officers and sergeant-majors on northwestern Somalia (BOI, L.Col. Morneault, 1402). His name was Matt Bridden (or Bryden). Another lieutenant whose name is not mentioned in the document acted as an expert in another briefing on Somalia given on 1 October 1992. I received a copy of this briefing through the Somalia Inquiry Liaison Team (SILT): it was not given by Lieutenant Bridden who appeared as an expert who could later answer questions from the audience, but by an officer of the CAR (unnamed in the SILT document). The briefing is very general and deals with the political situation in Somalia and with its social conditions.
- An RCMP team gave an afternoon's briefing to the CAR's officers and its master warrant officers on negotiating in crisis situations, particularly when hostages are involved. This briefing was theoretical and addressed any hostage-taking situation and not what might happen in a context like Somalia. The information received would be disseminated through the ranks, if it was found to be useful. (CIDCFS, Captain Walsh, 13: 2349-2352). There is some confusion about exactly who gave this briefing. Major Kyle believes it might have been given by a team from the Metro Toronto Police Department, but admits that his recollection of who gave the briefing is vague (BOI, Kyle, 524); in his testimony

before the Commission, Major Kyle said that he believed that this team came from the RCMP (CIDCFS, 21: 3896).

- Finally, L.Col. Watkin gave a lecture on the laws of war and international convention to the CAR's officers. This lecture was not specific to Operation Deliverance, its purpose being to give general information on the laws of war.

This appears to have been all of the training (briefings, actually), given on non-combat skills. Thus, except for L.Col. Watkin's lecture on the laws of war, which was not passed on to the troops, there does not seem to have been any significant training on the laws of war, the Geneva Convention or ethics.

> When I left, there had been no formal training conducted on ethics, or the law of war, at any level. (BOI, L.Col. Morneault, 1450).

> Specifically, for this operation, it was not included in the direction on the Geneva Convention or Laws of War, specific training for this operation, that I am aware of. (BOI, B.Gen. Beno, 269)

B.Gen. Beno added that this sort of training is part of all soldiers' training at one stage in their careers. This is generally the answer given by witnesses, when they were queried on this subject by the BOI or the Commission. There are, however, witnesses who claim to have had more training on the laws of war and the Geneva Convention. Such witnesses are, for instance, Sergeant Lehman (BOI, 654) and Captain Walsh (CIDCFS, 13: 2415). Sergeant Lehman was not a member of the CAR: the training might have been different in his unit, the Royal Canadian Dragoons. Captain Walsh's reference to the Geneva Convention was made in relation to the treatment of detainees and prisoners of war. Yet in testifying before the BOI, B.Gen. Beno specifically addressed the question of the handling of prisoners and detainees, but he mentioned neither the laws of war nor the Geneva Convention (BOI, B.Gen. Beno, 1016). I shall return to Captain Walsh's testimony at the end of this section.

- There was very little training on cultural awareness or sensitivity like that given to police forces such as the RCMP (CIDCFS, Major Seward, 30: 5812). In particular, the CAR had no training regarding the kind of reception they might expect when they arrived in Somalia (CIDCFS, Major Seward, 30: 5812). Interestingly, Captain Walsh was asked what kind of reception he was expecting from the Somalis to the deployment of Canadian troops:

Q.- Did you have any idea what the people would be like in that desert area, how they would respond to the Canadian presence?

A.- Very favourably. We expected to find women and children left. The males had gone off to join clans because if they didn't — there was certainly a lot of pressure to do that. (CIDCFS, Captain Walsh, 13: 2370)

The CARBG were to be bitterly disillusioned in this respect.

- No consideration was given to the issue of the presence of white supremacists or neo-Nazis within the force that was to be sent to Somalia (CIDCFS, Captain Koch, 23: 4231).
- There was no testing for racism within the CAR, such as might have been accomplished by provoking racist responses in the course of simulating incidents that might occur in Somalia (CIDCFS, L.Col. Macdonald, 12: 5010).
- The ROE were drafted late into pre-deployment (actually, they were not drafted until deployment had begun). Hence there was no specific training about when a soldier was permitted to shoot:

Q.- What I am driving at is what training in your calendars related to a soldier being trained when to shoot and when not to shoot?

A.- There was no — as I say, there was no training that I'm aware of that presented a soldier with that type of scenario. (CIDCFS, Captain Walsh, 13: 2374)

There are two general remarks that should be added to this description. First, Captain Walsh, who was the CAR training officer, was repeatedly asked to assess how many hours of training had actually been devoted to non-combat training. He dodged the issue and never came up with anything that approximated an actual time figure (CIDCFS, Captain Walsh, 13: 2368). Second, B.Gen. Beno testified to the effect that 2 Commando's weakness was that it had trained too much for counter-insurgency and not enough for peacekeeping (BOI, B.Gen. Beno, 266). This conclusion is echoed in many other testimonies.

The *Somalia Handbook*. The *Somalia Handbook* (Canada, Department of National Defence, 1992) was distributed to all troops who were to be deployed in Somalia. It is a revealing booklet, a direct reflection of the priorities of training. In all, there are two pages devoted to the social conditions, the political situation and the economic conditions prevailing in Somalia (pp. 7-8). There are nine pages devoted to "General tips on operating in the Somali terrain." They cover the desert and the effect of the climate on personnel, and include "safe sex," first aid in the desert and the effect of climate on equipment. The Handbook ends with ominous warnings about "Dealing with the locals":

> Unlike other UN missions, the different factions in Somalia have proven to be *very* unpredictable even day-to-day. Any locals with weapons must be considered as dangerous and potentially hostile on *every* encounter.
>
> Always remember, yesterday's allies can turn on non-vigilant groups if it is in their interest and they can get away with it. This is an unfortunate aspect of trust-building in Somalia. Never let your guard down. Good luck! (Canada, Department of National Defence, 1992, p. 21, par. 94-95)

In conclusion, I repeat my initial assessment that non-combat training in the pre-deployment phase and in-theatre was limited to officer briefings; the officers were at liberty to disseminate to their men the information that they judged useful. By any count (time devoted to it, perceived significance, integration of training), non-combat training during the pre-deployment and in-theatre phases was never more than a perfunctory exercise that did not go beyond briefings to selected groups. Since this training was undertaken at a time when the CAR was still to be part of a UN peacekeeping mission under Chapter VI of its Charter, we may wonder how much more aggressive the training might have been, if it had been known that the CAR would be deployed to perform a UN Chapter VII mission, where they did not have to wear blue helmets.

An Assessment of the Training

The training received is a reflection of the ambiguity of the Canadian position on the training of peacekeepers. This ambiguity is reflected in the *1994 Defence White Paper* (Canada, 1994: 34). It is declared on the one hand that: "The Government believes that combat training...remains the best foundation for the participation of the Canadian Forces in multilateral missions."

These multilateral missions are described as situations which are just short of war and where "such training equips Canadian Forces personnel with the complete range of skills that may be needed to meet the varied demands of the unexpected situations they will encounter" (p. 34). Yet, on the same page it is declared that Canada will support and contribute to the enhancement of peacekeeping training: "Recent experiences in UN operations have confirmed the value of cultural sensitivity, international humanitarian law, and dispute resolution training prior to deployment. Such training has always formed part of the preparation for Canadian peacekeepers sent abroad; it will be further enhanced."

The White Paper goes on to add that the Government has assisted in the establishment and funding of the Lester B. Pearson Canadian International Peacekeeping Training Centre at Cornwallis, Nova Scotia, under the auspices of the Canadian Institute of Strategic Studies. Two questions are raised by these declarations. First, can both objectives — general combat training for multilateral missions and specialized training for peacekeeping missions — be pursued at the same time? Second, assuming that both objectives can be pursued at the same time, which one should have priority? The reasonable assumption is that both objectives not only can be pursued at the same time, but that they must be. Whatever may be the case, combat training is an imperative that the CF cannot renounce without giving up at the same time the very reason for its existence, the protection of Canadian national security. The only real question is the setting and calibration of priorities.

Jockel (1994: 64-65), a keen observer of Canadian practices in relation to peacekeeping, wrote that Canada nearly always dispatched highly trained, professional troops on Cold War-era peacekeeping missions and that much of Canada's peacekeeping reputation could be attributed to this fact. However, Jockel goes on to say:

> But today all that is changing. International peacekeeping operations have become not only far more numerous but also more complex and often more dangerous for peacekeepers. Canada's protected place in the roster of peacekeeping countries is being challenged as more countries, including the United States, equip their troops with blue helmets or berets. And to the worry of many observers, the Canadian Armed Forces, squeezed by personnel shortages at a time of very tight defense budgets, are no longer able to deploy only highly trained, regular force personnel on peacekeeping operations. What are the government's options? (Jockel, 1994: 64)

Jockel outlines his own options for the Government of Canada one of them being to specialize in deploying "vanguard" personnel, that is, providing personnel such as communications, ground or air transportation, engineering resources or medical services. The other option would be an emphasis within Canada, and with support from the government, on peacekeeping research and training (Jockel, 1994: 65). With foresight, Jockel (1994: 65) adds that the Canadian military, in the short run, can be expected to be "cool" toward the prospect of specializing in training and research for peacekeeping missions.

There are three important sources of ideas with regard to cross-cultural training. The first one comes from the military peacekeeping experience, as it is recorded in the research literature on peacekeeping. As I previously noted, another important source, much lesser known, stems from the experience in the training of persons to be sent on humanitarian missions in foreign, and mostly, underdeveloped countries accumulated by governmental and non-governmental organizations involved in international development. The last source is found in police experience — now considerable — in cultural sensitization. I shall briefly review what springs from these three sources.

The Peacekeeping Experience. Instead of attempting to summarize one by one the books on peacekeeping that have devoted some space to the issue of training, I will formulate four general considerations which arise from the literature on peacekeeping.

1. There is first a growing acknowledgement that peacekeeping and peacebuilding involve a broader range of skills than the traditional military skills. This was very early recognized by Rikhye et al., (1974: 279, see also p. 267 ff.), who concentrated on the fact that "even the best-trained and most competent soldiers need training in peacekeeping skills if they are to fulfill effectively their tasks as peacekeepers." What ran against the current wisdom in 1974 is now the standard position in the literature and among high-ranking officers involved in peacekeeping (Fetherston, 1994: chapters 7-9; Cox, 1993; and Canada 21 Council, 1994).
2. This acknowledgement has led to the assertion that there was an imperative need to review the training of soldiers sent on peacekeeping missions. The need is encapsulated in what is called the "Experience/ Expertise Fallacy" (Fetherston, 1994: 228). Because you have previously participated in many UN peacekeeping missions, it does not follow that you can rise to the occasion created by the increasing complexity of peacekeeping. This requires expertise based on explicit

knowledge. In Canada, it was recommended by the Standing Committee on National Defence and Veterans' Affairs that DND conduct a "complete evaluation" of peacekeeping training (Canada, House of Commons, 1993: 25). There were numerous other Canadian bodies making a similar recommendation.[12]

3. The training which is called for should focus on the learning of "softer" skills or more flexible attitudes than those which are learned through traditional military combat training (e.g., International Peace Academy, 1984: 272-274 and 392). This is basically the leitmotiv found by Hannigan (1990) in his review of the literature on peacekeeping training. There are many words used to describe these skills; some of the most current are "mediation" and "negotiation" skills (Rikhye et al., 1974: 267). A more theoretical way of drawing the line between "hard" and "soft" soldiering is to make a distinction between the "constabulary ethic," characterized by the need to use the minimum degree of force, and military professionalism, which does not require such a limitation on the use of force (Moskos, 1975 and 1976; see also Miller and Moskos, 1995).
4. Hannigan (1990) and Fetherston (1994) list several items regarding skills and attitudes needed for peacekeeping. Building on Kealey's research, Fetherston (1994: 174) reaches the following conclusion:

> The outcome of this research has direct relevance to peacekeeping. This research suggests that effectiveness is not necessarily tied to technical competence, but instead has more to do with positive and effective communication. The critical point here for peacekeeping is the need to establish a definition of effectiveness which includes technical competence but which goes beyond it.

This point is closely related to the previous one: efficient communication is the only tool through which activities such as mediation and negotiation can be achieved.

I shall now briefly review that part of Kealey's research which is directly relevant to peacekeeping training. The recurring theme in the research is that despite the call for an alternative to traditional training in the development of basic soldiering skills, this alternative has yet to be defined. Thus Fetherston (1994: 219) points to "the training gap." There seems to be a consensus on why there should be an alternative to traditional combat training. How the armed forces should proceed with such training remains a largely unanswered question.

The Experience of International Development. In reviewing this vast body of literature, I will focus on Kealey's work. Kealey, a Canadian who did most of his work for CIDA, was concerned with sending Canadian advisors to developing countries. His work is, in consequence, of particular significance for my purpose. Kealey has published numerous papers, beginning in the late 1970s. His *magnum opus* is a Ph.D. thesis which attempts to develop a model for predicting which advisors are going to have success abroad and which are not (see Kealy, 1990 for a paper summarizing his thesis; see also Kealey, 1988).

As Kealey readily admits, there is a basic continuity in his work, which builds on previous landmark articles (Ruben and Kealey, 1979; and Kealey and Ruben, 1983). One particular conclusion of Kealey's and Ruben's work was already explicit in 1979 and can be found in all the later work:

> Six of the behavioural patterns observed during training in Canada were related to effectiveness at interaction with nationals and transfer of skills. Interestingly, the highest correlation between *effectiveness* and *task-oriented role behaviour*, was inverse (r= -0.544, p = 0.05). Apparently, persons who were exceedingly task-oriented (a pattern associated with success in Western cultures and therefore often an implicit criterion for selection for cross-cultural assignment) and "came on strong" in problem-solving in interpersonal and group contexts, tended to be less effective than less consistently task-oriented persons. (Ruben and Kealey, 1979: 42, emphasis in text).

The paper further states that persons who displayed self-centred role behaviour tended to be less effective in the field.

I have already discussed the implication of this research for the screening and selection of personnel. It also has immediate implications for training, because the traits which make for successful actors in deployment in foreign countries are not all inborn; a great number of them can either be acquired or reinforced through training. For example, Kealey (1989: 416) draws the profile of the effective technical advisor (the "winner"). The first trait of this profile is the ability to demonstrate *caring* behaviour as indicated by *sensitivity* (emphasis in text) to local realities — social, political and cultural.

This kind of research has numerous implications for training. I will not emphasize the need for sensitization to local realities, which is obvious and now recognized, at least in theory. The idea that too much task-oriented training may be dysfunctional is, however, more challenging. A possible misunderstanding must be dispelled here. Kealey's and Ruben's point is certainly *not* that training should *not* be task-oriented, as such a position is self-defeating. Their point is that a fixation on the technical accomplishment of a task, at the expense of all flexibility and also at the expense of all considerations which are not exclusively technical, is bound to be counter-productive.

With regard to task orientation, the nature of CF training is complex. It is, on the one hand, not task-oriented in the sense that there is a lingering belief that training peacekeepers in basic soldiering skills will be sufficient for the successful accomplishment of their mission. It must be remembered that several witnesses testified that there was no difference between the current Airborne regimental training and the training for Operation Cordon (redefined as Operation Deliverance). In this respect, it is desirable that CF peacekeeping training be more focussed on the *kind* of task expected of it — peacekeeping and its UN variants — while preserving the necessary flexibility which is a prerequisite of success in this *kind* of undertaking.

On the other hand, as I previously argued, it is far from certain that sending a specialized regiment such as the CAR on a peacekeeping mission was the right decision. In terms of training, the CAR's was highly specific and task-oriented.

Keeping the right balance between training that by definition must be focussed on specific objectives and training that incorporates the flexibility necessary to peacekeeping missions in developing countries is a daunting task.

The Policing Experience. There is much more expertise regarding cross-cultural training and sensitization in police forces than there is within the armed forces. It was stimulated several decades ago during police confrontations with demonstrators supporting the end of segregation in the United States (Bard and Zacker, 1976; for Canada, see Boyd and Bell, 1985). For instance, there is a Canadian Centre for Police-Race Relations [Centre Canadien pour les relations interraciales de la police] which has produced guidelines for (1) recruitment and selection of minority police officers; (2) police-minority community relations and liaisons; (3) intercultural and race relations education and training for police personnel; and (4) crisis management and conflict mediation involving police services and diverse racial and ethnocultural groups (Greater Toronto Region Working Group on Policing in Multicultural, Multiracial Urban

Communities, 1993). And more guidelines will be developed by this Centre. The Canadian Association of Chiefs of Police has issued three "how to" manuals on police race relations.[13]

Needless to say, there is an important body of literature on these subjects in the field of police research. Legal training is, by definition, integral to the training of any police force, since law enforcement and the use of certain specific powers constrained by the law are related to the core of policing. Cross-cultural sensitization is either integrated into the regular training or is part of refresher training which is often given on an ad hoc basis. In most large Canadian police forces both forms of training are given. The training tends to use the most advanced pedagogical techniques (e.g., role playing with credible actor participants and critical appraisal by the group of a trainee's performance in such situations, which are recorded on video for future discussion). The regular training is followed by all police ranks; since there are no commissioned officers in police forces, everybody begins his or her career at the bottom level. Refresher courses are addressed to police of various ranks. Despite all these efforts there still appears to be scope for improvement according to the findings and recommendations of the Report of the Commission on Systemic Racism in the Ontario Criminal Justice System (Ontario, 1995).

Merely quoting this literature without referring to its origins would only foster misunderstanding about the level of difficulty in solving the problems cross-cultural training faces. The only reason why the police have more experience in trying to solve these problems is that they have had to deal with many more *public* scandals, scandals which have driven them to take action or at least to claim that they have. The seriousness of the obstacles to reforming the police in this regard can be gauged by the number of times that public inquiries and commissions have had to make the same recommendations before finally getting some of them implemented. In three Canadian provinces (plus the federal government), the following bodies formulated in their respective reports what were essentially the same recommendations regarding the screening and training of police personnel: the Task Force on Policing in Ontario (Ontario, 1974), the Pitman Report (Pitman, 1977), the Carter Report (Carter, 1979), the Gerstein Report (Ontario, 1980), the Abella Report (Canada, Commission of Inquiry on Equality in Employment, 1984), the Bellemare Report (Québec, 1988), the Lewis Task Force Report (Ontario, 1989), the Marshall Inquiry Report (Nova Scotia, 1989), the Greater Toronto Working Group Report (1990), the Leighton/Normandeau Report (Canada, Solicitor General, 1990, *A Vision of the Future of Policing in Canada: Police-Challenge*

2000), and the previously mentioned Report of the Commission on Systemic Racism in the Ontario Criminal Justice System (Ontario, 1995).

As I have suggested, a great number of the recommendations made by these bodies are repetitious and have not yet been implemented. Why is this? There are no doubt many reasons, but I will elaborate on one answer to this question. Shearing and Stenning (1980) have conducted research on police training. Their design was fairly complex and I do not have the space to do justice to it. Basically, Shearing and Stenning asked police at different times of their career how frequently they were using a particular kind of skill and whether they had been trained enough in the exercise of this skill. The results of this inquiry were unexpected. It was found that there is actually a correlation between infrequency in the use of a skill and the demand for more training. For instance, 98 percent of the sample admitted that they either never used their riot equipment or only used it once a year; yet 54 percent wanted more training in the use of their riot equipment. With respect to the use of emergency weapons (shotguns, rifles and machine guns), 94 percent of the sample said that they used them less than once a year (82 percent said that they had never used their emergency weapons in the past several years of policing experience); 56 percent answered that they wanted more or much more emphasis on this kind of training. Last example: 85.1 percent of the police respondents answered that they never applied first aid for poison to a victim, yet 44.6 percent wanted more or much more emphasis on this kind of training (Shearing and Stenning, 1980: 43-45).

This study was in part replicated by that of a master's student under my supervision (Gimenez-Salinas, 1995). Gimenez-Salinas asked a cross-sample of police from Quebec and from Catalonia, Spain which courses they found the most and least relevant in their training. Among the courses ranked as not very pertinent by the Quebec trainees were courses in criminology, on the sociology of social differences, computer science and, oddly enough, judo and karate. Among the most relevant courses were physical education, first aid, crisis intervention and probationary periods in police forces. Interestingly, the course that was judged the least pertinent by the police recruits in Spain was a course on the history, geography and language of their native Catalonia (Gimenez-Salinas, 1995, 97-98 and 128).

These answers show that the recruits' demands for training are much more determined by *their perception of their trade — policing — than by the actual requirements of the actual jobs*. Having myself taught for several years in Quebec's police academies, I can vouch for the fact that these perceptions greatly influence the level of receptivity of recruits and even of seasoned officers to crash training in a particular field. The wider the discrepancy between their perception of their function and the kind of training imparted to them, the lower the level of receptivity. Pelletier (1990) conducted an evaluation of a two-day training session for Montréal police in cultural diversity awareness. The training was, to say the least, received with very mixed feelings by the police, although it did have some positive impact. The police deeply resented the assumption that they were racially prejudiced, an assumption they believed underpinned such training (Pelletier, 1990: 105-109).

I believe that these conclusions on training and perception apply even more strongly to soldiers who are part of units such as the CAR. They view their unit as aggressive and elite, geared to forceful intervention in crisis situations that involve conflict between warring parties (as opposed, for instance, to emergency situations created by natural catastrophes). Even if there had been more time devoted to non-combat training such as cross-cultural issues and the Geneva Convention, I strongly doubt that, given the culture and tradition of the CAR, this kind of training would have met with enough receptivity to be in any significant way effective. The closer one gets to the level of the soldiers who would be involved in actual fighting, if there was any, the truer is this inference.

Recent Military Experience: The Lester B. Pearson Canadian International Peacekeeping Training Centre. For reasons that will become obvious, I have delayed until now a description of the Lester B. Pearson Canadian International Peacekeeping Centre (PCIP). The PCIP was created by the Government of Canada in April 1994 and began its activities in April 1995. It is jointly funded by DND and the Department of Foreign Affairs and International Trade (DFAIT), which invited the Canadian Institute of Strategic Studies, chaired by Dr. Alex Morrisson, who is also the chairperson of the PCIP, to establish a private and independent peacekeeping centre. The centre is using the facilities at the former CF Base Cornwallis in Nova Scotia. The two federal departments just mentioned provided $5 million as start-up money, and they are subsidizing it at the rate of $1 million each year for five years. This contribution is to be renegotiated between the PCIP and the government after the initial five years. The PCIP is presently staffed by Dr. Morrisson, who acts as chairperson and tenured professor, a director of study, two military personnel seconded from the CF and support personnel. For its training seminars, the PCIP relies on external resources recruited by contract.

The mandate of the PCIP is to provide research and education concerning peacekeeping in all its forms. The centre serves as a point of contact for peacekeeping information, and aims to foster tolerance and cross-cultural relations. It is important to stress that *the PCIP does not conduct the military training required by the CF personnel assigned to peacekeeping duties.* According to information provided by its registrar, there is in 90 percent of the seminars a contingent of eight to 10 Canadian military (in a total group that varies between 14 and 33 participants, although the total group generally numbers more that 20 participants). The PCIP has formulated the concept of the "new peacekeeping partnership," which serves as a basis for all its activities.

This is the term applied to those organizations and individuals that work together to improve the effectiveness of modern peacekeeping operations. It includes the military; civil police; government and nongovernment agencies dealing with human rights and humanitarian assistance; diplomats; the media; and organizations sponsoring development and democratization programmes. (Anderson, 1995: unpaginated)

The activities of the PCIP, so far, are reflected in its courses and seminars. As I said, these are usually attended by more than 20 participants, who come from all continents (there were nationals from at least 10 countries attending each seminar). The subjects of these courses and seminars, which generally last 10 days but can extend to a month and which are regularly offered to different audiences, include:

- Interdisciplinary Co-operation: the New Peacekeeping Partnership in Action
- Creating Common Ground: Peacekeeping Negotiation and Mediation
- Refugees and Displaced Persons: the Humanitarian Challenge
- The Maritime Dimensions of Peacekeeping
- Myths and Reality: the Legal Framework of Modern Peacekeeping
- Peacekeeping Management Command and Staff Course
- Modern Peacekeeping Logistics

I asked the management of the PCIP whether the CF was taking full advantage of the services offered by the centre. The answer was that the PCIP enjoys the full and active support of the higher levels of DND and of the CF, but that additional guidance will be required before the middle and lower ranks of the CF take full advantage of what is offered by the PCIP.

This answer is an interesting one, because it puts the finger on where the most resistance is found when institutions such as the military and the police which use force and coercion, attempt to change. One of the keener observers of the evolution of policing towards what is called "community policing" and "problem-oriented policing," both which rest on the idea of a partnership among police departments and other non-police agencies to solve problems of domestic security, has noted that the idea threatened the role of middle managers, the lieutenants and captains (Skogan, 1990: 123; see also Skogan, 1994 and 1995). It is at this level that greater resistance to innovation is found, and more particularly resistance to the idea of a shared approach to policing that allows for the input of civilian partners in the co-production of security. That there is a similar resistance to change within the CF may eventually become clearer.

Recent Military Experience: Project SHARP and AETHRP. In 1994, the Canada Communication Group, a private firm of consultants acting on behalf of DND initiated a tendering process whereby private companies were invited to submit a multimedia training program designed to eliminate personal harassment, sexual harassment and abuse of authority. This project was eventually expanded to include Awareness Education and Training for Harassment and Racism Prevention (AETHRP). I will address the issue of training to eliminate racism in the next chapter.

Recommendations

My recommendations are dictated by my findings and my terms of reference. I did not examine all the training that takes place within the CF, nor was I required to do so. My recommendations will essentially focus on non-combat training for peacekeeping, which I divided between cross-cultural training and legal and human rights training.

Before proceeding, I wish to make one general recommendation with regard to the training of the CF.

RECOMMENDATION VI: **I recommend that a basic course on peacekeeping and related notions such as peacemaking, peace enforcing, peace and nation building be integrated in the general training of all members of the CF.**

In all probability, peacekeeping will not be a side task of the CF but will play an increasing role in future operations. This should be explicitly acknowledged in the training of the CF, where the specific demands of peacekeeping should also be recognized. For example, there is an urgent need to dispel the misunderstanding about the differences between missions authorized under Chapter VI and Chapter VII of the UN Charter. Missions of the latter sort are few and not necessarily tantamount to the waging of a full-scale war. Such training should pay special attention to the needs of junior and middle rank officers.

Although elements of such training already exist at different levels of the CF, they are scattered, unco-ordinated and unsystematic. Moreover, the belief that peacekeeping training is only a derivative of regular training and does not need to be specific is widely shared by CF personnel, particularly by the junior and middle rank officers and by the NCMs.

Cross-Cultural Training and Sensitization to the Specific Context of a Particular Peacekeeping Mission

RECOMMENDATION VII: **I recommend that cross-cultural training and contextual sensitization be much more elaborate than it was for Operation Cordon (later Deliverance).**

By more elaborate training I mean the following: (1) Cross-cultural training and contextual sensitization should be mentioned in their own right in the training schedule, and specific content should be explicitly determined and described. (2) Training should not only address issues of substance (e.g., the geography, climate, history, social conditions, politics, customs and culture of the country where the CF is going to be deployed) but also issues of process (e.g., whether this country is consenting to the deployment of UN-sponsored forces or is forced to accept these troops; what kind of reception the troops are likely to receive; what kind of reaction will be generated by performing the mission successfully — gratefulness, indifference, etc.). One particularly important issue is that personnel must be acquainted with the *normative framework* that permeates a society, i.e., what rules does it follow regarding right and wrong (the most sensitive aspect of this topic is divergent perceptions of "corruption" — the hidden reef on which many self-righteous humanitarian missions have been wrecked). (3) Training should be addressed to armed forces members of all ranks, who should benefit from direct contact with their trainers instead of waiting for information to filter down the chain of command. (4) Training should be interactive to the extent possible and not consist of perfunctory and tedious briefings. (5) The trainers for this kind of education should be selected on the basis of their belief in its usefulness because trainers who do not share in this belief only communicate their own cynicism.

RECOMMENDATION VIII: **The CF should look beyond its own personnel resources and tap into the experience and knowledge of other agencies and of civilians.**

This point is of crucial importance. Canadian troops may be deployed, sometimes on rather short notice, to a variety of countries all over the world. However competent its personnel may be, the CF cannot be expected to produce instant expertise on countries like Egypt, Cambodia, Somalia, the former Yugoslavia or any South American country that may call on the UN. A network of experts and agencies, such as CIDA, the RCMP and non-governmental organizations, such as CARE, that specialize in humanitarian missions, must be established and maintained.

RECOMMENDATION IX: **A special unit should be established either in the CF or at DND whose mandate would be to extract systematically from every Canadian experience in peacekeeping those elements — lessons learned, best and worst practices — that could be used to develop a knowledge base for the preparation for future missions.**

This type of undertaking is now current in the field of policing, crime prevention, civil protection and, more generally, crisis intervention. It is important that learning from the past be transformed from an intuitive and discretionary practice into a structured process for accumulating relevant knowledge.

Legal and Human Rights Training

RECOMMENDATION X: **All ranks should receive a refresher course on all legal and human rights matters pertaining to their peacekeeping mission.**

In its content, this recommendation sounds obvious. The real difficulty is to implement it in a way that is not counter-productive. Judging from my own experience in the training of police, people of action are less receptive to training about abstract notions, such as the Geneva Convention, which are presented as true limitations on their operations in the field. It must be realized that this kind of training should not consist merely in a transfer of information: its basic aim is to formulate norms of action that should become an integral part of all mission-related behaviour. The kind of teaching in which the temptation to proselytize is present, and must be resisted, requires instructors who are both highly competent and highly committed.

As such, the instructors should enjoy the respect of trainees and should be viewed as genuinely concerned about the success of the mission. What I mean here should be obvious: bringing in a high-minded outsider with no status in the eyes of the CF members going on a mission will, at best, have no impact and will, at worst, be counter-productive. This is perhaps one case where a selected group of officers endowed with special qualities of leadership should be trained to be instructors for the rest of the group (but still eschewing the imperfect procedure of letting the training trickle down through the established chain of command). The training should be strictly tailored to the needs of the mission and every effort should be made to avoid the perception that it is formal legal education imposed to satisfy external standards. Even if all CF members are exposed to such training at one time in their career, it should not be assumed that this training has left a lasting impression. A new round of training is necessary. And finally, this training should be directed toward all ranks.

RECOMMENDATION XI: **When taking part in a mission that requires the formulation of rules of engagement, the CF should make sure that these rules are formulated early enough for their meaning to be carefully explained to all persons who will have to apply them.**

The teaching of the proper significance of the ROE when such rules apply, is one of the most crucial parts of the legal or para-legal training for a peacekeeping or a peace-enforcing mission. For Operation Deliverance, the production of these rules was under so little control as to be chaotic, with dire consequences for training. Even when personnel do not serve entirely under their command, the CF should make absolutely sure that they are not deployed without having cleared all ambiguities with regard to the ROE.

There are two final remarks to be made about training. As I previously suggested, the real challenge with non-combat training is not only in its design but also in its implementation. It should be stressed that, even if properly applied, the recommendations that I make will have a perceptible impact only in the middle and long term. The need for persistence and patience is a lesson that can be clearly drawn from the police experience.

Although I believe that there is much to be learned from the research of Kealy and others concerning the counter-productiveness of being too rigidly task-oriented, it is difficult to translate these findings into precise recommendations. It is the spirit rather than the letter of the training that should be reviewed in this case. I cannot insist too much here on the necessity for the CF to team up with other knowledgeable agencies in reviewing its training policies for peacekeeping.

HAZING RITUALS

The question of hazing rituals has been treated thus far by the Commission as a pre-deployment issue. The main witness on this question was Corporal Joseph Christopher Robin, who testified before the Commission during its pre-deployment phase of investigation. Since these hazing rituals continued to take place after the return of the CAR from Somalia, this issue is connected both to the pre- and the post-deployment phases.

I shall address the issue of hazing rituals only to the extent that it is connected to my own terms of reference. I shall ask two basic questions. The first is whether these hazing rituals provide evidence that the CAR was the kind of elite unit that should have been sent on a peacekeeping mission. The second is whether these rituals reveal the presence of racism within the CAR.

Loyalty and the Development of a Bond Among Members of the CAR

Before addressing these two questions, I wish to discuss very briefly the general issue of the creation of a bond among members of a unit that perceives itself as being elite. There are many ways to create bonds among members of such a unit. One is through the sharing of experiences related to the mission of the unit. This way of bonding is natural and normally does not generate controversy. Another way is through exceptional experiences, such as hazing rituals, which are unrelated to a unit's mission. Though justified as bonding exercises, hazing rituals are artificial in the sense that they are exceptional, and they also can be very controversial.

There is one feature of the natural way of bonding on which I want to comment. Chief Warrant Officer (CWO) Raymond described this feature clearly when he was asked by L.Col. Watkin during a hearing of the BOI what was unique to a regiment such as the CAR. Here is how he answered:

Gee, that's a good question and the Commander and I were talking about that, and I said, "I hope they don't ask me that question." Okay, as we all know the three infantry battalions or regiments supply the 031s to the Airborne Regiment. What makes a soldier with experience in the Vandoos, the Royals or the Patricia's come to the Airborne Regiment and put that all aside? It is the mystique of the maroon beret and I'm quoting Montgomery, to bring it down to our terms, these soldiers, and I'm not trying to be dramatic, challenge death by jumping out of an airplane. And that's what binds them together. They hold more loyalty, I have seen more soldiers hold more loyalty, including the gunners that wear the maroon beret, that hold more loyalty and allegiance to the maroon beret than they do to their parent regiment. And that, in fact, is because together as a group, they face death and jump out of an airplane, and it's as simple as that. Because I've gone through that, I've grown up through that. (BOI, CWO Raymond, 1010)

CWO Raymond describes himself as being representative of "the old way" (BOI, CWO Raymond, 1012). This will help me to make my point. Without wanting in any way to disparage the CAR or to minimize the dangers involved in being a paratrooper, I believe that some critical distance should be taken from the "Airborne mystique," since it may, in certain of its aspects, be profoundly self-serving and little else than an escape into nostalgia. (It is no mere coincidence that CWO Raymond quotes Field Marshall Montgomery on challenging death by jumping out of an airplane.) The plain fact is that "challenging death" by jumping out of an airplane is now a popular sport practised by many — men, women and, sometimes, persons young enough to be called children (adolescents). In saying this, I am fully conscious that paratroopers jump in conditions that are much more dangerous and demanding than do people who practise the sport of parachuting. This is undeniable. However, I accept Captain Yuzichuk's point (BOI, 5565), also made by many other witnesses, to the effect that parachuting into combat does not make soldiers into an elite unit possessing the kind of special skills which the British SAS or the U.S. Navy Seals have. It follows from this point that the CAR members had to construct artificially their own uniqueness, since this did not automatically follow from the fact that they were airborne troops. The hazing rituals were an overcompensation for the lack of genuine uniqueness.[14]

The Hazing Rituals

My main source for what occurred during the hazing rituals are videotape records. There are three such tapes. One was made during a hazing ritual that occurred in July 1992, another one in September of the same year. A last video was shot in 1994, after the CAR's return from Somalia. The fact that hazing rituals occurred both before and *after* Operation Deliverance shows that not much had changed in the most outrageous practices of the CAR, after the incidents in Somalia. All rituals involved only members of 1 Commando, whose parent regiment is the Royal 22nd Regiment (the Vandoos). This regiment is composed mostly of French-speaking Canadians. The language heard in the videotapes is French.

Although such a description is lacking in good taste, it is important to report what can be seen on the three videotapes. There are basically three kinds of activities.

The first kind of activity is the hallmark of all hazing rituals and consists of testing the persons being initiated for their physical stamina. The recruits are made to do push-ups, jump blindfolded from a platform held a few inches from the ground and moved back and forth to simulate the experience of being airborne, and like exercises. One favoured ordeal is to have a recruit put his forehead to the end of a three-foot broomstick and then run circles around the broomstick, while keeping contact between his forehead and the broomstick's end (a few circles makes a person fall from dizziness). All this seems innocent enough, even if it is physically demanding. However, there is one exercise that borders on physical torture. Recruits have to hold on to a piece of metal, which is electrified, for approximately five to 10 seconds.[15]

The purpose of a second kind of activity is to humiliate and degrade the recruits. The recruits, naked down to their waist, first have their heads shaved. Then they are compelled to drink alcohol (mostly beer, bottles being pushed into their mouth) until they cannot stand on their feet (the broomstick exercise gets ugly when performed by persons who are completely drunk). They are also made to crawl and to walk on all fours, tied to a leash and called a dog's name ("Fido"; the latter treatment was reserved for the only Black recruit in 1 Commando). Some are made to do push-ups over excrement, their chests being in contact with it every time their bodies come down.

There is one particular ordeal that reflects what seems to be an obsession in all the hazing rituals that I watched on tapes. This obsession is having recruits vomit, and the "ancients" go to great lengths to achieve this result. In the 1992 rituals, recruits are repeatedly screamed at to vomit ("dégueule!", "dégueule!", which is the French equivalent of "puke!"). An infallible way of bringing this about was to press the white part of a piece of bread into a small ball, urinate over it, and then have every recruit chew on it until they spit it out with their vomit. The bread ball is caught by one of the ancients, who has made sure that the recruit vomited over it. The dissolving breadball is then pressed again into a solid substance and passed on to the next recruit who must chew on it until he also vomits; every recruit goes through the ordeal, the breadball [la boulette] becoming more disgusting as more people have vomited and urinated over it.[16]

The third kind of activity is sexual and mostly seen on the 1992 videotape. A few soldiers are seen masturbating in public or simulating masturbation. There are also simulations of sodomy and fellatio. Soldiers in drag are also seen.

There are two interesting features of these videos. At the end of the 1994 video, an officer — I believe he is wearing the bars of a major — is clearly seen chatting with members of the Commando while the hazing is going on.[17] This cannot be interpreted necessarily as meaning that officers condoned such hazing rituals, as the officer may have attended on his own and not as an officer. However, it certainly means that officers were not unaware that these rituals were taking place. Actually, Sergeant Gravelle of 1 Commando, who was present during the 1992 hazing, testified to the effect that it was impossible for the officers of this Commando to ignore that the hazing ritual was taking place (CIDCFS, Vol. 6: 1113). He furthermore testified that the permission to hold the party in the bachelors' barracks had to be obtained from officers (CIDCFS, Sergeant Gravelle, 6: 1125). Second the "Rebel flag," which was the insignia of 2 Commando, is seen displayed at the end of the 1994 video.[18] This shows that there may have been more communication among the three Commandos than is acknowledged. It also makes one suspect that such hazing rituals occurred in all the Commandos of the CAR.

Finally, it is interesting to note how the hazing ritual is described by Sergeant Gravelle, who attended it: it appeared to him as "youngsters who apparently wanted to enhance the sense of their own worth [des jeunes qui paraissaient à se valoriser entre eux-autres] (translation: kids who seemed to be proving themselves) (CIDCFS, 6: 1108 and 1117).[19]

What are we to make of these hazing rituals? In my selective review of the literature on rituals (among many titles, Eliade, 1965; Pistone, 1992; Mariel, 1974; Jamin, 1977; Young, 1965; Lafontaine, 1985; Lepper, 1932; Leemon, 1972; Lane, 1981), I have seen references to aspects of the CAR's hazing rituals (for instance, putting unsavoury mixtures into the mouth and having physical stamina tested). Yet, I have not found anything that was as degrading as what occurred during these CAR rituals. However, since such rituals are not well known (and researchers are reluctant to describe the obscene), the CAR's rituals may not be unique.

Whatever may be their frequency in other units, it is difficult to rationalize the CAR's hazing ritual as either a healthy or even an effective process of bonding among members of an elite unit. The kind of behaviour displayed in these rituals was depicted, albeit less excessively, by the film director Luchino Visconti in a celebrated picture entitled *The Damned*. His explicit purpose was to puncture the Nazi mythology of the ironhearted but idealistically pure storm troopers by showing them as drunken degenerates when they gathered together for "manly" celebrations. Visconti succeeded in his endeavour, and no one seeing his film would interpret these alcoholic gatherings as bonding exercises for elite warriors. In the year following the disbandment of the CAR, the French CBC (la Société Radio-Canada, in its news program entitled Le point) did a lengthy piece of reporting in Valcartier, where the Vandoos are based. It was obvious that the soldiers who appeared in the program were devastated by the broadcasting of the hazing rituals. What everyone seemed to agree on was that what occurred during these rituals was incompatible with military honour, no matter how it was defined (men of honour do not chew on human excrement and vomit). Thus it is difficult, if not impossible, to rationalize these rituals as the building blocks of an elite identity, as the word "elite" is understood.[20]

It is interesting in this regard to remember Sergeant Gravelle's description of how he perceived the hazing ritual (in translation: "kids who seemed to be proving themselves", CIDCFS, 6: 1108 and 1117). What resulted from these hazing rituals was a singular inversion of values: what is degrading becomes uplifting; toughness is paradoxically equated with losing all sense of pride and being willing to be abused as a "slave."[21] Such rites of passage are not meant to build character but to break it.

In the sections devoted to screening and selection, I referred to research that claimed that self-centredness was a personality trait associated with failure in cross-cultural postings. What the CAR's hazing rituals displayed went far beyond self-centred behaviour; they warped soldiers with a sham identity enforced through personal humiliation. Such rituals qualify as the worst prescriptions for the making of humanitarians and effective peacekeepers. A parallel can be drawn here to the sad truth that beaten children tend to become beating adults.

RECOMMENDATION XII: **All rituals of initiation allowed in the CF should be supervised by officers who would be accountable for any behaviour incompatible with military law and professional ethics.**

RECOMMENDATION XIII: **To the extent that CF authorities believe that initiation rites and similar activities are beneficial to the military esprit de corps and should therefore be allowed, they should write policy guidelines governing the conduct of such ceremonies.**

Hazing Rituals and Racism

As I previously said, my main source of information for what occurred during the hazing rituals that took place at the Airborne base in Petawawa in 1992 and 1994 are two videotapes presumably made by members of 1 Commando. In the first of these videos, Trooper (now Corporal) Joseph Christopher Robin[22] is the object of much hazing. He was the only Black member of 1 Commando taking part in the ritual. There is no member from a Canadian racial-ethnic minority who can be seen in the 1994 video. This, however, as I shall claim, does not make this second video any less relevant to our discussion.

Corporal Robin testified before the CIDCFS on 12 October 1995. His testimony is of course of prime importance in interpreting what is seen on the 1992 video of the hazing ritual. Sergeant Gravelle and Captain Collin also testified concerning the 1992 hazing ritual. I have already quoted the sergeant's interpretation of rituals such as these; for the rest, he does not remember anything specific that happened during the initiation of July 1992. Although he was rumoured to have been present, Captain Collin claims not to have been at the initiation, nor to have seen from a distance what was happening (CIDCFS, Captain Collin, 6: 1136-1137). I will first report what is seen on the video in respect to Corporal Robin, then provide an analysis of his testimony and finally, I will draw my own conclusions.

What Is Seen on the 1992 Video. In my previous description of the content of the 1992 video, I omitted certain details to avoid repetition in this section on racism. Although it is difficult to count them, more than 20 members of 1 Commando (CAR) submitted to hazing. In the course of the ritual, Trooper Robin had to endure the following, in addition to what everyone else was submitted to:

- The letters KKK were written on his back preceded by the shape of a heart (As in "I [heart shape] NY"; in this case "I [heart shape] the KKK").
- He was tied to a tree, showered with white flour, and taunted that he, following the supposed example of the singer Michael Jackson, might change the colour of his skin.

- He was called "nègre" (nigger) several times during the hazing ritual.
- At one point the following dialogue took place between two white participants in the hazing. They are talking about Trooper Robin (then tied to a tree): "Qu'est ce que tu pense des Nègres? Moi, j'ai rien contre les races inférieures, mais elles ont pas place dans le Commando 1." (Translation: Q. What do you think about Niggers? A. I've got nothing against inferior races, but they have no place in 1 Commando.)
- He had to walk on all fours, pulled by a leash held by his "master," who called him "Fido."
- When he was on all fours, Trooper Robin was led between two white participants, one simulating anal intercourse from behind him and the other one standing erect, his pelvis at the height of his face.
- He was urinated on.[23]

In all respects, except one, Trooper Robin is the only one to be submitted to such maltreatment. He testified to the effect that one other person was obliged to walk on all fours, but he could not identify who that person was (CIDCFS, 6: 1037). On the video, he is the only one seen in this position. In the 1994 video (in which there are no members from a racial-ethnic minority), no one is obliged to walk on all fours. The one contentious issue is whether he is the only person who was urinated on; I do not believe that he was.

Corporal Robin's Testimony. Corporal Robin was submitted to intense cross-examination when he testified in French before the CIDCFS. As the Chairman of the Commission noted, his memory was highly selective. In the course of a testimony lasting approximately one hour and a half, he said that he could not remember what happened during the hazing ritual 21 times and claimed that he had no idea ["aucune idée"] how to answer a question nine times.[24] He seemed at first forgetful of anything that might incriminate members of 1 Commando. The only thing that he remembered about the hazing ritual was having to perform push-ups and run around the tip of a broomstick, as I previously described.[25] However, once having been compelled to admit that he was made to walk on his knees, he remembered without any help that the persons pulling his leash were considerate enough not to have him walk over broken glass (CIDCFS, Corporal Robin, 6: 1052). Corporal Robin's hesitations only show the severe pressure under which he was testifying. However, I believe that despite the blanks and inconsistencies, his testimony is revealing about what he experienced in July 1992. It must be said that in being reluctant to testify against the members of his unit, Corporal Robin was following one of the most basic rules of survival in all organizations that use force. All persons familiar with public inquiries into policing organizations are familiar with this fact. In this respect, Corporal Robin does not deserve our scorn bur our understanding.[26]

Corporal Robin admitted several times during one of his cross-examinations that he was intent on protecting the reputation of his former Commando in the CAR (CIDCFS, 6: 1074-1075, 1095). He began his testimony by saying that he had never experienced racial discrimination in the CAR, neither from his fellow NCMs (CIDCFS, 6: 1045) nor from the officers (CIDCFS, 6: 1066). However, under more questioning, he qualified his initial answer and said that he was familiar with racial discrimination. He was born in Thetford Mines (Quebec), a Black in a white society, and things were no different in the CF (CIDCFS, 6: 1045-1046). There were racial incidents [accrochages] in civilian life as there were in his military life. The latter were not distinctive and on the whole no worse than what he had previously endured.

He initially viewed the initiation ceremony for what he believed it was: a hazing ritual limited to a tough but short period of time (two or three hours) which he, like his comrades, had to survive; racial harassment was, as it were, part of the game (CIDCFS, 6: 1050-1054). There is, however, something that stood out clearly in Corporal Robin's answers under cross-examination: he was always conscious of the fact that having KKK written on his back and talk about inferior races and similar comments were an expression of racial discrimination (CIDCFS, 6: 1062, 1064). To be accepted, he was ready to tough it out (CIDCFS, 6: 1050). Apparently on the condition that his racial humiliation not be made public, he was ready to make no fuss about it. When he was shown on television with the letters KKK on his back, he admitted in an interview conducted in English with the military police that he was "disappointed" (CIDCFS, 6: 1059). He reiterated before the Commission that he disliked seeing himself on a news broadcast being subjected to a racist action (CIDCFS, 6: 1062). He also admitted, poignantly enough, that he had discussed the content of the video that was shown on TV with his parents and that they were shocked (CIDCFS, 6: 1997).

Was the Maltreatment Experienced by Corporal Robin an Expression of Racism? The answer to this question flows unambiguously out of Corporal Robin's testimony, despite his efforts to protect his unit. It can only be yes. This can be demonstrated with almost syllogistic rigour.

Corporal Robin admits to being no stranger to racial discrimination in civilian society. He then states that what he experienced in the CAR was no different. But if it were no different, then it was the same. And if it were the same, it follows that he experienced racial discrimination in the CF and, particularly in the CAR, just as he had in civilian life.

Why is it, it might be asked, that Corporal Robin and many others like him, believe that they are not victims of racism in a particular environment, if the discrimination there merely proves to be no different from the discrimination they normally experience elsewhere? Part of the answer is provided by the formulation of the question. Within a prejudiced society, being discriminated against becomes part of the normal expectation of the victims of racial prejudice. When they move to a different environment which lives down to their fears, they tend to view this environment as corresponding to their usual expectation and hence as being normal. There is however a crucial difference between the two perceptions of normality. Something may be seen as normal because it does not deviate from an established pattern; something may also be said to be normal because it lives up to a moral standard. Racial prejudice may be current; that does not make it right.

There is another answer to the question formulated above. We tend to confuse descriptions with explanations. In a descriptive sense, admitting that a particular environment is no more racist than another one is tantamount to a recognition that it is, in fact, racist. The fact that there is, say, a certain amount of racial discrimination within the armed forces might be explained by claiming that such prejudice merely reflects what exists in the general context of society (from which it would follow that it is less the armed forces than society in general that should be faulted for racial discrimination). This explanation carries a certain amount of conviction and might guide us as to where we should start remedying racism. However, it cannot be used to deny flatly that there is, in reality, any racial discrimination within the armed forces. Describing a situation, explaining it and proposing remedies are different undertakings too easily confused. One ends up explaining away everything and remedying nothing.

I conclude this analysis of Corporal Robin's testimony with two quick comments. First, following the hazing that he went through in 1992 and his tour of duty in Somalia, Corporal Robin should normally have stopped being a new guy and been part of the old boys ("ancients") of the regiment (he was actually promoted to the rank of Corporal). He should then have been himself in the position of hazing the newcomers to the regiment at some later ritual. Is it conceivable that he could have done to a new white recruit everything that was done to him. Could he have walked this new recruit like a dog on a leash, called him a Pepsi frog, tied him to a tree with "I (heart shape) black power" written on his back, covered him with some blackening substance so that he resembled a re-embodied Al Jolson, and simulated anal intercourse with the new man down on all fours? I confess having difficulty conceiving this happening. In the 1994 hazing video, Corporal Robin is nowhere to be seen among the abusing "ancients."

Finally, if the word "elite" has any sense, is it justifiable to expect, from a unit pretending to such a title, moral standards which are as low as those prevailing in civilian society? This last question is particularly resonant for a unit about to be sent on a peacekeeping mission to an African country, and I take it up in the next chapter.

THE INFILTRATION OF THE CAR BY MEMBERS OF THE EXTREME RIGHT

I will deal very briefly with this issue in this chapter. The alleged penetration of the CAR by individuals with leanings toward the white supremacist movement began several years before the CAR was sent to Somalia. It can thus be hypothesized that the presence of right-wing elements in the CAR may have influenced the CAR's behaviour in Somalia. In this light, the issue of the infiltration of the CAR by members of the extreme right can be seen as a pre-deployment issue.

When it is treated as such, there is actually very little information gathered so far on this question. As we saw, there was no screening out of white supremacists before Operation Deliverance, as their presence was not perceived at that time as an issue.

When they are queried on the meaning of the Rebel flag used by 2 Commando, all witnesses, with one exception, either fail to see its significance other than as a rallying point for 2 Commando or perceive it as a disciplinary problem. For instance, Trooper Robin testified that he learned that the Rebel flag was used by the U.S. Confederates when he watched a television series ("The Dukes of Hazard"; see CIDCFS, Trooper Robin, 6: 1084). The NCOs charged with enforcing the discipline (for instance, MWO Mills) and officers who commanded the CAR, such as Colonel Holmes, L.Col. Morneault and L.Col. Mathieu, viewed the flying of the Rebel flag as a gesture of defiance. Its displaying was banned by Colonel Holmes. With one possible exception, all witnesses — including Trooper Robin, Sergeant-Major (S.M.) Jardine, majors Kyle and Seward, and the former commanders of the CAR who testified — denied that it had a racist significance. According to the respective testimonies of Colonel Holmes, Colonel (ret.) Joly and Major Kyle, the Rebel flag was used as an insignia for PPCLI's hockey team in Germany, which was called "The Rebels." Colonel Joly (CIDCFS, 18: 3355) claimed all PPCLI sport teams were christened "The Rebels." It came to be adopted by 2 Commando as their insignia after a training period with U.S. troops either at Fort Bragg (CIDCFS, MWO Mills, 23: 4365) or in some other U.S. base. In a video made during the CAR's tour of duty in Somalia, a Rebel flag is shown to be present in the gear of one soldier. It is safe to assume that he was not the only one to have the forbidden insignia.

The one possible exception to the chorus of like-saying witnesses is Trooper Grant, who testified before the BOI. His testimony in this regard deserves to be quoted, because it is the most explicit.

Q.- LCOL WATKIN: Is your — what is your understanding of the basis for the rebel flag? Is it racist supremist or is it something else?

A.- PTE GRANT: No, it was a flag that was picked, like...they tried to make our own flag. They wanted ...what the guys all got together, big 2, folk songs, everything else, they shot it down. So the guys picked the rebel flag, I mean they like that flag. Then they got kind of carried away I guess for a couple of years, I am not aware of how carried away they got, but a lot of guys that I met who are ex-2 Commando, had the webs on the elbow, and the screwdrivers on the forearm and that's ... if you're not aware are like white supremist type tattoos, and when I was going there and I met a couple of these people and yeah, they were racist, I mean, if you ask them, they would say, yeah, they hate niggers. Well that's fine, that's their own opinion, that's how they value the rebel flag, that's not me.

Q.- LCOL WATKINS: Is this prevalent in the Commando?

A.- PTE GRANT: No. It was.

Q.- LCOL WATKIN: When was it?

A. PTE GRANT: This is a guess, because I don't know, because I haven't been there. I'll say maybe about eight years ago. There was probably a few people who were dominant in the commando, and had that opinion. (BOI, Private Grant, 884)

This testimony is highly interesting because Trooper Grant is the only witness, whose testimony I have so far read, who is candid about racism within the CAR. Even he hesitates to make a narrow connection between the Rebel flag and racism. As he says, the flag meant different things to different persons and its meaning also varied over time, depending on how numerous and dominant white supremacists were (or supremacists in the CAR).

As I have noted, Trooper Grant is the only witness to be forthcoming on the presence of White supremacists within the CAR. When he testified before the BOI, he estimated that there were approximately 11 to 12 racist right wingers in the CAR in the late 1980s, half of them now gone (BOI, Private Grant, 1146-1147). His testimony is open to criticism in connection to his estimation, because as he himself admitted, he was not in the regiment when the greatest number of White supremacists were there. Private Grant is also the only witness who seems to know about White supremacist tattoos (the web and the screwdriver). The sole witness who referred to racist tattoos before the Commission was S.M. Jardine (CIDCFS, S.M. Jardine, 25: 4693). These were apparently swastikas, the meaning of which is well known.

In sum, the sources for estimating the extent of the presence of members of the extreme right in the CAR are few with respect to the testimony of witnesses who took part in the operation, Private Grant being the main — and almost the only — such source. Furthermore, there is no extensive documentation on the penetration of the extreme right produced before 1993. For example, the Hewson Report does not raise the issue of the penetration of White supremacists. There was one document dated 17 December 1991, which consists of a memorandum relating to neo-Nazi organizations and skinheads, to which there is attached another document called SIROS DND report and a confidential list of 23 individuals alleged to have some kind of affiliation with right-wing extremist groups. However, in the wake of the incidents that occurred in Somalia and of the courts-martial, this question became a major public issue. The report of the Security Intelligence Review Committee (SIRC, 1994) on the Heritage Front affair raised the spectre of a deep penetration by persons affiliated with neo-Nazi and White supremacist organizations. The SIROS program was expanded, and more attention was paid to this issue by the CF, in part in response to what happened in Somalia.

The examination of these documents and of the general answer given by the CF to the problem of the alleged penetration of right-wing extremism within the armed forces raises issues reaching much further than what happened in the pre-deployment phase. They involve what happened in-theatre and perhaps more important, they raise questions of CF accountability and transparency in the post-deployment phase. They also imply an assessment of the adequacy of the response to racism within its ranks as well as related problems. This is why I believe it is better to treat this whole issue of the penetration of right-wing extremism as a post-deployment issue, granted that I may have to draw conclusions that apply in retrospect to the pre-deployment and the in-theatre phase.[27]

CHAPTER FOUR

In Theatre

INTRODUCTION

In this chapter, I will address two issues. The first is racial prejudice and the second is the use of force. In addressing racial prejudice, I will look for indications of the influence of right-wing extremism in theatre but will postpone the definitive discussion of this question until my final chapter. Racial prejudice will be treated as an attitude of members of the Canadian Airborne Regiment (CAR) potentially directed against other members of the Canadian Airborne Regiment Battle Group (CARBG); it will also be examined as attitude and behaviour directed against Somalis. Our review of the use of force will include all incidents that were marked by the use of force, with a specific attempt to account for the context in which these incidents took place. To the limited extent permitted by presently available information, I will focus part of my discussion on the policing duties which the CAR had to assume and on the question of whether, in performing these activities, it observed its rules of engagement (ROE).

The analysis of two issues raises methodological problems,[1] particularly in connection with the issue of racial prejudice. I have already mentioned in this regard that any intimation of racism in any form is bound to be forcefully denied by almost everyone (unless there is an explicit profession of racism as is the case with White supremacist groups). The consequence of this denial is that any conclusion about the presence of racism in a given organization is usually made on the basis of the implications of what is observed, heard or read. It is a rare occasion when such a conclusion can be reached on the basis of a direct observation or on the hearing or the reading of an admission. Even in a case where the presence of racism could be directly asserted, as in the case of the videotape showing Corporal Robin's ordeal, there is always scope for contention since the victim himself denies having been the target of racism.

The second problem is more difficult to describe. I will present it through an illustration. Shidane Arone was beaten to death, mainly by two people. One of these was Master Corporal Matchee and the other was Trooper Kyle Brown.[2] Master Corporal Matchee is Aboriginal; Private Kyle Brown has testified that he was partly Aboriginal (he may have been labelled as Aboriginal by members of the CAR). Hence the question: is it not possible that because they were part of a small minority within a racially prejudiced outfit, both Matchee and Brown tried to gain acceptance by being tougher on a Somali prisoner? (This prisoner would have been part of a racial-ethnic minority, if detained in Canada; in Somalia, he was of course only detained in a Canadian compound).

It is very difficult to answer this question. The difficulty has to do with the limitations of reflecting on a single incident. On the one hand, there is no general conclusion that can be drawn from this single incident (unless it is widely corroborated, no theory can be built on the thin basis that once upon a time two Aboriginal soldiers from a nearly all-white Canadian regiment beat a Somali national to his death; this may be good material for a play but not for the social sciences). On the other hand, one might try to apply to this case a theory already in existence, which was built from similar cases. The problem with this procedure can be formulated in a single question: how similar were the original cases on which such a theory was elaborated? Were these cases connected with persons from one minority being "tougher" on people from their own ethnic group or did these cases involve cross-minority relations (e.g., a Black person in an all-White unit being harder on Aboriginal soldiers or the reverse)? These questions become ever more intricate, nor is it correct to assume that there is a huge amount of research that incontrovertibly fits the case at hand. As we shall see, unfortunately there is not.

RACIAL PREJUDICE

This section is divided into four parts. First, I describe the incidents that may be interpreted as being motivated, wholly or in part, by racial prejudice. This description contains all the publicly known or Canadian Forces (CF)-recorded incidents where armed force was used. Second, I present the perception of these incidents by those who were directly or indirectly involved in them. I then propose my interpretation of these incidents in the light of present research. I finally make recommendations for action to avoid the repetition of these incidents.

The Incidents

Any incident that implied the use of force by the CAR in Somalia may be seen hypothetically as motivated by racial prejudice. The reasoning would be as follows: as the CAR may have been racially prejudiced, every recorded action that it took should be examined from the perspective of its possible motivation by such prejudice. I will not adopt this approach, as I find implausible the notion that every action taken by the CAR might have been motivated by racial prejudice.

This having been said, I shall provide a brief description of the incidents involving either the use of force, verbal reference to the use of force or alleged violations of the law, such as theft. Our sources are (1) the Canadian Forces Board of Inquiry's (BOI's) investigations, (2) a synopsis of in-theatre incidents and the investigations of these incidents prepared by Mr. Robert LeBlond (CF document not publicly available), (3) press reports and (4) the recent book by Corporal Michel Purnelle.[3] There is one last source which is not devoted in itself to the record of incidents, but which presents an informative picture of life in Somalia during Operation Deliverance; this source consists of several videotapes shot in Belet Huen by a crew of Canadian soldiers. Sources (1), (2) and (3) will be used for this section; source (4) and the videotapes will be used in the next section entitled "the view from the field."

It is imperative to bear in mind that my enumeration of incidents that occurred in Somalia is bound to be incomplete, since at the time of writing this last part of my report, the Commission of Inquiry into the Deployment of the Canadian Forces to Somalia (CIDCFS) had not begun its hearings on the "in-theatre" phase of Operation Deliverance.

Incidents Mentioned by the BOI Report and in the Press. I will start with the incidents involving the use of force that were mentioned in the BOI Report and in the press. Several members of the CF faced charges before courts-martial for these incidents and were, in a number of cases, found guilty.

- 17 February 1993: One or several members of the CARBG Mortar Platoon fired into a rock-throwing rioting crowd at a Bailey bridge, killing one Somali and wounding two others. The BOI's terms of reference precluded it from examining the incident, because it was the subject of a separate investigation (BOI, Phase I, Volume XI, Annex A, par. 34)

- 4 March 1993: Two Somalis trying to infiltrate the ENGR and the MEDEVAC HEL compound were shot. One was shot dead and the other was wounded. This is the incident that was described by medical officer Armstrong as murder. What he meant is that the two Somali casualties were part of a set-up. This version is confirmed by Purnelle (1996: 166-167). Purnelle tells how a truck carrying food rations was left partly unloaded as bait. Two Somalis fell into this trap set by the CAR's Captain Rainville.[4] One was killed and the other wounded. Major Seward declared several times after the death of Shidane Arone that he issued his infamous order to abuse captured Somalis trying to infiltrate Commando's compound in Belet Huen in order to intimidate potential thieves without having recourse to such traps.
- 16 March 1993: Arrest and death in detention of a Somali national, named Shidane Arone (BOI, Phase I, Volume 11, Annex A, par. 38). This is the event that eventually led to the Somalia inquiry. It is important to stress that in the statement that he made to the military police, Master Corporal (MCpl) Campbell declared that he had heard on other occasions the kind of howls Shidane Arone gave voice to while he was being tortured. This is a revelation that only the CIDCFS can follow in its hearings and its investigations.
- 17 March 1993: A section of troops of 2 Commando was assigned to escort some trucks to the International Committee of the Red Cross (ICRC) compound. When they arrived, a disturbance involving some Somalis and the Somali guards at the compound was taking place. One of the guards (not wearing a uniform but possibly displaying an ID badge) was firing warning shots. As he wrestled with one of the Somalis, the guard's weapon discharged into the ground near a 2 Commando corporal. The latter fired at the guard and killed him (this description is adapted from that of LeBlond).
- 19 March 1993: Master Corporal Matchee (believed to be responsible for the beating death of Shidane Arone) was arrested. Later he was found hanging unconscious in his cell. At the time of the writing of this report, he has not recovered from the brain injuries suffered in his suicide attempt.
- 3 May 1993: Master Corporal Smith accidently discharged his weapon and hit Corporal Abel in the forehead, killing him instantly. Charged with criminal negligence, he pleaded guilty to negligent performance of duty and was sentenced to four months in jail.

This listing is by no means inclusive and will be complemented by the findings of the CIDCFS. At the time of writing this report (August 1996), the CIFCDS has not begun its hearings on the operations conducted in-theatre. As I previously said, there are other sources for the incidents that happened or are alleged to have happened during Operation Deliverance. The Montréal daily *Le Devoir* in June 1996 published photographs of young rope-bound Somalis. Signs proclaiming the youngsters were thieves had been hung around their necks.

Incidents Mentioned in the LeBlond Synopsis. One important source is the synopsis of these incidents prepared by LeBlond. I have tentatively classified the incidents described in his synopsis:[5]

Verbal abuses:	3
Abuses of prisoners in detention:[6]	7
Being AWOL/drunk:	1
Motor vehicle accidents (MVA):[7]	6
Possession/destruction of evidence (photographs):	4
Shootings:	2 (in addition to those already described)
Thefts:	6
Unreported loss of ID:	1

There is an interesting feature of this classification. Much has been made of the looting of Canadian material and goods by Somalis. There were six cases of theft reported from December 1992 to May 1993 (five months). The evidence against potential suspects was judged inconclusive in two of these cases (although the violations did happen); in four cases, there was lack of evidence against suspects.

The CIDCFS received a summary, prepared by Colonel Wells on behalf of the Chief of the Defence Staff, of military police files involving members of the Airborne Regiment in Canada — in Petawawa — from 1988 to April 1993. The summary is dated 23 May 1993. According to this summary, there were no cases of theft in 1993 (January to April); there were two cases in 1992, one case in 1991, two cases in 1990, one case and one attempt in 1989, and two cases in 1988. *In other words, there were three times as many cases of theft during the five months of Operation Deliverance as there were in any year in Petawawa from 1988 to 1993.*

It is difficult to interpret these figures, because the number of cases is so small, regardless of the increase in percentage. The least that can be said is that they are indicative of a relaxation of discipline. However, there is a more troubling feature. In an incident that occurred on 15 February 1993, ex-Major-General Mise Ismail Arale complained that a Canadian soldier (in fact, Master Corporal Matchee) stole a copper sword from his residence. The complainant was paid compensation by the military. Some Canadian troops seem to have been obsessed by the idea of bringing back trophies from Somalia and behaved as if they were occupying a conquered territory. The most excessive examples of this kind of behaviour are the photographs of smiling members of the CAR beside prisoners who they had — in some cases grievously — abused, even though these photographs provided incontrovertible evidence that the soldiers were party to a crime or a serious breach of regulations.

The Perspective from the Field

I shall present the perspective that flows from the BOI hearings, from Purnelle's book (1996) and the videotapes made by members of the CAR during their tour of duty. At the time of the writing this report, there are no other sources, nor have the hearings of the CIDCFS on the in-theatre phase been conducted yet.

The BOI Hearings

- With only one exception, all the witnesses heard by the BOI deny that anyone in the CAR, and particularly in 2 Commando, ever displayed racially motivated behaviour. The witnesses who denied any racist behaviour, when queried on this subject, are B.Gen. Beno (BOI, 261), Colonel Holmes, (CIDCFS, 4: 750), L.Col. Morneault (BOI, 1376), L.Col. Mathieu (504), Major Magee (BOI, 322), Major Seward (BOI, 1241), Corporal Jones (BOI, 989), Trooper Reddick (BOI, 7336) and Trooper Gilbert (BOI, 966). CWO Raymond testified to the effect that the only expression of racial-ethnic antagonism that he was aware of was in connection to Anglophones and Francophones in the CAR (Anglos vs Frogs; BOI, 1002).
- Other witnesses did not flatly deny any expression of racial prejudice but tried to account for it. Captain Reinelt admitted that there may have been pejorative language used in connection to Somalis but believed that it was only an expression of temporary frustration with the difficult situation faced by the CAR (BOI, 738); Captain Blackman and

Corporal Sprenger interpreted racial slurs as only said in jest (jokes) (BOI, 1137 and BOI, 1188). Troopers Hodgson and Gilbert denied that the use of words such as "niggers" expressed any real racial prejudice (BOI, 1096 and BOI, 969). Trooper Hodgson stated that the CAR was hostile to any non-CAR member, whoever that person might have been (BOI, 1091).

- Captain Blackman and Trooper Hodgson testified to the effect that officers would caution the rank and file against the use of racial epithets (BOI, 1120 and BOI, 1096).
- Several members of the CARBG testified that members of the CAR were victims of Somali racism (BOI, L.Col. Mathieu, 503; BOI, Captain Blackman, 1120; BOI, Trooper Gilbert, 962). These witnesses said that Black members of the CAR, who were morphologically different from Black Somalis, were called "niggers." On the contrary, Trooper Reddick (a Black member of the CAR, who was part of a vehicle repair unit and did not wear the maroon beret — his beret was green) testified that Somalis "looked up to him" (BOI, Reddick, 808).
- As I have said, the only explicit testimony with respect to all questions surrounding racism within the CAR was from Trooper Grant. (His testimony is important and deserves as careful consideration as that of Corporal Robin; I postpone my analysis of his testimony, for he may be called to testify by the CIDCFS.)
- Finally, there were allegations about the use of racially prejudiced language by officers of the CAR. Trooper Grant alleged that MWO Mills and Major Seward had used racial epithets (BOI, Trooper Grant, 888). Captain Blackman testified that he heard a rumour to the effect that Colonel Labbé, the CARBG Commander and his PA "had a bet, a case of champagne, over who would shoot the first Somali" (BOI, 1137-1139). Captain Blackman later said in his testimony that he did not believe this was a credible anecdote.

Incidents Mentioned in Purnelle's Book. In his book, Purnelle describes various reprehensible actions of members of CAR (Purnelle, 1996: 150). However significantly blameworthy these incidents may be, there is even more significance in the way in which he presents them. To illustrate this, I shall translate one paragraph from Purnelle (1996:150), where he lists examples of "irresponsible behaviour":

There was irresponsible behaviour, such as pouring the dirty oil from a Grizzly on the ground near a well, trying to exhume the body of a dead person by removing the stones covering the corpse, putting gas into orange juice given to a Somali, showing pornographic magazines to adolescents, or running over a goat or a turtle [translation]. (Purnelle, 1996:150)

The significance of this paragraph is that Purnelle mentions the exhumation of a human body *and* running over animals in the same breath and without any indication of his awareness of the disparity between these two acts. His apparent insensitivity to the overwhelming difference in the degree of culpability that should be associated with the two can be underscored by calling to mind that one of the greatest tragic plays in Western culture — *Antigone* by Sophocles — is precisely about the absolute duty to bury the corpse of the dead. (In the Greek tragedy, Antigone chooses to die rather than to leave the corpse of her brother without burial.) Purnelle, it needs to be said, appears to be one of the most morally conscious members of the CAR. Similar lack of perspective can be found in other accounts in his book.

Another significant feature of Purnelle's book is his discussion of the abuse of alcohol and drugs, an abuse he suggests is widespread (Purnelle, 1996: 146-147).

The Videotapes. The videotapes vary greatly in their visual and acoustical quality, parts of them being blurred or barely audible. Yet, they have great value as corroborating evidence.

- The off-quoted statement to the effect that the members of the CAR "haven't killed enough niggers" can be heard. Master Corporal McKay has been identified as the speaker. Other racial slurs can also be heard on the tapes.
- The videos confirm beyond any doubt that alcohol (particularly beer) was freely available and consumed in considerable quantities.
- Expressions of racist stereotypes, such as remarks about Africans being lazy and deserving their current misery, abound. A typical comment is made by a trooper who, when asked about what he thinks about the mission, answers "I haven't done much thinking, I just rearrange my prejudices."

What most strikes the viewer about these videos is similar to what strikes one about the trophy-photographs of abused prisoners and the hazing ritual videotapes. These troops viewed themselves as being so detached from all standards of ethics and decency that they did not seem to be bothered in the least — indeed, appeared to be proud — that their misconduct was recorded on tape and in photographs. Their indifference to what is generally regarded as acceptable human behaviour, an indifference bordering on provocation, is a source of deep puzzlement for the author of this report.

Interpretations of the Findings

One of the purposes of this section is to provide an answer to the question of whether members of a minority recruited in a unit would tend to be harder on other members of the same minority in order to gain acceptance from the majority.[8] As I previously said, this question is complex and admits of many variants.

It is not however the only question that I intend to address. The section is divided into four parts: (1) the Gurkha syndrome, (2) minorities policing minorities, (3) racially prejudiced language and (4) racially prejudiced behaviour.

The Gurkha Syndrome. I have discussed this theory, initially elaborated by Cynthia Enloe (1980) and widely influential in the literature. It implies the use by a colonial power of a peripheral ethnic group within a colonized country to police the rest of the society, as the British used the Gurkhas from Nepal to help enforce their domination in India. Dreisziger (1990, ed.) discusses several cases of the use of minority groups in a policing or in a soldiering capacity. White's essay, "The American Army and the Indian" in this collection describes the uses of American Aboriginals in the U.S. Army as a manifestation of the Gurkha syndrome.

Could this theory be applied to the situation of persons belonging to different Canadian ethnic minorities in the CAR and provide an explanation of their behaviour in Somalia? For several reasons, I believe that it cannot.

First of all, a peacekeeping mission is not a colonial venture, although ironically it may have been perceived as such by the armed factions and the war lords in Somalia. More important, however, is that the number of CAR members recruited from Canadian racial-ethnic minorities was very low, as the figures quoted in chapter 2 of this report show. By January 1995, there were 25 members of the CAR who belonged to racial-ethnic minorities. Eight minorities were represented. The more numerous groups were the Afro-Canadians (12 members) and Aboriginal Canadians (six, plus one "Amerindian"). Whatever may have been the composition of the

CAR from October 1992 to May 1993, it cannot have been much different in terms of the number of persons from racial-ethnic minorities serving during this tour of duty. All testimonies confirm that persons from such minorities were few, were from different backgrounds and were sprinkled in different units of the CARBG. When members of racial-ethnic minorities are used as Gurkhas, they are generally numerous and grouped together in special units, the Gurkha phenomenon being essentially a group phenomenon and not an individual one. Those two operationally defining conditions were completely absent in Somalia. They are also absent from the Canadian Forces and its traditions. The only attempt to recruit minorities in a more systematic way is the recruitment of Aboriginal Canadians as Rangers deployed in Northern Canada. Even in this case, the numbers remain relatively low, and the Rangers are not used in a special policing capacity.

Aboriginal Canadians are used in a policing capacity by the RCMP. This RCMP program of organizing Aboriginal units to police certain reservations was created at the demand of the First Nations themselves. They are used as community policing agents and not as Gurkhas.

Even if it does not apply to our case, the Gurkha syndrome theory is interesting. It serves to explain why people — generally White people — are so inclined to believe spontaneously that members from minority groups recruited by majority-dominated organizations will display harsher behaviour toward persons who either share the same racial-ethnic background or share their status as being part of a minority (even if it is not the same). They are expected to show more harshness to gain recognition from the majority.

On reflection, this belief is neither particularly natural nor reasonable: one would more reasonably expect that potential Gurkhas would in fact feel solidarity for their own kin or for people sharing a similar status as a minority. The other source of this spontaneous belief comes from prison literature and particularly from literature describing living conditions in concentration camps. The use of inmates as "kapos" to police their fellow inmates has been widely acknowledged in the memoirs of ex-prisoners, in the research literature and in written, film and television fiction. As far as I can see, this is the most important source for the belief that dominated individuals or groups are harder on other persons or groups in a situation similar to their own.

The literature on prisons and concentration camps is far from being as simplistic as the belief described above — few "kapos" actually outdid their masters (Hilberg, 1983; Langbein, 1975). I cannot review this huge body of literature, which is very tangentially related to my topic. I shall instead review the policing experience.

Minorities Policing Minorities. Police forces are the main type of organization that can be compared to the armed forces in that they also have the power to use legitimate force. Can it be said that police from racial-ethnic minorities overidentify with the norms of the group forming the majority (the Whites) in the police forces of Western countries? I gathered evidence on this question by interviewing police officers from minorities.[9] I have also interviewed colleagues who are involved in research projects on the behaviour of police recruited from minorities (in France, the United Kingdom and in Canada). Finally, I reviewed the research already published in this regard. The evidence strongly favours the conclusion that police from minorities do not overidentify with the norms of the majority group in the organization. I shall briefly present the evidence.

Present research: The researcher presently conducting the most systematic research in this regard is Dr. Simon Holdoway (1979, 1983, 1988, 1991 and 1995) from the United Kingdom, who has published extensively on the police. Dr. Holdoway is involved in a multi-site exploration of policing by officers belonging to racial-ethnic minorities (in the United Kingdom and Canada). The following is his response to my query (11 January 1996).

My research has included long interviews with black and Asian (respectively from an Afro-Caribbean and Indian sub-continent background) police officers serving in constabularies throughout England and with black and Asian resigner. I asked them about the ways in which they adapt to the racial prejudice and discrimination of their colleagues. My preference is to use the terms "racialised prejudice" and "racialised discrimination" because this signifies that I am working with the analytical-framework that accepts that "race" is socially constructed and articulated through organisational and occupational processes.

Officers respond to the racialised categories of their white colleagues in various ways — there is an element of choice here, which may be important to your inquiry. What I can say is that within the English Police black and Asian officers face a highly racialised working environment. The experience of racial prejudice and, to a lesser extent, discrimination from colleagues is a daily experience. There are strong pressures to conformity amongst the work group, which includes the acceptance of the use of racialised categories within the work force.

Responses to this work context lead to a number of responses amongst black and Asian officers. I have not found that they over-identify with the norm of the work group — which is your specific interest.

Holdoway's conclusion reflects accurately my own experience in interviewing Aboriginal and Black police officers. There is however an important reservation made by Holdoway with respect to the "element of choice," which also stems from my own research. The hub of the issue is that there is no general answer that would apply in all cases to the question of overidentification. The weight of the evidence, as I said, disconfirms the overidentification hypotheses. It may however still be applicable to particular cases. In my interviews with Aboriginal police, I found out, for instance, that an Aboriginal police officer was more likely to identify with the professional culture and values of his or her White Sûreté du Québec (SQ) colleagues when that person was policing an Aboriginal community that did not belong to his or her own tribe.

The research literature: The main theme of the research literature on police officers from minorities focusses much more on the racial discrimination from which they suffer than on their responses to this discrimination. They are the object of racial prejudice both from the community they police and from their colleagues (Alex, 1969; Campbell, 1980; Kuykendall and Burns, 1980; Walker, 1983; Leinen, 1984; Wilson et al., 1984; and Cerky, 1985). This latter perspective is called the "double marginality hypotheses." As identified in the general literature, minority officers do not overidentify with the values of the dominant group.

There is, however, some support for the overidentification thesis. In his studies on police brutality Reiss found that *both* Black and White police officers are more likely to exercise undue force against members of their own community (Reiss, 1967 and 1968). Jacobs and Cohen (1978) found that minority officers responded with great ambivalence to their double marginalization: they tended on the whole to be either more lenient with Black offenders or particularly diligent in the prosecution of offenders which they see as threatening to Black communities. Particularly disturbing are the results of studies that show that Black officers are overrepresented among police shooters (Geller and Karales, 1981). Fyfe (1981: 370) in his study of the New York Police Department even found that the rates of involvement of Black and Hispanic officers in fatal interracial shootings (Black officer shooting Black person; Hispanic officer shooting Hispanic person) were more than twice as high as the rates of fatal shootings by White officers involving victims from the Black and Hispanic communities.

The analysis of this data, however, indicates that this minority officer overrepresentation among police shooters is closely associated with racially varying patterns of assignment, socialization and, most important, residence (Fyfe, 1981: 381; Geller and Karales also came to the same conclusion regarding Chicago police). Black and Hispanic officers are assigned to ghetto precincts, and they live within the inner city milieu where there is much greater violence. It is significant in this last regard that their rate of involvement in off-duty shootings is more than three times higher than their White counterparts, who mostly live in non-violent New York suburbs. Hence, on closer examination, these findings provide a much weaker confirmation of the overidentification thesis. According to Fyfe's own interpretation of his research, the most significant finding is that there is a relationship between the race of New York Police Department officers and police shootings, but that apparently it is a spurious one in which many other variables intervene, such as place of deployment and residence. Walker (1983) found that Black officers had values that were more progressive and community oriented; their patterns of value were in fact much more similar to the Black population and more oriented toward social goals as opposed to the personal goals that were more strongly pursued by White officers.

In sum, the research does not support in general the overidentification hypotheses. The most significant finding in this regard (police shootings by minority officers) is only weakly, if at all, supportive of this hypotheses. The research does show, however, that there is an element of choice at play. As I said, minority officers may respond to their double marginality by being either more lenient or harsher toward members of their own communities.

Contextualization: I will now try to interpret the previous findings and put them in the military context.

- It might be said that the armed forces demand an even greater degree of loyalty than the police, and that the consequences of not affirming group norms and becoming a loner may be catastrophic in a military setting (Holdoway, 11 January 1996). This is entirely possible, but it remains speculative. Miller and Moskos interviewed U.S. Black and White combat and non-combat personnel who were sent to Somalia concerning their opinion of the local population. Their opinions were much more positive than those of the White personnel, as can be seen from Table 4.1.

Black soldiers were offended by constant anti-Somali remarks from Whites, including jokes about intentionally running down Somalis on the road (Miller and Moskos, 1995: 629). Overall, the attitude of the Black military was much more favourable to the Somalis, although they resented even more than Whites the violence used by Somalis against U.S. soldiers.

Table 4.1
Soldiers' Opinions of the Local Population, by Gender, Race and Combat Status*

	Men					
	Black		*White*		*Other*	
	Combat Status %	*Non-Combat* %	*Combat Status* %	*Non-Combat* %	*Combat Status* %	*Non-Combat* %
Mostly Positive	15	19	4	4	4	6
Mixed Positive/Negative	58	71	40	56	57	72
Mostly Negative	27	10	56	40	39	22
	100	100	100	100	100	100
N=	(33)	(92)	(195)	(179)	(23)	(36)

	Women		
	Black %	*White* %	*Other* %
Mostly Positive	20	9	–
Mixed Positive/Negative	74	64	82
Mostly Negative	6	27	18
	100	100	100
N=	(76)	(58)	(17)

Note:
*Women were not eligible for combat status at this time.
Source: Miller and Moskos (1995, p. 629).

- There is one feature of the murder of Shidane Arone that has not really been researched so far. The two persons who murdered him were from a minority — they were Aboriginal soldiers — but they were not Black. Is it possible that the overidentification hypotheses receives more confirmation when relations between individuals belonging to different races are involved? Again, this is possible and plausible enough to be researched. However, we do not have much police research on this subject. Generally, minority officers are deployed within their own community: in addition to equality of access to a well-paid job, this is the usual justification for hiring them. Hence we have no systematic data on cross-cultural relations between police belonging to one racial-ethnic minority and members of other racial-ethnic communities. This research is worth pursuing, but the nature of its findings cannot be anticipated one way or the other.
- One variable in cross-cultural policing is important to note. This is the number of persons from a racial-ethnic minority who are present at the same time in a policing organizationand its subunits. The more that minority members are isolated, the more likely it is that they will overidentify with the norms of the dominant group. And on the contrary, the more numerous they are and the more they work together, the more they can resist the norms of the majority. In this respect, it must be noted that the Aboriginal soldiers were quite isolated, like other minorities, in the CAR.
- One last consideration is that racial prejudice on its own may have played a crucial part in the murder of Shidane Arone. Master Corporal Matchee was identified by Trooper Grant (BOI: 886) as the most virulent racist of 2 Commando. He was reported by a witness interviewed by the military police to have made a racial comment in connection with Shidane Arone: "at approx 2230 hrs, MCpl Matchee came into the CP to bum a cigarette from him [MWO Mills, my note] and before he left MCpl Matchee said, 'the white man feared the Indian, so will the black man'." (Voluntary statement of Corporal MacDonald to the military police, 1 April 1993). The evening referred to by Corporal MacDonald is that of 16 March 1993; during that evening and the following night Shidane Arone was tortured and beaten to death by Master Corporal Matchee and Private Kyle Brown.

Racially Prejudiced Language. When we summon all the evidence, it becomes obvious that racial slurs were current in the CARBG, while it was in Somalia and also in Canada, before and after its departure for

Somalia; the hazing rituals make clear the CAR's behaviour on their Canadian base. This behaviour raises two kinds of question: (1) is racist language conducive to racist behaviour (other than in words)? and (2) is racist language *perceived* as an indication of unqualified racism? For reasons of simplicity, I shall address these questions in reverse order.

Racist language and perceptions of racism: The answer to the question "Is racist language perceived as an indication of unqualified racism?" is easy to give and quite straightforward. This answer is "yes." There may have been a time when it could be claimed that a person using racial slurs such as "niggers," "spicks," "frogs" or "Jewboys," was not a racist but only verbally indelicate or flexing verbal muscles. That time is over, a fact that owes very little to so-called political correctness but much to a growing sensitivity to the power of language.

I quote in this regard a paragraph from Allport (1954/1979: 478-479):

> In Chapter...we discussed the immensely important role of language in building fences for our mental categories and our emotional responses. This factor is so crucial that we shall return to it again — as it bears on childhood learning. In Goodman's study it turned out that fully half the nursery school children knew the word "nigger." Few of them understood what the epithet culturally implies. But they knew that the word was *potent* [emphasis added]...Not infrequently in a temper tantrum a child would call his teacher (whether white or coloured) a "nigger" or a "dirty nigger." The term expressed an emotion — nothing more. Nor did it always express anger — sometimes merely excitement. Children wildly racing around, shrieking at play might, in order to enhance their orgies, yell "nigger", "nigger", "nigger." As a strong word it seemed to vocalize the violent expenditure of energy under way.

Needless to say, since the first publication in 1954 of Allport's book, the emotional potency of words such as "nigger" has grown so much that they are now only referred to in an oblique way ("the n-word"). The power of these racial slurs is such that anyone uttering them is immediately branded as a racist. I believe that it is a complete delusion to claim that language can be separated from the rest of behaviour, such as saying that someone is using racial slurs *only* as a *way of talking*, but that deep inside that someone is not racially discriminating. Whatever its past credibility — if it was ever credible — this position is now thoroughly discredited. More important, it is not going to become legitimate again if we just we wait long enough. That in the future Whites will "descriptively" call people of African descent “niggers,” while claiming to be racially unprejudiced, is unlikely.

The CF has commissioned an internal study entitled *Racism and the Canadian Forces* (Canada, Department of National Defence, 1993; this study was not made public). One of the conclusions of this study is “that members of the Canadian Forces, like most Canadians, will carry with them in some measure racist attitudes derived from their cultural and family heritage” (5/35). A similar point has been made repeatedly with respect to the police: racial discrimination is present among them to the same extent that it is in the wider society.

This point may be true *de facto* but it is also quite wrong *de jure*. It is not the general public that will police our streets or be sent on peacekeeping missions. Holding organizations such as police departments or the CF to standards which are no higher than those which are displayed in public opinion polls is unjustifiably complacent. If such organizations have any professional meaning, then they should be held to higher standards than those of the general public. Would it be any defence for the Red Cross to say that it was no less and no more aware of blood tainted by the HIV virus in the 1980s than the general public? The very least that can be expected of organizations claiming any degree of professionalism is precisely that their standards be higher than those of the general public. Consequently, it is no excuse whatsoever for Canadian peacekeepers to say that they use racial slurs no more and no less than the average Canadian. Canadian peacekeepers should be held to standards developed in relation to their task and not in relation to Canadian public attitude polls. Not only do peace-keeping and related tasks not allow for any expression of racism, they oblige peacekeepers to avoid any appearance of racism, indeed, to eschew any behaviour which would make them vulnerable to accusations of racism. To accept any other course is to imperil the success of peacekeeping missions.

Racist language and racist behaviour: We previously saw that the use of racial slurs is in itself a manifestation of racism since it expresses an attitude of racial prejudice. Hence, Allport (1954/1979: 14) classifies such verbal behaviour in the category of “antilocution,” which is the first degree of negative action motivated by racial or ethnic prejudice. The other forms such negative action can take are, in increasing order of gravity, avoidance, discrimination, physical attack and extermination.

The existence of increasing degrees of gravity in the expression of racial or ethnic prejudice naturally raises the following question: does indulging in behaviour of lesser gravity lead to the more extreme forms of racism? In other words, does one start by antilocution, to borrow Allport's terminology, and end up with individual and systematic physical attacks? I addressed this question in part in chapter 2 of this report. I now analyse it in greater detail.

This is a difficult question, and one that has been forcefully debated, particularly since the end of World War II. Without pretending even to approach the beginning of an answer, I propose to make a distinction between two different ways of framing this issue.

The first way is through what I call the "individual causation" perspective. A classic illustration of this perspective is the debate on whether someone using soft drugs will eventually become addicted to hard drugs. As with this analogy, would an individual using racial slurs behave with increasing degrees of hostility toward the persons that he or she demeans through language? Generally speaking, this question can be answered in the negative: it is simply not true that everyone using antilocution against certain ethnic groups will eventually resort to physical attack.[10] However, even if there is no predetermined path leading from verbal to physical aggression, it is not unlikely that people who verbally abuse a particular group will also tend to avoid and discriminate against members of that group.

There is, however, another side to the individual causation perspective, where the analogy with drugs completely breaks down. The question of whether smoking marijuana in the presence of an individual who never used soft drugs before might prompt this person to use hard drugs is felt to be spurious in drug research.[11] This is far from being the case in racial relations: one can easily imagine that a person who was always careful to avoid using racial slurs toward others would feel such outrage in being submitted to such abuse that he or she might react by escalating the degree of violence of his or her *response*. Being the target of racial slurs, a person may respond by physical attack. The use of racial slurs is the object of an increasing taboo in society precisely to avoid such escalation in the response. Needless to say, the same prohibition should apply in situations of peacekeeping in multi-ethnic contexts, where tensions are already running high.

Barring the response to antilocution by people who are its object, the use of derogatory language does not inevitably cause the individual using it to increase the violence of the expression of his or her racial or ethnic hostility. However, it may be asked if the pervasive use of racial antilocution could not collectively facilitate an escalation in aggressiveness. Viewing the use of racist language as "collective facilitation" rather than as individual causation can lead to the formulation of a different answer to the question of escalation. The widespread use of verbal racist stereotypes may create within a group a psycho-social environment that would be supportive or, at the very least, be perceived as supportive, of an increase in the degree of aggressiveness in the expression of racial antagonism. It is important to stress that there is no strong determinism operating in this case, and that there may be no overtly active pressure to act out one's racial prejudices. However, even if it be passive, general acceptance of the expression of racial prejudice may facilitate escalation in aggressiveness. It may even incite a weak individual, seeking what he expects to be a higher level of approbation and acceptance, to resort to physical coercion against members of the group discriminated against. These remarks do not contradict what I previously said on minorities policing minorities. Research does not reveal a general trend whereby minority police would be tougher on their own. However, this does not preclude the possibility that weak individuals anxious to gain acceptance in a racially prejudiced group would discriminate more aggressively against racial or ethnic minorities regardless of whether they themselves are members of a minority.

I believe that collective facilitation of the expression of racial prejudice is more of an issue that individual causation. The two issues are also quite distinct. Because it can be argued that the link between word and deed is weak on the level of individual causation, it does not follow automatically that it is also weak on the level of collective facilitation.

Logics of Racism. Previously I made a distinction between two logics of racism, the logic of domination, which generated a particular kind of racism characterized by exploitation of the dominated group, and the logic of exclusion, which generated another kind of racism, grounded in systematic segregation. I made several points with respect to these different logics of racism. First, I stressed that the logic of exclusion was no less offensive than the logic of domination, since it ultimately resulted in excluding the "racialized" group, to borrow Holdaway's terminology, from the human kind, thereby potentially legitimating its extermination. I also said that because of the nature of their mandate — peacekeeping rather than conquest and subjugation — peacekeepers were more threatened by exclusionary racism than by the racism of domination.

Although it was in great part justified as protection against looting, there may have been more than reasons of security at play in the Canadian obsession with keeping Somalis out of their compounds and in their desire to segregate themselves almost completely from the local population. Purnelle (1996: 150) wrote in this respect:

One day, after having been in charge of security on the road to a village where a conference of Somali leaders was to take place, we made contact with the French military. Our first surprise was to have seen one military jogging ten kilometres from his base without any weapon. Second surprise, the French were cohabiting with the Somalis. There were several huts in the midst of their camp, and their dwellers freely circulated through it. As for us, we had requested the evacuation of two huts for reasons of security. The French commander also explained that he had solved the weapons' problem by issuing permits to their possessors; he would seize the weapons that were not registered. [our translation]

This account and testimony given at the BOI's hearings raise the possibility that Canadian efforts to segregate themselves as much as was feasible from the local population may not have been justified solely by considerations of security. In my interviews with the French military, I found that the French did not perceive the problem with looters as particularly acute.

Whatever may have been the reasons for the Canadians to seclude themselves in their compound, this is not the most disquieting aspect of their performance in Somalia. A much more problematic feature of their behaviour was their taking of photographs, particularly photographs of the swollen face of Shidane Arone, when he was being beaten to death. The testimony of Corporal Bibby during one of the courts-martial is, in this regard, particularly revealing. First, Corporal Bibby acknowledges that it was common for all the troops to take pictures of almost everything that they came across during the course of their tour in Somalia:

Q. So the point being that it would be appropriate then to take pictures of the good, the bad and the ugly here, is that fair to say?

A. Yes. sir.

Q. Including disturbances in the town perhaps, actions of the troops and perhaps searching out for weapons, that sort of thing, if time permitted?

A. When you're doing your job you wouldn't have time to take pictures, searching houses. On the other hand, you know, something like the meat market downtown, the butchery where they threw all the bones into the river and you wondered why there is so many diseases there, that would be a good example, taking a picture of that.

Q. Why do you say that's a good example? Can you tell us a little bit more about that, the butchery?

A. It was fairly grisly. You could see where all the diseases was coming from. There was literally a mountain of camel bones clearly right in the river where they bathed and washed. Maybe three hundred metres down the stream there would be kids playing in the water.

Q. So the fact of taking photographs of the prisoner in the situation that he was in that night would not have appeared to you to be particularly remarkable then?

A. No, it wouldn't, sir. [Note: the prisoner referred to here was Shidane Arone.] (Transcripts of the courts-martial, Volume 6, 1118-1119)

It is truly striking that this man does not show any consciousness of the difference between taking pictures of practices harmful to the environment and taking pictures of a Somali adolescent being tortured to death. The bones of freshly butchered camels and the bleeding face of Shidane Arone all blend into the good, the bad and the ugly of the Somali landscape and were equally fated to enliven the saga of the CAR. As was apparent in the corporal's testimony and confirmed by later testimony, a good number of Canadian solders participated in the Somali "picture safari." The Canadians were not alone. Soldiers in the Belgian contingent — the contingent singled out by African Rights (1993) for its particular disregard for human rights — also took pictures of Somali prisoners being abused. It was these pictures, passed around among friends at home, that eventually led to a public scandal and courts-martial in Belgium.

How do we explain photographs of abused people appearing in collections of travel pictures? The possibility exists that the abused became just one photographic opportunity among others because they had been dehumanized, made into objects, by the Canadian soldiers. If so, what I characterized as exclusionary racism may be some part of the explanation for this apparently brutish voyeurism.

Conclusions and Remedies for Racism

Attempts to eradicate racism in the military have been spearheaded by the U.S. effort to integrate their armed forces. Essentially, this effort is internally directed: what is to be remedied is racist behaviour within the military. In the context of peacekeeping, and particularly the context of Operation Deliverance, the problem is not primarily internal but external since the number of members from racial and ethnic minorities serving in the Canadian Forces is nothing like that in the United States. Even if the internal problem does not have the same magnitude as in the United States, the CF could not even pretend to be fair to external minorities encountered in peacekeeping operations, if it was not fair to its own minority personnel. However, tackling this issue falls outside the mandate of the CIDCFS and, consequently, of this report.

The issue that does have to be addressed is how to ensure that CF behaviour during peacekeeping missions will be free of racial and ethnic prejudice. Because this issue has surfaced only recently, there are very few useful examples to look at.

As we have seen, the most readily available examples are in the field of policing, where, in a number of instances, police have shot at and killed suspects allegedly because of their race. I have reviewed the recurring, and finally redundant, recommendations made for reform: (1) recruiting a higher proportion of members from racial and ethnic minorities, (2) screening out candidates whose profile suggests they may be troublesome at a later stage, (3) improving training and (4) developing cross-cultural sensitization through varied approaches.

All these remedies are commendable, although none in itself seems to be the answer. For example, each new shooting incident results in a recommendation to increase cross-cultural training and sensitization. This recommendation is made without any acknowledgement that not only has it been made before but it has been *applied* before without perceptible improvement in the relations between the police and racial and ethnic minorities.

The CF has asked for submissions to develop and implement a program of training and sensitization for the prevention of sexual harassment and racism. Despite my efforts, I could not find out what stage this project has reached. As the call for proposals was made in August 1993, it is much too early to attempt any assessment of such a program. If it has been developed, it has not yet been implemented, according to my information.

Despite the shortage of models, there are some points about preventing racial prejudice, before it gets deeply rooted, that can usefully be made.

First, there is a great difference between developing an ethos and implementing tailor-made programs. The crucial difference is that developing an ethos — bravery in the face of danger, for example — is not achieved through selected training programs but through the co-ordination and transformation of a whole value system and the corporate image of an organization. It is achieved only as the result of unrelenting conditioning.

Let me try to suggest what I mean here by the difference between developing an ethos and implementing specialized programs. As we previously saw, it is alleged in documents such as *Racism and the Canadian Forces* (Canada, Department of National Defence, 1993) that it is a reasonable expectation that "members of the Canadian Forces, like most Canadians, will carry with them in some measure racist attitudes derived from their cultural and family heritage" (p. 5/35). I have already shown the general fallacy which is implied in this apologetic argument. I would now like to use it to illustrate the nature of ethos.

Does it follow from the statement quoted above that we should be satisfied that the proportion of racially prejudiced persons is not higher in the CF than in the general society? In trying to answer this question, let us consider an analogy drawn from what is the core of the military ethos. Suppose the proportion of soldiers that could be expected to turn coward in combat is the same as the proportion of people in the general population who would flee in the face of danger. Would the armed forces be satisfied with this state of affairs? The answer provided by history to this question is an emphatic "no." Indeed, the whole structure of the military was and still is devoted to ensuring that all its men and women are prepared to confront danger; the raison d'être of armed forces is precisely to stand in harm's way.

The mandate of the CF is not solely nor primarily peacekeeping. I have, however, shown that there are strong indications that participating in such operations will be an increasing role for the CF. It is as unthinkable that such a role could be played by a racially prejudiced force as it is that a cowardly army would adequately protect Canadian national security. To the extent that the Canadian military views its contribution to international peacekeeping as crucial, it should endeavour to develop among its personnel an ethos for the acceptance of racial, ethnic and cultural diversity. This result cannot be achieved through cosmetic programs which are immediately recognized as such by those undergoing them.

There are two conditions for developing an ethos. The first condition is that the effort to develop and implement such an ethos be *systemic*. The second is that this effort be self-consistent. In this regard, we can only rejoice that Canadian Forces Administrative Order (CFAO) 19-43, the recent CF policy on racist conduct, which is devoted to curtailing racism, did not follow the lead of *Racism and the Canadian Forces* (July 1993). Its devisers did not see as justified the formulation of a new regulation prohibiting CF members from associating with any racist organization (p. 22/35). Their startling conclusion would have radically undermined the cogency of any effort to develop a new peacekeeping ethos. Fortunately, it was not followed by CFAO 19-43 which includes in its definition of racist conduct "participation in the activities of, or membership in, a group or organization that a CF member knows, or ought to know, promotes discrimination or harassment on the basis of race, national or ethnic origin, colour or religion." This move toward consistency can only be seen as a step in the right direction. Whether it goes far enough remains to be seen.

THE USE OF FORCE

This second part of chapter 4 is divided into three. First, to provide the general context of the Canadian deployment in Somalia, I discuss the most recent findings in the research literature concerning the different UN peacekeeping operations there. Second, I will discuss the ROE with which the troops were provided. Finally, I focus on the use of force by members of the CARBG, particularly with regard to the detention of prisoners.

The Wider Context of the Canadian Intervention in Somalia

As I have repeatedly stressed, at the time of writing, the CIDCFS had not yet begun its hearings on the in-theatre phase of the CARBG's deployment to Somalia. However, the three UN peacekeeping operations in Somalia, First United Nations Operation in Somalia (UNOSOM I), United Task Force (UNITAF) and the Second United Nations Operation in Somalia (UNOSOM II), have generated a large amount of interest among scholars; some key participants in this operation also wrote about it. Since 1990, the following publications all discussed aspects of Somalia, its culture and history and the different UN peacekeeping missions undertaken there since 1992: African Rights, (1993); Africa Watch, (1990); Brauman, (1993); Drysdale, (1994), Hirsch and Oakley, 1995; Lewis, 1993 and 1994; Marchal, 1992, 1993a, b and c; Prunier, (1990-1991; Sahnoun (1994); Schraeder (1992); Smith (1993); Torrenzano (1995); and Willame (1994). There were also numerous article in journals such as *Horn of Africa* and *Politique Africaine* and in magazines such as *Jeune Afrique.* It is impossible to summarize all this literature within the scope of this report. I shall only briefly discuss points that bear directly on the issues raised here.

The Police State of Syad Barre. Somalia was created as a country in 1961, but it did not recognize the sacred nature of the former colonial frontiers, as most of the other African states had. On 16 October 1969, President Abdulrashid Ali Shermake was assassinated and temporarily replaced by Daud Abdulla Hersi. On 21 October, General Syad Barre was appointed president by a military junta in a bloodless coup. From 1969 to 1977, he tried to establish a variant of "scientific socialism" patterned after the Kim Il Jung regime in Korea. In 1977, he invaded Ethiopia with the view of annexing the Ogaden, claimed for a long time to be Somali territory. Syad succeeded at first, but when the Russians decided to support Colonel Mengestu, and Fidel Castro sent some 15,000 Cuban troops to help Mengestu, the Somalis retreated precipitously to their own country. That defeat was the turning point in Syad Barre's rule. From then on, he resorted to increasingly totalitarian means to remain in power. The title of the Africa Watch Report of 1990, *Somalia: A Government at War With Its Own People,* is significant in this respect. Furthermore, Syad Barre tried, for his own advantage, to revive clan loyalty, after having tried previously to eradicate the influence of clans in Somalia. The members of the different UN missions sent to Somalia, with the possible exception of Mohamed Sahnoun, who was forced to resign by the UN Secretary General, did not even begin to understand the complexity of the clan issue. The main warring parties in Somalia's civil war were those of General Aideed and of Ali Mahdi. Yet they were from the same clan, the Hawiye family.However, General Aideed belonged to the Habal Gebir subclan, comprised mostly of country people who raised cattle, whereas Ali Mahdi belonged to the Abgal subclan, whose members were, in the main, city businesspeople.

The authoritarian nature of the regime, particularly of its police, is important to bear in mind when studying the mistreatment of detainees by their Canadian keepers. This mistreatment was in line with the way Somali police dealt with looters and other violators of the law. Actually, excepting the murder of Shidane Arone and other incidents that the CIDCFS may bring to light, as we shall see, the Canadian authorities generally refused to go as far as the Somali police would have let them.

The General Failure of the UN in Somalia. This second general point cannot be emphasized too much. I alluded to it in chapter 2 when I discussed the international dimension of the intervention in Somalia. As the titles of several books devoted to the action of the UN in Somalia, *Le crime humanitaire* by Brauman; *Somalie, La guerre perdue de l'humanitaire* by Stephen Smith (1993), *Somalia: The Missed Opportunities* by Mohamed Sahnoun; and *L'imbroglio somalien* by Torrenzano, tend to indicate, the UN/U.S. intervention in Somalia was at best controversial and generally perceived as a dismal failure that cast doubt over the whole notion of humanitarian intervention.

Actually, few of the parties involved in the different interventions in Somalia remained unscathed by their experiences.

- Mohammed Sahnoun was the Special Representative in Somalia of the UN General-Secretary, Boutros Boutros-Ghali. Although he believed that he was making progress, he had to resign in October 1992, after having been disowned by Boutros-Ghali for his efforts to bring together a group of Somali intellectuals for a conference in the Seychelles islands (Sahnoun, 1994).
- Hirsh and Oakley (1995), who were two high-ranking civilian officers in the UNITAF, wrote a book in praise of its success. Actually, several participants in the UNITAF, most notably the Belgians, the Canadians and the Italians, were denounced by organizations, such as African Rights, for their violations of human rights in Somalia. Some of these violations were very serious and resulted in courts-martial, public inquiries and in a lasting blot on the image of the armed forces in Belgium and in Canada, where the crisis has taken on major proportions.
- UNITAF could be claimed to be an American success by Hirsh and Oakley (1995) only because it lasted no more than five months. UNITAF was succeeded in May 1993 by UNOSOM II, which was under the sole command of the UN. UNOSOM II ended in a full-scale war between the troops of General Aideed and the UN contingents. The U.S. Rangers and the American Quick Reaction Force (QRF) were particularly active and 13,000 Somalis lost their lives, according to figures given by General Aideed.[12] On 3 October 1993, U.S. elite forces (Rangers and Delta Force) conducted an operation against a Mogadishu hotel

where the principal lieutenants of General Aideed were supposedly holding a meeting. The Americans were trapped at this hotel; 18 Americans were killed and 78 wounded, the highest number of casualties suffered by U.S. troops since the terrorist attack against the Marines in Lebanon. Following this disaster, President Clinton announced that all U.S. troops would leave Somalia by 25 March 1994. The number of Somali casualties during this battle was very high (between 500 and 1,000 dead and wounded). With the exception of the Korean and Gulf wars, which were not under UN command although sponsored by the UN, UNOSOM II is the only peacekeeping operation that resulted in a full-fledged war between the forces of the UN and the local population. Resolution 885 of the UN Security Council created a commission of inquiry into the violence that erupted during UNOSOM II. Its conclusions were highly critical of the whole operation, whose mandate to impose peace and to reconstruct the political institutions of Somalia was characterized as "pretentious."

Whether we refer to UNOSOM I, UNITAF or UNOSOM II, it would seem that all interventions in Somalia were ill-fated. This fact does not excuse the behaviour of the elements of the CARBG who were involved in gross violations of human rights. It, however, places these violations in the context of an intervention with several phases where nothing went right and where a certain kind of anomie, to use the famous concept developed by the French sociologist Durkheim, seems to have prevailed.

One of the most knowledgeable experts on the humanitarian interventions in Somalia and one of its severest critics, Rony Brauman (1993: 26), offers the following scathing description of the peacekeepers' actions:

To the feelings of revulsion experienced previously by the members of the humanitarian teams as they witnessed atrocities daily, was added the feeling of despair before the utter insensitivity of the UN knights-errant, who, by order of the UN, had replaced their Somali security guards. The sad situation was made so much worse by the paratroopers' drunken behaviour. Having settled themselves in the offices of the NGOs, the "peace soldiers" gave to dumbfounded humanitarian volunteers the spectacle of men urinating from the top of roofs, whistling at women, brutalizing kids, beating up people here and there for no apparent reason other than their fancy or bad humour. So much so that the doctor from Medecins sans frontieres [Doctors without frontiers, a French humanitarian agency] was not unduly surprised when one of the paratroopers confessed to him, one day, that he could not bear hearing his colleagues bragging about the number of Somalis that they had killed anymore ... Sexual aggression, acts of violence, overtly racist, had become the daily lot of Kismayo under the "protection" of these "Blue Helmets" [it was the Belgians who were deployed in the Kismayo area during the UNITAF operation].

The Canadians then were not the only troops to behave badly in Somalia. And it must be said, in all fairness, that the Canadian government was actually singled out in the report issued by the African Rights organization for how it reacted to the deplorable behaviour of its forces.

The Canadian government deserves full credit for its willingness to treat allegations of homicide by its soldiers with the seriousness that the charges warrant. No other government contributing soldiers to UNITAF or UNOSOM has shown comparable concern for accountability. It is revealing that of the four cases of killings of civilians by the Canadians, half of them have been considered worthy of criminal investigation. This level of scrutiny puts the other UNOSOM forces to shame. (African Rights, 1993: 30)

The Ambiguous Character of the Mandate. In chapter 1 of this report, I referred to the certain degree of ambiguity in the UNITAF's mandate. This ambiguity was highlighted in the Report of the UN Commission of Inquiry, which insists that only UNOSOM II was explicitly mandated to enforce the disarmament of the warring parties.

Hirsh and Oakley (1995) reveal that these ambiguities were in fact much deeper (Oakley was the U.S. President's special representative in Somalia for both President Bush and Clinton). There were two main ambiguities, which stem from the UN Security Council Resolution 794 (3 December 1992) creating the UNITAF. First, paragraphs 8, 9 and 13 accepted the offer of several countries to participate in UNITAF, but recognized that the United States was the "first among equals," although the co-ordination of the whole operation seemed to remain in the hands of the UN. In plain words, it was not altogether clear who was in charge of the operation, the situation being characterized by the classic split between leadership in theory (the UN) and leadership in fact (the United States).

The second ambiguity concerned the nature of the mission. According to paragraph 10 of Resolution 794, the mission's main purpose was to provide a secure environment for the distribution of food and other humanitarian aid in Somalia, which had been devastated by drought and famine. However, paragraph 16 of the same resolution asked all UN states to take all the necessary measures to apply paragraph 5 of UN Resolution

733. This resolution, approved by the Security Council on 23 January 1992, was the first UN attempt to deal with the Somali crisis; its fifth paragraph requested of all member states the application of an embargo on the provision of arms to the warring parties in Somalia. By referring explicitly to this paragraph, Resolution 794 actually raised the issue of disarming the factions at war in Somalia.

Hirsh and Oakley (1995: 102 ff.) argue that there was a serious disagreement between President Bush, who was the prime champion of the UNITAF, and the Secretary-General of the UN, Boutros Boutros-Ghali, who had actively tried to resolve the Somali crisis since the beginning of 1992. Boutros-Ghali urged the enforcement of disarmament of warring parties, whereas President Bush insisted that the only purpose of the intervention was the provision of security to the convoys that were to distribute food and other needed assistance to the Somalis dying of hunger. The President made this point very clearly to Boutros-Ghali in a letter, quoted by Hirsh and Oakley, that was written at the end of 1992 (1995: 103, 2).

Neither President Bush nor Boutros-Ghali won the argument. On the one hand, some degree of disarmament had to be enforced to create the conditions of security necessary for the distribution of food. The U.S. troops proceeded to put the Somali "technicals" in cantonments and also attempted to seize heavy military equipment.[13] Thus disarmament was enforced to some extent. It was not, however, systematically enforced. For example, Ali Mahdi surrendered most of his technicals to the Americans, while General Aideed moved his outside Mogadishu, where he hid them for future use. The U.S. troops did not undertake extensive searches for arms in the course of the UNITAF. Such searches were conducted by the UNOSOM II troops, which included U.S. troops.

This lack of a clear mandate with regard to the enforcement of disarmament helped to undermine the action of the UNITAF. For example, *The Somalia Handbook* that was distributed to all Canadian troops being sent to Somalia was basically written for Operation Cordon, where Canadians were to be deployed within the context of UNOSOM I, a Chapter VI operation. The role of the CF was not redefined according to the new UNITAF mandate and is characterized thus: "The role of the military force is to protect the people who are providing food aid. The Canadian battalion group is a security force for the civilians who are operating the aid agencies" (Canada, Department of National Defence, 1993, p.3, par. 5).

This definition of the role of the military is actually quite close to President Bush's view of the role of the UNITAF, as it was set out in his correspondence with Boutros Boutros-Ghali. It is thus not surprising that such ambiguities would generate confusion in the field. For example, the BOI's inquiries revealed that the Canadian troops went searching for hidden weapons during their patrols. Purnelle (1996: 161) claims rather that the CAR's commandos mainly seized weapons from villagers who needed them to protect themselves from armed thieves. He suggests they avoided a systematic search for weapons where they were really hidden, that is, in Belet Huen. It must be borne in mind that the issue of disarmament and of the possession of weapons by Somalis is at the heart of the problems experienced by the Canadians in dealing with the persons who were looting their compounds. When it was believed that Somalis were stealing weapons, the orders to fire on them were changed. Following a controversial order given by Lieutenant-Colonel Mathieu, a soldier could aim between the loincloth and the flip-flops if it appeared that the looter was carrying away something that seemed suspicious and that might be a weapon; the identification of the object did not need to be positive, the object only had to *look* suspicious. It was in the same spirit that Major Seward, the OC of 2 Commando issued his own order to "abuse prisoners."

The lack of unity in the command of the operation also resulted in other uncertainties in the field. The CARBG was first to be deployed in the northeast of Somalia near the seaport of Bosaso. Purnelle (1996: 124) describes how the troops learned only after their arrival in Somalia that they had been assigned a new sector (Belet Huen) by the Americans. He also relates that the Canadian and American definitions for imposing peace were different and that this added to the confusion in the field. As we shall see in my discussion of the ROE, Corporal Purnelle is quite right on this point.

The effects of the unclear mandate and the shortcomings created by a lack of cohesion in the command of the operation are numerous, and I could provide further examples. Instead of doing so, I shall quote two comments, one made by a French colonel and the other by a Canadian corporal, on the results of operation "Restore Hope," as it was called by the Americans, or "Operation Deliverance," as it was called by the Canadians. The French colonel is quoted by Stephen Smith (1993: 180), the correspondent of the French daily, *Liberation,* in Somalia:

> Basically, even when things are going well, we ask ourselves what is the point... There are no local authorities who can relieve us; we only deal with fake leaders who are totally corrupted. So, on the day when we leave, everything will collapse again because our politicians did not do their jobs. [My translation]

The Canadian corporal is quoted from Purnelle (1996) [my translations]:

There is a war here, but where? Why are we here? We were increasingly aware that those who had us sent knew nothing of this archaic world. Politics are lived here according to the rhythm of the sun, and what is true now is no longer true an hour later. Our soldier's dream will not be realised here. The only glory will consist in going back in one piece as quickly as possible. (Purnelle, 1996: 148)

Our feeling was that everything would start again as soon as we left and that the internal problem could only be solved by itself. We came too late. Famine had already destroyed and killed what it could and our aid was in reality only fictitious and very costly. It is hard to live here and to realize that humanitarian agencies are only stealing from the poor. (Purnelle, 1996: 159)

Although separated by country and by rank, the French colonel and the Canadian corporal both concluded that the mission was at best a temporary remedy and that it lacked a long-term purpose. This feeling of being part of an enterprise that serves no purpose is highly destructive of morale, and the collapse of morale is an important cause of wrongful behaviour by military in the field.

Once more I would like to stress that the mandate's ill-defined character is not used here as an excuse. But also once again, I wish to stress that the CARBG operated within a general context that had an adverse effect on members of the different contingents that took part in the intervention in Somalia.

The Lack of Military Intelligence. I cannot assert confidently that all participants in the UNITAF lacked intelligence on crucial aspects of their mission as much as the CARBG did. I have not thoroughly researched the action of the other participants in the UNITAF; what I know I have learned through a review of the literature.

There are, however, important aspects of the situation prevailing in Somalia that it would have been useful for the Canadians to have known in advance. I believe that the Canadian troops were disconcerted by the fact that many of the Somalis who were caught (or not caught) looting their compounds were children or young adolescents; Shidane Arone was only 16 years of age.

Although difficult to assess, there was a phenomenon, which field researchers and journalists described, that seemed to have played a part in the Somali conflict. Some of the most active combatants, particularly among the persons manning the technicals, appear to have been very young men referred to by Somalis and non-Somalis alike as "mooryaans." They are described by Stephen Smith as a cross between freedom fighters, thieves and dispossessed people inclined to riot [our translation].[14] Marchal (1993a), who studied them, reports that when the Mogadishu mooryaans heard of the Los Angeles riots from 29 April to 2 May 1992 that followed from the acquittal of the police who had beaten Rodney King, they started to dream about a crusading trip to Los Angeles to join forces with their Afro-American brothers. Apart from being very young, the mooryaans were daring, uncontrollable even by the war lords whom they served (according to their fancy), and unpredictable, at times terrorizing the Somali population.

Mooryaans are described by Marchal (1993a) as a typically urban phenomenon. Were there mooryaans in Belet Huen? We do not know. What we do know, however, is that young people played a significant part in attempts to loot the Canadian compounds and engaged in hostile behaviour against the Canadian troops. Had the Canadians been aware of a phenomenon like the mooryaans, they might have been in a better position to devise a strategy to deal with the youths.

There are numerous kinds of information that might have been useful to the CARBG that it did not possess. For example, due to the very high level of infant mortality in Africa, the life of an adult is significantly more valued than the life of a child. Life expectancy increases once an individual survives to adulthood. And adults can always beget more children as the high rate of fertility demonstrates. In Western countries, where infant mortality is quite low, the life of children is valued above everything else. Hence the culture shock for Canadian troops witnessing the lack of value put on children's lives. Such shocks could have been avoided if accurate intelligence had been disseminated by briefing the officers and non-commissioned officers (NCOs), who would then have been in a better position to command and to inform their men. But in Operation Deliverance, there was scant intelligence for any other topic than the physical health of the troops.

THE RULES OF ENGAGEMENT

Before discussing the ROE, I must reiterate what I have already said in chapters 1 and 3 that paratroopers do not make the best peacekeepers. This is fully confirmed by entries in Major Seward's diary and letters to his wife. Major Seward voluntarily turned over these documents to the military police shortly after the death of Shidane Arone. In one significant entry in his diary, Major Seward asked himself whether paratroopers/ commandos are suited for peacekeeping.

[Diary entry dated 6 Jan 93 - " 3.] It's difficult to say whether soldiers, especially Commandos are required for the task that we're doing. Capt Dan Alvin has said to me that my Commando is blood thirsty. He admires the soldiers aggressive patrolling techniques moving from fire position to fire position but advises that a more relaxed approach is required. He's suggesting that we move with the weapons at the ready, bounding from shadow to shadow, remaining alert at intersection and gaps but observing from a standing position vice [sic: the meaning is most probably "versus"] a fire position. I've already conveyed this to the Platoon Commanders but I intend to reinforce the concept again today and to view the rehearsals.

However, a Record of Reproof dated 16 January 1993, issued to Major Seward and signed by both Major Seward and L.Col. Mathieu, said:

Aggressiveness. Despite repeated direction by the Commanding Officer to reduce the level of aggressiveness exhibited by his Commando while conducting patrols in Belet-Wen [sic], Major Seward continued to permit his Commando to act aggressively toward the population. This was in complete contradiction to the Hearts and Minds policy being implemented by the Unit.

The "Hearts and Minds" policy referred to in this quotation was a strategy for winning over the Somali population by peaceful means, such as mediation and humanitarian assistance, rather than intimidating them with a show of force. Not only did Major Seward not pay heed to the advice given to him on January 6 by Captain Dan Alvin, but he was fined C$2,400 by Colonel Labbé after a summary trial for having accidentally discharged his weapon. In an entry in his diary dated from the same day that he was fined (19 January 1993), Major Seward wrote:

Lieutenant-Colonel Mathieu issued me a Record of Reproof for my lack of Control of the Commando and my personal conduct. Personal conduct was of course related to my accidental discharge. Lack of control was related to the following: Failure to follow direction; failure to ensure weapons control and aggressiveness — *if I hear any more "hearts and minds" bullshit I'm going to fucking barf!* [emphasis added]

These entries take on their full significance in the light of one of the first entries in Major Seward's diary (31 December 1992, mistakenly dated 31 January in Seward's original diary):

> Early this morning Sgt MaCauley had contact with a technical — a Toyota ½ t [half ton] mounting a machine gun, probably a 12s ... Sgt MaCauley had the machinegunner in his sights and was prepared to seize the vehicle. Rules of engagement however limited our actions to observation only. We should have had our first kill!

These entries in Major Seward's diary — he was the OC of 2 Commando — illustrate the paradox that surrounds the ROE that governed the UNITAF's action in theory. For reasons that I shall discuss, the ROE had many shortcomings, the main one being that they did not cover all the situations in which members of the CARBG might potentially have to use force (or to refrain from doing so). In other words, the ROE had to be supplemented by direction coming from the officers. If officers disagreed with the basic thrust of the ROE — the use of minimum force — the kind of direction they provided their men might well have been inconsistent with the letter or the spirit of the ROE.

It might be objected that the paradox is weaker than I make it sound, because Major Seward was not typical among the officers of the CARBG. There is no doubt that he stood apart because of his aggressiveness. As we have seen earlier, there was doubt expressed on several occasions as to the wisdom of sending him to Somalia, his attitude having been the object of severe criticism during the training of the CAR. However, even if we grant the fact that Major Seward was indeed, along with Captain Rainville, among the most aggressive officers of the CAR, it does not follow that officers who were less aggressive than he were in fact more moderate or no more aggressive than the average officer on a peacekeeping mission. All indications lead to the conclusion that however excessive the attitude of Major Seward, it was not inconsistent with the kind of esprit de corps that prevailed in the CAR and, in particular, in 2 Commando. Although the question of whether the CAR was truly an elite regiment is moot, it was clear that its members were viewed as more physical and more aggressive than members of other units. This point is explicit in the Hewson Report and comes out clearly in the BOI Report. The kinds of initiation rites that were held in the CAR were not designed to foster the values of mediation but rather those of aggression and resistance to physical pain. The number of CAR members who were later court-martialled for different incidents involving the unjustified use of force is significant. Finally, even the BOI Report, which is not overly critical of the CARBG's performance in Somalia, recognizes that there was a real problem with the frequency of accidental discharges of weapons. These accidental discharges are a sure sign of a kind of over-eagerness for combat. Purnelle (1996: 81) captures the spirit that prevailed in the CAR in one pregnant sentence: "we are taught war and made to improvise peace" (our translation)" [on nous apprend la guerre et on nous fait improviser la paix]. To conclude, the problem, which was exacerbated in the case of Major Seward, was nonetheless present in varying degrees with the other officers of the CAR, whether they were commissioned members of this regiment or not.

Let us now discuss the reasons why it was vitally important to provide more direction to the men than was given in the ROE.

First of all, the ROE were distributed to the troops after they had begun to deploy. They deployed between 20 December 1992 and 3 January 1993. The memorandum signed by M.Gen. Addy, which provided the terms of reference and ROE, was dated 24 December 1992. The BOI Report addressed this problem in the following terms:

> ...the issuance of the mission specific ROE occurred as forces were being deployed. For training purposes, the Canadian Airborne Regiment used the ROE which were at that time, and still are in effect in Bosnia-Hercegovina. Since these ROE were more restrictive than those issued for Somalia, this did not, in the opinion of the Board, present a significant problem. (BOI, Phase I, Volume XI, Annex G, par. 3)

The BOI's reasoning is quite clear and also plausible: it was easier to adapt to more permissive rules than it would have been to more restrictive ones. There is, however, another side to this coin. Persons moving from greater restrictions to lesser ones may be prone to taking greater liberties with the newer and more permissive rules than they really allow, thus "letting off steam" or "feeling more loose." This later interpretation is not without factual bases. In chapter 1, I quoted the court-martial testimony of Captain Hope who declared that since it was authorized by Chapter VII of the UN Charter, Operation Deliverance was "a straight military operation" rather than the usual peacekeeping mission under Chapter VI provisions. Captain Hope was nothing less than the intelligence officer of the CAR. What was believed at his level, that Operation Deliverance was a straight military operation, was likely to be exaggerated at the level of the troops and the non-commissioned members (NCMs) of the CF.

The need for delineating, with all due precision, the stringent limitations on the use of armed force that were embedded in the new ROE was greater because the transition from the requirements of an operation conducted under Chapter VI of the UN Charter to another one pursued under the authority of Chapter VII was at risk of being misinterpreted. Not only was it misinterpreted by Captain Hope, the CAR's own intelligence officer, but it was also crucially misunderstood by the men and the NCMs, as is clear from their testimony at the courts-martial, before the BOI and the CIDCFS.

Second, the ROE distributed to the men were inaccurate in important ways. The Canadian ROE stated that riot control agents might be used in the case of unarmed harassment or threat, before resorting to deadly force. However, the Commander of the UNITAF forbade the use of tear gas as this was against U.S. policy (BOI, Phase I, Volume XI, Annex G, par. 7). Purnelle (1996) bears witness to the fact that this kind of contradiction severely undermined the will of troops to follow ROE which were, in fact, shown to be at odds with the policy of the higher command of an operation.

Third, there is a wide discrepancy between the ROE issued to officers and their summary distributed to the troops. The ROE in M.Gen. Addy's memorandum of 24 December 1992 cover seven pages. The substance of these rules was communicated to the troops on a plastic card. This discrepancy is quite normal. One would not expect soldiers in the field to carry with them a seven-page memorandum. However, this situation demonstrates the need for higher ranking officers to supplement the sketch provided to the troops in the field with more explicit directions.

This need is all the more obvious when we look at the plastic card — field aide-mémoire of Operation Deliverance — that was distributed to the troops. Since I will comment on this aide mémoire, I will excerpt from it:

FIELD AIDE-MÉMOIRE OF OP DELIVERANCE
RIGHT OF SELF DEFENCE: Every soldier has the right to take all necessary and appropriate action for self-defence.
MINIMUM FORCE: Only the minimum degree of force required to deal with the situation shall be used.
DEFINITIONS ...
[HOSTILE FORCES, HOSTILE ACT, HOSTILE INTENT, TERRORIST ATTACKS]

GRADUATED RESPONSE. (The level of response to be exercised initially will depend on the situation, as common sense dictates.)
a. warning (verbal, visual);
b. minimum force (manoeuvre, physical restraint);
c. warning shot; and
d. deadly force (use of firearms or crew-served weapons).
USE OF DEADLY FORCE PERMITTED FOR
a. Self-defence;
b. Defence of Coalition member;
c. Defence of relief personnel and supplies; and
d. Defence of Noncombatants.
THREAT/RESPONSE
a. Threat - Unarmed harassment or threat (should first take the following procedure)
Response.
(1) verbal warning
(2) show of force
(3) warning shot
(4) deadly force (use as last resort) [Note: Riot Control Agents may be used earlier if approved by Commander of CIFS].
b. Threat - Armed Individual
Response [the same as above, except that "minimum force" is substituted for "show of force" in "b."]
c. Threat - CREW-SERVED WEAPON
Response (same as in b.)
d. Threat: AIRCRAFT (there is no identified threat)
Response.
(1) identify aircraft
(2) use deadly force only if the aircraft is committing a hostile act.
DETENTION OF PERSONNEL. Personnel who commit a hostile act, or demonstrate hostile intent, interfere with the mission may be detained when ordered by the commander.
UNATTENDED MEANS. No booby traps or mines are authorized.

In contradiction to the expectations of the members of the CAR who believed that the switch from Chapter VI to Chapter VII meant that the CAR was involved in a "straight military operation," the guiding principles of the field aide-mémoire are completely consistent with a "straight peacekeeping operation." Its cornerstone is the right of self-defence. The right of self-defence is coupled with two other key principles, namely the use of minimum force and the resort to a graduated response. However, the principle of self-defence is the only basis for the use of deadly force, which corresponds to the use of maximum force and is at the end of the spectrum of a graduated response.

There is another feature of these ROE, as they were communicated to the troops, that deserves to be mentioned. The rules for responding to threats are, to a large extent, perfunctory: the graduated response is the same for unarmed harassment or threat, armed individual and crew-served weapons, namely the sequence of verbal warning, minimum force, warning shots and deadly force. That perfunctory sequence gets even less instructive when it is remembered that the use of riot control agents was prohibited by the American command in the case of unarmed harassment or threat. What were the troopers expected to do then? Fire into an unarmed crowd or use combat helicopters to disperse rioting mobs? It is precisely in situations such as these that the troops needed additional guidance. Did their officers possess the required will and peacekeeping competence to provide this needed guidance? Our hypothesis, which might be explored by the CIDCFS, is that they did not.

There is one topic on which guidance was particularly needed, because the ROE were noticeably deficient. This topic is the arrest, detention and treatment of detainees. It was in this part of the operation that the most serious abuses took place, most notably the torture and killing of Shidane Arone.

First, as can be seen by reading the field aide-mémoire, there is almost no guidance on this subject: the aide merely declared that a person behaving in a hostile manner can be detained when the commander orders it. Nothing is said about the place and condition of detention.

Second, the longer version of the ROE is also very brief on this subject. It states that persons hostile in act or in intent may be detained and that "detained personnel will be evacuated to a designated location to appropriate military authorities." This rule raised more problems than it solved, as the BOI Report makes abundantly clear. Within the Canadian Joint Forces Somalia, there existed no clear legal authority to detain Somalis on a long-term basis. Canada not being at war with Somalia, there was the additional problem that Somalis were not war prisoners. UNITAF's headquarters were aware of the problem, and took the easy way out by suggesting that no Somalis be detained for any length of time. The solution of handing over the detainees to Somali authorities or clan elders was potentially risky for Canada. According to the BOI Report, there is an obligation under the Geneva Convention for a detaining power to ensure that a transferred prisoner will be properly treated by an authority into whose custody that prisoner is transferred. Actually, even the obligation is not clear, since the Geneva Convention applies to prisoners of war, which the Somali detainees were not, at least technically speaking (BOI, Phase I, Volume XI, Annex G, p. 3, par. 8). The BOI concludes its discussion of the issue of detention by saying that in the case of a new type of operation, such as peacemaking, the necessary structures and policies to support the tactical commanders do not exist within the UN. Therefore "these circumstances, in the opinion of the Board, call for a high degree of judgement, initiative and innovation by the in-theatre commanders (BOI, Phase 1, Volume XI, Annex G, p. 3, par. 8)."

Such "a high degree of judgement, initiative and innovation" was obviously not exercised in the course of Operation Deliverance. On the contrary, it is the very want of these qualities that had a dire impact on the success of the mission.

ROE are not the only guidance that can be provided to commanders in the field and to the troops. What are called Standard Operating Procedures (SOPs) have been developed in peacekeeping operations in the past to supplement the ROE. According to the analysis of an interview given by Major Seward to the military police on the morning of 28 May 1993, there were no SOPs in regard to the handling of prisoners, and the functional responsibility for dealing with prisoners was passed to him by L.Col. Mathieu. In view of his record for aggressiveness, it was not the wisest decision. It ultimately had disastrous consequences.

I end this brief discussion of the inadequacy of rule-governed guidance in the case of the detainees with a quotation from a recent book, on the psychological cost of learning to kill in war and society by U.S. L.Col. Grossman (1995). Grossman devotes numerous chapters to war atrocities defined as "the killing of a noncombatant, either an erstwhile combatant who is no longer fighting or has given up as a civilian" (Grossman, 1995: 193). Grossman suggests pertinently that modern war and, particularly, guerilla warfare blur such distinctions. The definition of atrocity developed by Grossman applies directly to the torture and killing of prisoners (an erstwhile combatant who is no longer fighting because he has given up). In the course of his analysis, Grossman makes the following point, which is immediately relevant for the situation that I have been describing.

> The most important point here is that nobody has ever pointed out to me the potential repercussions of improper POW handling. No leader of mine has ever

> stood up and clearly stated this position to me and defended it (that the mishandling of prisoners is one of the worst things to do in trying to win a war). In fact the opposite has occurred. As a private and a sergeant, I have had enlisted superiors strongly defend the execution of POWs whenever it was inconvenient to take them alive and at the time I accepted it as reasonable. But they never made me understand the vital importance and the deadly ramifications of POW handling (or mishandling) on the battlefield because I think they themselves did not understand. (Grossman, 1995: 205)

As I previously remarked, the Somali detainees were not prisoners of war in the technical sense of the term. They certainly qualified as prisoners and as non-combatants whose mishandling and killing equalled atrocity. The crucial point made by Grossman (1995), in a book written long after the enactment of the Geneva Convention and other protocols protecting the rights of detained non-combatants, is how little attention is paid to this subject either in training or in the guidance provided by officers in the field. One very clear conclusion that comes out of my analysis of the ROE, particularly field instructions regarding detention, is that the CF cannot rely on the fact that there are legal texts and conventions protecting detainees taken into custody in the course of military or peacekeeping operations. These rules must be supplemented when they are incomplete or inadequate and, most important, their content must be communicated to the troops with firm orders regarding the obligation to respect them.

Elements of the Situation In Theatre

What happened in Belet Huen was exactly the opposite of the conventions I have been discussing. Major Seward, the person holding the functional responsibility for dealing with prisoners, issued an order to "abuse the prisoners." How could this have happened and what led to this aberrant order being carried to an extreme? This is not a question that I am now in a position to answer. The CIDCFS will no doubt shed much light on this question. In this report I explore what is known of the facts surrounding the violent incidents that I have already briefly described. I remind the reader again that the purpose of this exploration is to reflect on elements that may provide an understanding of what happened. It is not to look for excuses. Finally, my discussion is selective rather that exhaustive, focussing on the broader factors. Donna Winslow's report (1995) to the CIDCFS provides an extensive description of the in-theatre phase. The factors that I discuss were generally widespread and had an impact on a significant number of the members of the CARBG, if not on all of them.

Alcohol and Drugs. I have already mentioned that large amounts of alcohol were consumed by CAR members in Somalia. The presence of considerable quantities of alcohol, in particular beer, is often referred to in the transcripts of the interviews of the military police. In one of these interviews, Private Furdal claimed that he had been drinking too heavily on the night when Shidane Arone was killed to have any precise recollection of what happened.

The prevalence of alcohol consumption was also mentioned during the BOI's hearings, although not as clearly. The videotapes shot by members of the CAR in Somalia offer incontrovertible evidence that a significant amount of alcohol was consumed during Operation Deliverance. In one of these tapes, a corporal is heard saying that the only good thing about this mission is that the troops have free access to a kind of beer that seems to be particularly appreciated by the troops. One of the most damning indictments comes from Purnelle (1996: 147, 169) who not only asserts that beer was flowing freely, but that one warrant officer, generally responsible for enforcing discipline, was a known alcoholic. According to Purnelle (1996: 146), the CARBG higher leadership often indulged in drinking bouts. Finally, Purnelle (1996: 147) claimed that drugs were imported to Belet Huen by soldiers returning from holidays in Nairobi. The drug problem was so significant in 1 Commando that its OC, Major Pommet, ordered a search of all tents for hidden drugs. Although some of these claims should be confirmed by testimonies given at the future hearings of the CIDCFS, the sum of evidence and the variety of its sources are such that it can be safely assumed that there was a problem with the consumption of alcohol and drugs among the Canadian troops in Somalia.

It was asserted by different witnesses at the court-martial of Private Kyle Brown that Corporal Matchee had consumed a large amount of alcohol during the night he tortured and killed Shidane Arone. Several witnesses interviewed shortly after the incident by the military police also said that Corporal Matchee had had several beers that night (see, for instance the interview with Corporal McKay). Because Corporal Matchee is incapable of testifying, it is unlikely that we will ever get to the bottom of this affair. Whatever the truth is, it cannot be doubted that freely flowing alcohol accentuated whatever potential for mischief was already present. It remains to be seen to what extent it undermined the respect of the men and NCMs for their officers.

The Situation Confronting the CAR. Another factor, which is probably more important in that it affected all the men, is how much what they faced in Somalia contradicted what they had been led to believe from various sources including the avidly watched the Cable News Network, before they left Canada. The principle contradictions, as gathered from testimonies given by members before the BOI, courts-martial and from the videos and Purnelle (1996) include the following.

Famine: What was called the famine triangle lay to the south of Mogadishu and was a figure roughly circumscribed by lines joining the three cities of Baidoa, Berbera and Kismayo. The CARBG was deployed around Belet Huen in the south central part of Somalia, outside this triangle. Although Canadian soldiers witnessed some famine (for instance in the village of Caagaboy, as reported by Purnelle (1996: 142), the overwhelming majority of testimonies by members of the CARBG on this subject suggest that they were deployed in a zone where there were many more black marketeers profiting from international relief than people dying of hunger. This consideration greatly undermined their sense of purpose.

War: It also appears that the Belet Huen zone was fairly peaceful and that little combat occurred in it. There were incidents that involved a measure of danger, but the CARBG was faced on the whole by hostility, aggressive curiosity and would-be looters. Purnelle (1996: 149) writes in this regard that what the troops were secretly wishing for was to be attacked, so they could defend themselves — that something at last would happen to make them feel like soldiers. Whether they took out their frustration for the lack of action on the few looters they could catch will always remain a plausible hypothesis. Even if this hypothesis were confirmed by witnesses, which is highly unlikely, it would only constitute one part of the explanation.

Corruption: In contrast with the expected famine and civil war, the troops witnessed a high level of corruption. To their dismay, goods that were taken into villages under UNITAF protection found their way back into the black market. This is not the place for a digression on what was perceived — and what was real — as corruption. There are, however, wide cultural and religious differences about what constitutes acceptable trade practices. For Moslems, for instance, charging an interest rate on loans is a sin. The point here is not to wave away with some airy notion of multiculturalism the "corruption" plaguing African (and other) countries but to emphasize how little prepared the members of the CARBG were to deal with the situation into which they were plunged.

Hostility: Instead of meeting clearly identified enemies, the CARBG were met at first with indifference and increasingly with hostility. This hostility was not the kind of hostile intents, acts, forces and terrorist attacks with which the ROE were replete but by an insidious animosity omnipresent in the general population, and particularly in demonstrating crowds and looting youths.

The end product of the failure of reality to fit expectation — no famine, no war, no gratitude — was threefold. First, there was the oppressive boredom, which is depicted vividly in the CAR's videotapes of their mission. Then this boredom begot a craving for action that was never really satisfied and that found a deviant expression in the incidents I have described. Finally, there was an overwhelming confusion as to the purpose of the mission and the deployment of the CAR in Somalia. It is this sense of confusion, expressed in the testimonies of the troopers and NCMs before the BOI, which was the strongest catalyst for the most serious abuses of power.

The Youth of the Looters and of the Detainees. A strong element for destabilization was the youth of the Somalis displaying hostile behaviour toward the Canadians. This hostile behaviour took different forms, the most irksome apparently being acts of looting. The members of the CAR captured a number of looters, mostly youths, and abused them in various ways. The most frequent kind of abuse was putting captives on public display with their hands painted white or with signs around their necks proclaiming that they were thieves. The most serious incidents involved ambush and the wounding or killing of looters. Of course, most reprehensible of all, was the torture and killing of Shidane Arone. In his enumeration of incidents that occurred in Somalia, LeBlond found no less than seven cases of abused prisoners, including the Arone incident. There may have been more incidents of shaming prisoners by putting them on public display; LeBlond reports numerous instances in which photographs were ordered to be destroyed. These might have revealed other cases of abused prisoners. There was at least one other incident of abusing detainees. In a letter to his wife dated 27 January 1993, Major Seward tells how five Somali teenagers were caught stealing from Service Commando, adding that: "The troops are however taking advantage of the situation to put on a demonstration. They're pretending that their intentions are to cut the hands off these kids with machetes. It sounds awful but if you were sitting here you'd be laughing too. Soldier's humour is infectious."

Major Seward acknowledges in his letter that he is indeed responsible for co-ordination of police matters "and hence [I have] the responsibility for the care of the teenagers." Yet as the officer bearing the responsibility for the treatment of these teenagers, he condones the behaviour that is described in his letter. How can we account for this apparent lack of judgment?

In his interview with the military police (28 May 1993), Major Seward reported that on approximately 14 March 1993, Colonel Yussuf, who was the Police Commissioner of the region of Hiran, came to his office to request that three looters be shot to demonstrate to the local population what would happen if they were caught infiltrating the Canadian compound. That Colonel Yussuf made such a request is not entirely implausible, given what I previously said of the Syad Barre police state and its systematic abuse of the rights of Somalis. Mr. Abdi, a native Somali who was acting as interpreter for the Canadians, stated in an interview granted on 26 March 1993 to the military police "that looters should be killed for their action against the Canadians." Furthermore, a CIDCFS note from C. Bouchard to D. Pomerant — dated 18 July 1995 — referred to the fact that, on 21 January 1993, two Somali youths were caught attempting to steal military equipment inside the Service Commando Camp. They were placed on public display with a sign stating that they were thieves. According to Captain Jeff Alexander DiLallo, this measure was recommended by a Somali civilian lawyer. However, Captain DiLallo did not give the investigator questioning him a written statement in relation to his testimony.

Major Seward said that he refused Colonel Yussuf's request. He furthermore testified on several occasions that he wanted to avoid another incident like the one that had taken place on 4 March 1993, when one Somali was killed and another injured in what was described as nothing less than an ambush set up by Captain Rainville. Major Seward claims to have given his notorious order to abuse the prisoners in an attempt to deter the looters by roughing them up at the time of their apprehension. Then when they were handed back to the Somali authorities, the population could see that infiltrating the Canadian compounds carried a stiff price.

This explanation may well be an apologetic rationalization after the fact. Furthermore, Major Seward's orders actually resulted in one more killing — Shidane Arone was abused to death. This only raises, in another and more dramatic way, my original question. Why was it, I asked, that officers showed poor judgment in condoning and even ordering the abuse of prisoners? It can also be asked, why was it that an order to rough up prisoners was interpreted as licence to torture?[15]

There is no unequivocal answer to this question, which reaches into the darkest places of the human soul. Yet, I submit that the fact that a great number of the looters were either children or young adolescents was an important factor in their abuse. We profess to respect and cherish persons who are more vulnerable (children, women, the elderly). Yet it also seems that they bring out the worst in us: when we feel justified in inflicting punishment on them, we tend to over-react. This is especially true when young teenagers are perceived to be offenders. In many countries now, they are the object of extreme abuse.

Not all this abuse stems from a diabolic desire to hurt. The incident where five teenagers were led to believe that their hands would be cut off by machetes was regarded as an infectious joke by Major Seward and his troops. The risk that the "joke" would result in lasting trauma for the youths who were its object did not occur to them. The equivalent measure of punishment for adults is the staging of a faked execution. Faked executions are generally perceived as a particularly sadistic form of mental torture, which no one in his right mind would mistake as a joke. Another indication of the ambiguous nature of the CAR's treatment of young offenders is the fact that in the photograph published in the Montréal daily *Le Devoir*, an army chaplain is standing close to a group of young prisoners. They are tied up and sign-posted as described above. Why would a chaplain condone such treatment unless he did not believe that it was really mistreatment? The fact that it was youths who were submitted to such treatment may have generated confusion as to its real nature. We have all seen pictures of naughty schoolboys, wearing dunce caps, perched on stools at the front of their class.

The Lack of Military Police. In his study of the consequences of the limitations imposed by ceilings on the number of personnel, Eric Myles of the CIDCFS research staff found that there were only two military police in the Belet Huen area (see CIDCFS, Final Report). They were to play a crucial role. The only precise guidance given by the ROE on detainees was that they were to "be evacuated to a designated location for turn-over to appropriate military authorities." Such a guideline made sense in locations such as Mogadishu, where the Americans had provided a designated area for evacuating prisoners. The situation was drastically different when the CARBG was deployed in the region of Belet Huen, much too distant from Mogadishu for two military police to maintain by themselves an adequate liaison with the American military police company stationed there. If two military police were too few to maintain an adequate liaison with the Americans, they certainly could not adequately assume all the duties related to the apprehension and detention of the looters and other troublemakers in Belet Huen.

The consequences of this shortage of military personnel were much more drastic than appear at first glance. One of the key principles of the criminal justice system — indeed one reason for its very existence — is called "Montero's aim" (cf Walker, 1971: 12). According to Montero, a Spanish jurist, the criminal justice system exists among other reasons to protect the offenders from unfair retaliation. In other words, Montero's aim is an anti-lynching device. The offenders are at the greatest risk of being unfairly retaliated against when their victims or those close to their victims have the responsibility of apprehending and punishing them. Hence, the general need for a justice system that would process offenders independently from the pressures of victims (which may, of course, be taken into consideration as one factor among others). On a smaller scale, the lack of an adequate force of military police in Somalia had the disastrous consequence that the looters were caught, detained, and in certain cases, sanctioned by their victims, that is, the soldiers from whom they had been stealing equipment and even weapons. This situation, where the victims were also metting out the justice, at least in part, was dangerously close to that which leads to lynching. In a way, it might be said that Shidane Arone was lynched. The mob — in this case the CAR members within earshot — were not standing around ready "to string up the rope," but they could not have helped hearing Arone's cries; they had to know what was happening to him, and they did nothing to stop it.

Conclusions and Remedies

The first conclusion of this examination of the use and abuse of force in Somalia is that any explanation of what went wrong in-theatre has to be based on a discussion of both internal factors — factors which were specific to the CARBG — and of external factors — the broader context. By the latter I mean not only that factors such as the physical nature of the terrain, the heat and the living conditions should be taken into consideration, as they must. I mean that the performance of the CARBG should be assessed in comparison to that of the other contingents. If this is properly done, it will be found that the Somali intervention is not only a blemish on the reputation of the UN, but that there are few countries who participated in this intervention who did not experience situations similar to those of the Canadians and who did not get into serious trouble. Consequently the relevant question is not only why did the CARBG get into trouble, but, why did the Americans, the Belgians, the Canadians, the Italians, the Pakistanis and so forth experience considerable difficulties during this mission? Asking this question does not mean that the CIDCFS must pursue an inquiry into the behaviour of all these other contingents. It means, however, that there are elements of explanation that necessarily rest outside an investigation exclusively focussed on the CF. Among others, these include the considerable ambiguity of the mandate, the behaviour of the Somali population and its lack of consent to the intervention. Certainly, not all contingents had commandos who revered the Rebel flag. Yet racism was rampant among the troops who operated in Somalia. The overall explanation must rest on elements that were common to the troops deployed in Somalia; to these common elements specific factors should then be added to account for the behaviour of the Canadians.

The second conclusion is that there was a really dismal failure of CF intelligence in respect to the Somali exercise. Whatever handicap the speed of the switch from Operation Cordon to Operation Deliverance may have represented, both operations were to occur in Somalia. It might then have been assumed that the CARBG would be much better informed about what to expect. Instead, the extent of the ignorance of the troops and their commanders about situations in which they were to be deployed is nothing less then astonishing. They were told a lot about the reptiles and insects of Somalia and almost nothing about its people.

RECOMMENDATION XIV: **If the CF is to continue being involved in peacekeeping operations, it should develop an intelligence capacity which it completely lacked at the time of the intervention in Somalia. Let us remember in this regard that Captain Hope, the CAR's intelligence officer, rushed to the *Encyclopedia Britannica* when he learned that Canadians were to be deployed in Somalia under Chapter VII of the UN Charter. Such shortcomings must be remedied.**

Hirsch and Oakley (1995: 151) believe that Somalia very soon became a laboratory for testing out the application of new theories of peacekeeping. I believe this is an accurate assessment. Whether anything positive came out of the Somali laboratory, however, remains to be seen.

The main task in assessing this enterprise and the possibility of applying the concepts that it embodied is, I believe, the following. The BOI described very lucidly the kind of operation in which the CF was engaged:

> The Board noted that the Battle Group in Somalia experienced a profoundly different operational environment from that previously encountered by Canadians during other peacekeeping missions. This was peacemaking, which might informally be described as falling somewhere between peacekeeping and low-intensity war. (BOI, Phase I, Volume XI, Annex G, p. 5, par. 4)

I tend to agree with this description, with two reservations. The first is that the conditions in the field varied greatly from region to region. There tended to be more of a low-intensity conflict in Mogadishu. In comparison, it was much quieter in the region where the Canadians were deployed (the French were also deployed in a more peaceful region). The second reservation is that during the last months of the UNITAF and the first months of the UNOSOM II, peacemaking degenerated into a full-scale war.

Those reservations made, the general description of the operation as proposed by the BOI retains most of its validity. One of the basic features of UN and UN-sponsored operations is that they are governed by rules, with ROE, SOPs and other kinds of procedures. I have tried to show in detail to what extent the ROE were deficient in Operation Deliverance. This deficiency, which became more acute as the intervention developed into an open military conflict in 1993, is actually such that it raises at least four general questions. (1) Is there such a thing as a situation that falls between war and peace? (2) If there is such a situation, are the recently developed concepts of peacekeeping, which rest in great part on the *imposition* of a solution on a nation *without its consent*, viable concepts, or are they but mirages created by armchair strategists? In reflecting on this question, it must be remembered that the consent of a nation to an intervention on its territory was the cornerstone of peacekeeping as developed by Lester B. Pearson and Dag Hammarskjold. (3) If the answers to the preceding questions are positive, to what extent is it possible to constrain operations conducted in the twilight zone between war and peace by explicit rules? In other words can a low-intensity conflict be rule-governed in any significant way? (4) What rules would be applied, if there is any ground for their application?

These questions are complex and hard to answer. Let me briefly illustrate the difficulties. The Department of National Defence has recently issued on authority of the Chief of the Defence Staff a work on military doctrine entitled *Use of Force in CF, Joint and Combined Operations* (B-GG-005-004/AF-005, 1995-04-13). Chapter 4 of this document, entitled "Use of Force in International Operations," has a section on the ROE during international operations (section 403). ROE are discussed in relation to the "Spectrum of Conflict." This discussion bears on ROE in peacetime, this subsection including ROE during non-combat operations and ROE during periods of tension and crisis. The discussion also tackles the issue of ROE during international armed conflict and takes into account political considerations, safety and operational considerations. This discussion is followed by a section on rules of engagement at the outbreak of sustained hostilities. The rules which are proposed are of an extremely general nature.

What is interesting about this doctrine is that it is still based on the dichotomy between peacetime and conflict. For example, "nation-building missions and international enforcement regimes" are discussed under the heading of "ROE in Peacetime" and the subsection (1) entitled "ROE During Non-Combat Operations." Preservation of this dichotomy is particularly surprising in view of the fact that Operation Deliverance was precisely a nation-building and peace-enforcement operation, and experience has shown that this kind of operation tends to drift toward low-intensity war, if not all-out conflict.

A "Supplementary Direction During International Operations" including sections on the authority to stop and search and on the treatment of detainees is part of this doctrine. Nevertheless, none of the questions that I raised above are systematically discussed. This remark is not meant as a criticism of the document, presently available only in its "Advance Copy" non-final form. The real purpose of this remark is to emphasize the difficulty of genuinely tackling the questions that I have raised above. It is my conviction that these questions must be answered to avoid repeating the mistakes that marred the operations in Somalia.

CHAPTER FIVE

Post-Deployment

INTRODUCTION

This chapter addresses two basic issues. The first is accountability; the second is the relationship between the Canadian Forces (CF) and the press. These two issues are closely linked: pressure from the media played a crucial part in making the CF account for the behaviour of its members. The discussion of each of those issues will be followed by conclusions and suggestions on how to remedy the problems that have been identified.

Because the Commission of Inquiry into the Deployment of the Canadian Forces to Somalia (CIDCFS) did not deal with post-deployment issues in its hearings, my sources will mainly be of a documentary nature. Whatever is relevant in the Board of Inquiry's (BOI's) hearings and also in the hearings conducted so far by the CIDCFS will naturally be used.

ACCOUNTABILITY

This section is divided into five parts. I begin by stating the main conclusions that came out of my previous review of the research literature on accountability, and I also elaborate on these conclusions. However, my main purpose in this first section is to discuss the conception of accountability that prevails in the CF. Second, I want to discuss a specific issue that will illustrate the problems the CF is facing in connection with the issue of accountability. This issue is racism and, more particularly, the penetration of the CF by organizations that are racist and linked to the extreme right. In the original sketch for this report, I was supposed to discuss the issue of racist and extreme right organizations in chapter 4. However, since most of the material that deals with this issue was produced during the post-deployment phase, I decided to postpone my discussion of certain aspects of the extreme right's penetration into the CF until now. Third, I will pursue, in a more focussed way, a comparison of the structures of accountability in the police forces and in the CF. Fourth, I will compare the mechanisms that promote accountability in the Canadian security intelligence agencies with that of the CF and ask whether some of these mechanisms could not be borrowed to enhance the CF's accountability. Finally, I will propose conclusions that stem from the previous analyses and suggest remedies for the deficiencies that have been uncovered.

The Theory of Accountability Revisited

In this subsection, I briefly formulate again the main conclusions that I arrived at in the first chapter of this report and I elaborate on them.

I shall state, as briefly as possible, the main conclusions that I came to in the first chapter. This section was structured according to three questions: who should be accountable, what should be accounted for and to whom should an organization be accountable? The answers to these three questions were preceded by a discussion of definitional issues and preliminary remarks. They were followed by a presentation of different models of accountability, the policing model being compared to a potential CF model.

In my preliminary remarks, I tried to distinguish between accountability and responsibility, while recognizing that the notions overlap since they are but different forms of "answering for" something. The basic distinction that I found between the two notions is that accountability is an *information process* through which an accurate description of individual, collective and organizational behaviour is presented and through which justification for the behaviour thus described is offered (particularly when the behaviour described is potentially objectionable). Responsibility is an *action process* through which known behaviour is answered for and sanctioned either positively or negatively. This distinction implies that the attribution of responsibility is dependent on the success of the accountability process: if there is no knowledge of what happened in a given situation, it is impossible to determine who was responsible for what.

In my preliminary remarks, I also used the concept of the thickening of government, which I borrowed from Light (1995). The thickening of government has at least three kinds of implications for our purposes. First, because present government agencies are now characterized by the multiplication of bureaucratic layers, it is increasingly difficult to get an adequate picture of what happened in a situation, the perspectives presented often being in conflict. Second, this difficulty has an impact on the determination of responsibility, which varies according to the different accounts of the situation. Finally and crucially, the development of structures to enhance accountability must not add significantly to the thickness of government structures. They should be modest, simple, flexible and geared to maximum effectiveness (results) and efficiency (a good cost-benefit ratio).

The first question I asked after these preliminary investigations was who should be accountable. Based on the premise that accountability was an information process, I argued that an organization as a whole was the primary subject of accountability. Making only individuals accountable would generate incomplete, fragmented and often conflicting accounts. Using the words accountability and responsibility in their colloquial rather that their theoretical sense, I could say that accountability is a collective/organizational responsibility rather than an individual one.

The second question was what should be accounted for. The answer follows, in great part, from the answer given to the first question: what should be accounted for is the organizational behaviour of an agency rather than the individual behaviour of its members considered independently. The difference between individual and organizational behaviour is particularly easy to illustrate in the case of an agency, such as the army, where members act, at times, individually and, at times, under the orders of authorities within the organization.

A second way to tackle the same question is to use Light's analysis of the different aspects of accountability (Light, 1993: 14; see Table 1of this report). Light distinguishes between three approaches to accountability, which he respectively designates as the compliance, performance and capacity building approaches. The compliance approach aims at making individual behaviour conform to rules; the performance approach seeks to enhance efficiency and effectiveness through incentives; and capacity building is an attempt to achieve the same result at the level of the whole organization rather than at the level of its individual members.

For reasons that will be stated later, but which can now be intuitively grasped, the aspect of accountability that is most characteristic of an organization, such as the army, is compliance. There are, among others, three problems with this approach

- The compliance view of accountability often results in a trade-off between rigid conformity and the values of innovation and inventiveness. In a context such as peacemaking, for which there are few accepted rules, such a trade-off may yield adverse results.

- As I repeatedly argued, the compliance approach is crucially incomplete as it targets the individual rather that the organization.
- Finally, compliance is generally taken for granted by organizations which give priority to accountability. It is taken for granted at least in the sense that it is assumed that the rigid enforcement of a system of regulations will secure compliance and that few assessments of the degree of compliance thus achieved are actually made.

The third question that I raised was to whom an organization should be accountable. Again, the answer is determined by the answers given to the preceding questions. A basic feature of accountability is that you account to *another.* When self-accountability is the only form of accountability, it is equivalent to a denial of accountability. It follows from this feature of accountability that an organization considered as a whole — the proper subject of accountability — can only be accountable to another body. All internal accountability is only a form of self-accountability, when the organization is considered as a whole, as it should be in the true perspective on accountability.

This conclusion does not mean that there is no place for internal processes of supervision. On the contrary, I stressed the crucial need to avoid external interference in the conduct of an agency's operations. Such interference may result from a confusion of management and control. Control should be handled internally, but with oversight, is better exercised from the outside.

Finally, as I have said, I explored different models of accountability which are applied within police forces or intelligence security agencies, with a view to finding elements that could be borrowed from these models to develop a framework for accountability that would be more suited to the CF. I will not, at this point, repeat the main conclusions that came out of this exploration, because I intend to focus explicitly on the results of comparing different models of accountability, as they are found in other organizations.

Before leaving this summary of previous analyses, I want to elaborate briefly on their result. I have repeatedly emphasized that accountability is a process that applies as much, if not more, to whole organizations as it does to individuals in these organizations. There is one consequence of this perspective that I would like to develop through the tentative concept of "organizational agency." The word "agency" is taken here to mean the capacity for acting or for exerting power; it does not refer to an establishment engaged in doing business for another party, as in "advertising agency."

This notion of organizational agency takes on a particular significance when it is applied to organizations which have strong hierarchical structures, as is the case with an army.

Although one cannot, of course, deny that individual behaviour affects the environment and other people, there can be, and there generally is, great variation in the impact on others of different lives and of each individual in his or her own particular circle. Many people live unnoticed except by a rather small circle of persons. In the case of hierarchical organizations, such as an army, individual behaviour takes on an increasingly important organizational dimension as one moves upward in the hierarchy. After some threshold (e.g., colonel or general), almost anything that one does professionally ripples through part of the organization, if not throughout the whole of it. For instance, there are times when a rank and file soldier will dispatch his or her professional duties with little consequence for the organization considered as such (notice that these soldiers are appropriately called "privates"). However, it is difficult to imagine a single instance of a general's performance of professional duties that will not affect in any way, or at the very least, in a small way the organization to which he or she belongs.

The upshot of this notion of organizational agency is that if accountability refers more to organizational than individual answerability, it follows that the notion of accountability applies to persons in proportion to the importance of their organizational agency. There is a point when a person's agency meshes almost completely with that of his or her organization. In other words, the higher one is in a hierarchy and the more power one exerts within that organization, the more one should be accountable for his or her actions. Notwithstanding this, what is striking in highly hierarchical organizations, such as the armed forces or police organizations, is that the lower one is in the hierarchy, the more one appears to be accountable. The basis for this inversion is that the notion of accountability that prevails in such rigidly structured organizations is generally punitive, being applied from above through the process of discipline. Such a conception of accountability fosters the confusion between accountability and responsibility, as I have tried to articulate these notions. According to my view, penal responsibility is a top down process whereas accountability is a bottom up process. If almost everything that persons higher up in the hierarchy do professionally carries consequences for the organization, this is all the more reason to require that these persons be made accountable, the target of accountability being the organization rather than the individual.

There is one final aspect of accountability that I want to mention before leaving the subject. I have stressed that accountability was an information process by which an individual provided an account of his or her conduct and by which an organization accounted for its activities. It would be naive to believe that these accounts are only going to be used by those to whom they are presented as a means to enrich bureaucratic or academic knowledge. Accounts are followed by assessments and assessments by directions or instructions. In other words, whoever provides an account of himself or herself also makes himself or herself open to control. Quoting Goldring and Wettenhall (1980: 136), Stenning (1995b: 51) stressed rightly that accountability and control were "two parallel and interlocking mechanisms." Control is a notion that contains a very wide spectrum of practices, ranging from retribution, administration, instructions and directions and policy guidelines. If external accountability necessarily fostered the use of all these practices by the external body or person requesting the account, the unescapable conclusion would then be that accountability is the door to interference. But as I have argued before, accountability does not in any way imply the use of all these kinds of control. Some, such as administration and retribution, must remain internal practices whereas others, such as policy guidelines and directions with regard to objectives or compliance with legal constraints, can come at least in part from outside an organization.

Accountability as Viewed by the CF

The closing remarks of the preceding section are general and are meant to apply to organizations which are highly structured by hierarchy. I will now examine whether they apply to the CF, which certainly qualifies as a hierarchical organization. My sources for presenting the CF's view of accountability are essentially documentary, the CIDCFS not having held its hearings on the post-deployment phase. Among other documents, the CF has submitted a brief to the Commission on Chain of Command and Military Accountability (henceforth, CCMA). It is important to realize that "the chain of command and accountability measures within the CF are largely based on the need to operate in *combat conditions*." (CCMA, p. 9, par. 31, emphasis added). This raises, immediately, the issue of the adequacy of these measures in non-combat situations, which characterize most peacekeeping operations.

Before discussing the features of this conception of accountability, it is important to dispel a possible misunderstanding that this conception may generate.

Basic principle. In principle, each commander or individual in the chain of command is accountable to the next superior for their actions and conduct as well as the well-being, performance and conduct of their subordinates. Technical accountability is achieved through observation, inspections, audits, exercises and routine reporting. Formal review agencies exist within DND to analyze and to report on the effectiveness of internal procedures and rules. These agencies work directly for the DM and CDS and can be tasked to investigate any area within the Department or the CF. (CCMA, p. 8, par. 26, emphasis in text)

The Department of National Defence (DND) is headed by the Minister of Defence, assisted by the Deputy Minister (DM) and his or her functionaries on the one hand, and by the Chief of the Defence Staff (CDS), who is charged with the control and administration of the CF (as distinct from DND). The structure which is headed by the Minister of Defence can be viewed as hybrid, being staffed both by civilian and military personnel. Because of the specific combat functions of the CF and of historical traditions, the difference between the military and civilians is quite sharp and even dichotomous. In consequence, there is a temptation to interpret the dual structure of the Ministry (military and civilian functionaries) as a blend of the internal and the external, what is deemed internal referring to the CF and what is, in a sense, external to the CF referring to the civilian component of the Ministry. From this it might be inferred that in being accountable to the Minister and to the formal review agencies existing within DND, the CF is actually accountable to external bodies.

Although this is literally true in terms of the difference between the military and civilians, it is specious in relation to accountability. This can easily be shown by referring to the structure of a ministry such as that of the Solicitor General. Both the Royal Canadian Mounted Police (RCMP) and the Canadian Security Intelligence Service (CSIS) answer to the Solicitor General. Although it is not as deeply rooted in historical tradition as the military-civilian dichotomy, the difference between police and civilian and the difference between security intelligence agents and non-agent civilians is also quite sharp. Yet, when there were debates in Canada about the need for external accountability in relation to the RCMP and the Security Service (later CSIS), it was never seriously argued that since the Ministry of the Solicitor General was staffed by civilians it provided this external counterpart to which the RCMP or CSIS was accountable. What was meant by external accountability was the creation of reviewing bodies that were independent of both the Ministry and the agencies that they were overseeing.

I believe that the same reasoning applies in the case of the CF and DND. It would be stretching the meaning of words unduly to argue that since the Ministry is staffed in part by civilian functionaries, it is in a position to act as an external body to which the CF would be accountable.

This possible confusion hopefully having been dispelled, it is rather easy to present the conception of accountability which can be found in the CCMA brief and in other documents provided, for example, by the Judge Advocate General (JAG). The CCMA begins by describing the unified defence structure, which I previously discussed. The brief is then almost entirely devoted to a description of the chain of command (par. 8 to 14) and the concept of military discipline (par. 15 to 25). Accountability as such is covered in five short paragraphs which can be printed on one page (par. 26 to 30; the brief is nine pages long). The section on accountability begins with the basic principle, which I previously quoted and discussed. The basic principle is followed by a paragraph entitled "CF accountability," which is immediately succeeded by a paragraph on sanctions, reminding us that the *National Defence Act* (NDA) still contains the death penalty. The section ends with two further paragraphs devoted to formation level accountability and command accountability, which summarize material already presented in the sections on the structure of DND and on the chain of command.

This view of accountability has several features, which I will briefly discuss. Most of these features logically follow from one another.

Traditional: Accountability is almost exclusively viewed from the perspective of compliance. The word is actually used in paragraph 28 of the CCMA ("the sanctions and penalties available to encourage compliance are many and far exceed, in both number and magnitude, those available in civilian life"). This conception of accountability is, in fact, a rather narrow application of the compliance model described by Light (1993), presented in chapter 1 of this report (see Table 1). This view of accountability is historically the oldest and is very traditional. Such a model was applied for instance in police forces several decade ago, when accountability was confused with penal responsibility. As I pointed out earlier, the movement toward accountability was essentially a successful attempt to go beyond the narrow compliance model.

Punitive: In the CCMA, the discussion of the chain of command is largely devoted to a lengthy description of the power to enforce discipline and to apply sanctions, which is exercised by officers at the different levels of the hierarchy. There is actually a large amount of overlap between the briefs on accountability and military justice. ("The hallmark of the military justice system is the chain of command," Brief on Military Justice, p. 5, par. 15)

Reactive: This feature is a consequence of the previous trait. One of the most fundamental principles of justice is that there can be no punishment if no law or rule was violated. Consequently, penal justice can only be triggered by an act of deviance and takes place after the fact. Such a conception of accountability has few beneficial effects in the prevention of violations.

Individualized: This feature also follows from the fact that accountability is here conceived as penal justice. Penal justice essentially targets the individual. This aspect of penal justice was dramatically illustrated in recent attempts to hold whole organizations criminally responsible for very serious instances of polluting the environment. These attempts generally failed. There is, however, an aspect of the power to discipline and punish that is embedded in the military chain of command and which is at odds with civilian criminal justice. It is the application of collective sanctions, as when Colonel Morneault threatened to bar the whole of 2 Commando from Operation Deliverance, if the persons responsible for setting fire to the vehicle of a sergeant did not come forward. There are other instances of Colonel Morneault informally applying collective sanctions. I do not know whether collective punishment is applied either formally or informally in the CF. Whatever the case, it must be said that collective sanctions are very controversial in the field of civilian criminal justice and that a systematic recourse to collective sanctions in the CF — which I regard as highly improbable — would not be a good instrument for improving accountability.

Internal: I have already addressed this point in my discussion of, what is called in the CCMA, the basic principle of accountability and I will not repeat my argument. I shall add to it by saying that the conception of accountability articulated in the CCMA and the brief on military justice is essentially a compliance-through-the-application-of-sanctions model. The few generalities offered on accountability to the Minister, his colleagues in the Cabinet and to Parliament are standard variations on themes that are developed in any textbook on government theory. As is well known, the application of sanctions to military personnel when they violate the rules of the organization is not achieved through the civilian systems of criminal, civil and administrative law but through a separate system with its summary trials and courts-martial, referred to as military justice. Not only is this system internal to the CF but it is actually separated from all other civilian justice systems, as we just noticed. It could then hardly be more internal.

Non-specific: The attribution of this feature to the CF concept of accountability may be viewed as controversial. What I mean is this. The CCMA understands the word accountability in the same sense as we might say that every citizen is accountable for his or her respect of the Criminal Code and the criminal courts. Military justice is different in an important respect from civilian criminal justice, but it embodies the same fundamental idea of generating compliance to a set of legal rules through the use of penal sanctions. Not only is this idea very general but it is also one of the oldest ideas of humankind, one that was formulated literally thousands of years before there was any talk of accountability as the term is presently used. In other words, what is offered as a conception of accountability is only a martial variation on the antique notion of criminal responsibility. With regard to the current debates on accountability, it fails to address any of the relevant issues and lacks, almost completely, any specific content in relation to accountability.

There is one last feature of this version of accountability that I have not listed among the others, because I intend to demonstrate the extent to which it exists in the CF. I have argued that accountability in the CF embodies the same fundamental idea that we find in any model of penal justice. There is a trait of civilian criminal justice which is presently the focus of a sustained debate. Criminal law is applied by persons, such as police, judges and members of parole boards, who exercise a certain amount of discretion in making their decisions. The discretion enjoyed by judges in imposing a sentence is fairly wide. Discretionary powers are necessary to ensure that the criminal justice system is flexible enough to deal with the great variety of cases that it must process. It can also be exercised in a way that defeats the purpose of a particular piece of legislation. For instance, if there is an informal but widely shared consensus among the different professionals involved in applying the law that persons should not be prosecuted for simply possessing small quantities of soft drugs for their own use, this lack of enforcement is nearly equivalent to a de facto legalization of the use of soft drugs. Discretion in the application of the law can encourage compliance to the rules, when the law is actually enforced; it can also foster disobedience when there is a perception that one can violate the law with impunity.

With regard to the perspective on accountability offered by the CF, I would like now to raise several questions. Some are typical of the debate on accountability as it is presently pursued. For instance, does the CF provide the Minister with an accurate account of the situation that prevails with respect to particular issues? Other questions are more specific to the type of accountability requested in the CF, i.e., compliance to rules and regulations under the threat of sanctions. In this regard, I will explore whether the officers who play a crucial role in the enforcement of these rules and regulations exercise discretion in the way they enforce them. I will also ask myself if the way in which discretion is used defeats the purpose of certain policies as it sometimes does in the criminal justice system.

Testing Accountability: The Issues of Racism and of the Penetration of the Extreme Right in the CF

Because of the focussed nature of this report and the drastic limitations of my sources of knowledge, I cannot hope to provide general answers to the questions I raised at the end of the last section. I have chosen racism and the penetration of the extreme right in the CF as the testing ground for the theory of accountability developed by the CF.

Accountability as an Information Process: Knowing the Facts. I collected a certain amount of CF documents on the issues of racism and of the extreme right. I extensively reviewed them in order to verify whether they presented an accurate picture of racism and the penetration of members of racist organizations in the CF. I now propose to give the results of this examination. There is one important reservation that I must formulate before presenting the results of my analyses: the documents I used were transmitted to the CIDCFS by the CF. They are significantly incomplete. In particular, several of these documents are followed by annexes. With very few exceptions, all these annexes were missing and had not been transmitted to the Commission when we collected them.

A review of documents: I will begin by reviewing part of the content of a selected number of documents, i.e., those documents in which the CF tries to assess how many members of the CF also belong to racist right-wing organizations. These estimates, as we shall see, vary a great deal over time.

- *Memo 4 February 1993:* Typically, this is a memo that was written after a Canadian Broadcasting Corporation (CBC) radio broadcast on the presence of right wingers in the CF. The memo, which is classified as very highly confidential, makes a reference to Project SIROS on the identification of right wingers in the CF. This project was begun in 1991. According to data produced by Project SIROS, there were apparently 19 cases of CF members belonging to racist organizations, *these 19 cases involving 60 members of the CF (23 regulars and 37 reservists).* The memo outlines an explanation of the presence of right-wing extremists in the CF that was to be repeated for at least the next two years. The presence of such persons in the CF only reflects their growing number in Canadian society at large. Their number in the CF is proportionally very small and most of them are just curious followers looking for an alternative lifestyle as opposed to serious activists. Finally, the problem concerns more the Primary Reserve than the Regular Force.
- *Report on Right-Wing Extremism in the CF (Executive Summary), 11 May 1993:* This is also a very confidential report. It claims that the *sum total of allegations of right-wing extremist activity is 58* in the last two years, not all of these allegations having been substantiated. The number of members of the Canadian Airborne Regiment (CAR) involved in these activities is said to be six, two persons from this group of six serving at the time in Somalia. The significance of right-wing extremism is greatly diminished by comparing it, as the report does, with other cases of delinquency in the CF (58 allegations of right-wing extremist activity as compared to 457 cases of sexual assault, 121 deaths by suicide, homicide and natural causes, and 7,061 thefts and robberies over the last two years). Whether or not such comparisons are completely spurious is not discussed.

- *Briefing note for the Minister, 13 May 1993:* The number of CF members involved over the last three years in right-wing extremist or racist activities is assessed to be 65 (notice that it was assessed at 58 for the last two-year period on the 11 of May, two days before).
- *Briefing note for the CDS and DM, 22 June 1993:* This is also an interesting document. It begins by an admission that:

> It also became evident that our lists of involved personnel were not as complete or comprehensive as they could have been. DG Secur and SIU staffs have conducted an extensive review of their files to develop as comprehensive a data base as possible. As a result, the list has now grown to 91 names and four new investigations which have just commenced and may involve an unknown number of persons. (p.1, par.2)

- *A Draft Document entitled "Personnel Policy Review, Racism and the Canadian Forces," July 1993:* Annex A of this document provides a detailed review of racist organizations and activities in Canada as well as a summary of the extent and nature of involvement by CF personnel. This annex was not provided with the document itself, so we must rest content with the assertion that "while exact numbers remain classified, they are said to be very small, with activities ranging from attendance at rallies by some individuals in the regular force, to more active participation and incidents of criminal activity on the part of some reservists" (pp. 11-12). This document concludes that the limited nature of the problem does not warrant the prohibition of simple membership or association with certain organizations. It is further argued that such a prohibition would limit the freedom of association guaranteed in the Canadian Charter of Rights and Freedoms.
- *BOI, Report, Phase I, Volume XI, July 1993:* Annex I of the BOI Report addresses the issue of racism and the presence of individuals associated with right-wing extremist groups. Its conclusion deserves to be quoted:

> The Board did find some evidence that there may have been *one or two individuals* in the Battle Group who were believed to hold views that might be classed as white supremacist in nature. However, after extensive questioning, the Board concludes that this problem was, at worst, confined to this *tiny minority.* (BOI, Report, Phase I, Volume XI, Annex I, p. 5, par. 18, emphasis added)

This conclusion is so strongly worded that it is somewhat surprising to learn in the next paragraph that the BOI will raise again the issue of racism in Phase II of its inquiry, when it will investigate specific incidents.

- *Professor Harriet Critchley, interview with CBC Newsworld program, "This Country," 31 August, 1993:* Professor Critchley is one of the four members of the BOI and its only civilian member. When asked about the extent of racism in the Canadian Airborne Group Battle Group (CARBG), she first repeated the Report's assertion that it was limited to a "tiny minority," but she added that this conclusion may not have been definitive and that the issue might be revisited during Phase II of the inquiry (this phase did not take place, the CIDCFS having been appointed). I shall quote Professor Critchley's answer to a query about why the BOI would not say that racism was a systemic problem in the CARBG. The journalist's rather long question ended on this sentence: "So are you telling me that it's isolated?" ("it" refers here to racism.)

> This is, we haven't been able to finalize our conclusion in that regard, but I can tell you that because of media reports concerning racism in the Canadian Airborne Regiment, this was one thing which the Board examined very carefully, with all of the witnesses from that Regiment that we saw, and any other evidence that we could gather. We were constantly looking for that. Our general conclusion was that there may have been a tiny minority of people who were, for example, white supremacists in their leanings, in the Regiment at some time during the deployment to Somalia. *We can't be more definite than that now, because we have not been able to look into certain of the incidents. We will be able to look into those, and then we will be able to reach a much more definitive conclusion in part two of our report.* (Professor Harriet Critchley, interview with the CBC Newsworld, 31 August 1993, emphasis added).

The reservations of Professor Critchley on the definitive character of the BOI's Phase I Report are all the more interesting because they contrast with the general tone of the Report and of Annex I on racism. The tone adopted is that these conclusions are the last word of the BOI, although it is stated that the BOI will address the issue of racism again. The conclusions are actually followed by recommendations which do not appear in any way to be tentative.

- *Briefing Note for the Minister, 13 September 1993:* the number of members of the CAR in Somalia that may have been involved in racist activities and associated with the extreme right wing is now believed to be *six*. The total number of persons being investigated throughout the CF is said to be *45*. There is intelligence on an additional *35* persons, but it is insufficient to warrant opening a formal investigation. The conclusion reached in this briefing note will again be formulated word for word in other briefing notes: "The above information gives the impression that there is a major problem in the CF and while the seriousness of it should not be underestimated, it should be noted that the concluded investigations to date do not support media allegations of a serious problem." (p. 3, par. 7)

- *Briefing note for the CDS 24 March 1993:* The number of extremists under active investigation by the SIU is assessed to be *49 throughout the CF* (not all these investigations are conclusive). In addition, the SIU has intelligence dealing with a further *37 persons*, although this intelligence is insufficient to warrant the opening of a formal investigation. The conclusion reached in the briefing note for the Minister dated 13 September 1993 is formulated again: "The above information gives the impression that there is a major problem in the CF, and while the seriousness should not be underestimated, the concluded investigations do not support media allegation of a serious problem" (p. 5, par.11).
- *Confidential memo, "Racism and the Canadian Forces," 6 April 1994:* The number of files held by Director General Security (DG Secur) from 1991 to the above date on individuals in connection with membership in right-wing extremist organizations is now said to be *140.* They belong to different categories:

– Intelligence (insufficient at the time to warrant a formal investigation)	37
– Active (investigation tasked to SIU)	49
– Concluded - Non conclusive (allegations not confirmed)	15
– Concluded - Conclusive (allegations confirmed)	17
– Released/retires/deceased	22

Sixty-six files concern regular members of the CF and 70 concern reservists. It is no longer assumed that the problem of racism and right-wing extremism is confined mainly to the Primary Reserve. Part of the information contained in the briefing note for the CDS dated 23 March 1994 coincides with the figures given in this memo.

- *Briefing Note for Minister, 7 April 1994:* This briefing note was given to the Minister before his meeting with the National President of the Jewish Congress on 15 April 1994. With very minor variations, this briefing note is exactly the same as the note prepared for the CDS on 23 March 1994. The conclusion is naturally the same; it is also the same as the statement formulated in the briefing note for the Minister on 13 September 1993: "The above information gives the impression

> that there is a major problem in the CF, and while the seriousness should not be underestimated, the concluded investigations do not support media allegations of a serious problem." (p. 6, par. 12)

Conclusions: One thing must be clear from the outset. There is nothing in the documents that I just reviewed that shows any wilful intent to deceive and to depict knowingly the facts other than they were. This being said, these documents reveal that if accountability is to be understood as an information process through which the CF and DND were to give a true and complete account of the extent of racism and of right-wing extremism in the CF, then it leaves much to be desired.

First of all, the account is entirely reactive. It is not a self-initiated effort on the part on the CF to shed light on racism and right-wing extremism in its midst, but a follow-up to media pressure. It is highly doubtful that in the absence of pressure from the media, the greater part of the previous documents would have been written. Professor Critchley explicitly declared that the BOI's canvassing of the racism issue was fostered by media reports on racism in the CF.

Second, the assessment of the scope of right-wing extremist penetration varies greatly over the two years. It begins with 65 files; then it grows to 91 files, with the admission that the lists previously produced were not comprehensive. In the end, the number of files has grown to 140. In a similar way, the number of White supremacists in Somalia is reduced to one or two individuals by the BOI, and later expands to six individuals. However, throughout these changing assessments, there is one thing that remains absolutely the same. It is the estimation of the scope of the problem: it appears to be serious, but despite media allegations it remains insignificant (although, paradoxically, its seriousness should not be "underestimated"). This interpretation of the DG Secur and SIU findings is recited throughout the documents that I reviewed.

Third and foremost, the accounts I reviewed are defensive, complacent and have a marked tendency to *down play* the problem on which they are reporting. There are many strategies which are used to produce this effect. I shall single out two.

1. The first one is to treat all numbers as if they were equivalent. This gets us into "tiny minorities" and into the time honoured "rotten apple(s)" explanations of organizational deviance, explanations which have been discredited for at least a decade in other fields. Numbers are actually not at all equivalent and take on a different signification

depending on their context. For instance, one mole infiltrated into a security intelligence agency is not a "tiny minority of spies," it is a catastrophe. One may in the same sense wonder how many White supremacists in a contingent of peacemakers in an African country it would take to make their presence a serious hazard to the mission. DND and CF policy statements (DCDS 9-83 and Canadian Forces Administrative Order, CFAO 20-53) do not allow Jews and Muslims to participate in Middle East peacekeeping operations. Accordingly, one Jewish officer was not permitted to take part in the Gulf War. Why is it that one Jewish Canadian was seen as a threat to the Canadian intervention in the Gulf War, while up to six White supremacists in a peacemaking mission in Somalia are not seen to pose a "systemic problem?"

2. A second strategy is to be overly subtle in the description of participants in right-wing extremist groups. Thus a DG Secur Input on the Policy paper on racist organizations, dated 8 July 1993, makes a distinction between three kinds of skinheads. The first category is described thus: "...non-racist skinheads who are only marginally organized. They believe in the original asocial and violent aspect of the movement and consider the Skinhead movement to be an alternative lifestyle" (p. 12, par. 62a).

Until recently there was a zero tolerance policy in the CF toward homosexuality. It would have been unthinkable when this policy was in force to make finely shaded distinctions between homosexuals who associated with the gay and lesbian movement just because they found it an "alternative" lifestyle and real gay activists. At that time, any declared homosexual in the CF found himself or herself without any career prospects. Again, why all the subtlety when it comes to declared racists?

Playing problems down is not peculiar to the CF, although in my experience with investigating boards, there are few examples of such self-induced blindness as that of CAR 2 Commando. Such insensitivity is greatly facilitated when all assessment of the seriousness of a particular problem is done internally. The internal analysts tend to gear their assessment to what they perceive to be the wishes of the organization rather than to the significance of the problems they are investigating.

There was a recent confirmation of this analysis, which I shall quote before leaving this subject. General Jean Boyle, testified before the CIDCFS in August 1996. The newspapers reported the nature of General Boyle's testimony before the CIDCFS on 12 August 1996. All newspapers that I reviewed agreed on the content of this first day of testimony. I quote a Canadian Press release under the byline of John Ward.

Boyle testified at the Somalia inquiry yesterday that *in the fall of 1993* he found discrepancies between military police reports and the results of an internal Defence Department inquiry that filed its first report that summer.

At the time, Boyle was a major-general and an associate assistant deputy minister. Today he's a full general and chief of the defence staff, the country's senior military officer.

Documents tabled yesterday by the commission show Boyle briefed Collenette after the minister took office in *November 1993*, warning that there would be troubles ahead when the public learned the grim details of what happened in Somalia.

Boyle wrote that the closed-door internal inquiry under Maj.-Gen. Tom de Faye had missed the mark on some issues.

"One of the things I realized is that what was in the de Faye report did not coincide with what I was reading in the military police reports," Boyle told the commission, which resumed hearings after a short summer break. (Ward, *The Gazette*, 13 August 1996, p. A7)

Let us remember that despite what it found in its investigation, the de Faye Inquiry — BOI — still recommended that the members of the CAR be awarded the medal which they apparently deserved for their tour of duty in Somalia. Whatever were the incidents that occurred in Somalia, they were of such a limited nature that they did not reflect on the whole Canadian intervention in Somalia. This attitude, according to General Boyle, led the de Faye Inquiry to miss the mark on some issues and he had thus forewarned the Minister of Defence in November 1993.

The Minister was not the only person to be briefed by General Boyle. On 20 April 1995, several members of the CIDCFS staff (François Daviault, Legal Counsel, David Pomerant, Director of Research, Eloise Arbour, auxiliaire juridique, Jennifer Oulton, future auxiliaire juridique, Barbara McIsaac, Legal Counsel and Stanley Cohen, Secretary of the Commission) met with the JAG. These persons were briefed by Colonel Leclerc on the structure and the organization of the CF. They were also briefed by Major General Boyle on Operation Deliverance and the CARBG in Somalia. Here is the line followed by Major General Boyle in his briefing to six persons from the CIDCFS' staff.

The performance - I have been dealing with Somalia for about 1 and a half years. You are going to read documentation that is quite negative. 95 percent of the activities conducted, maybe even higher than that, is exemplary and heroic. These guys were the best in theatre, they were the best unit in UNITAF, they provided the greatest and most visible humanitarian assistance of any kind in UNITAF. There are letters of commendation from his Excellency and everyone. It is eye watering the positive mail. 5 percent of what they did was bad and it is gathering 99.9 percent of the press. When you go through the documentation and say "why was DND so ambivalent?", Serge Labbé was sending reports everyday, as Commander of the force. Everyday there was a situation report sent.

In theatre we had six major problems. These are the ones that we are aware of...[the general then gives a description of these six problems]. (General Boyle, transcripts to a briefing to members of CIDCFS' staff, 20 April 1995)

Although in a later part of his briefing, General Boyle actually refers to the discrepancies between the BOI Report and the reports from the military police, it is somewhat difficult to reconcile the gist of his testimony before the CIDCFS in August 1996 and the line that he pursued in the April 1995 briefing. In his August 1996 testimony, General Boyle stresses the discrepancies between the BOI Report and the military police reports and his action to forewarn the Minister of the impending problems that were going to be caused by these discrepancies. In his initial briefing to the CIDCFS' staff, the General seems much closer to the line pursued by the BOI: to belittle the significance of the negative aspects of Operation Deliverance and to stress the positive. When he refers to "eye watering" mail in his April 1995 briefing, General Boyle sounds more like the cheerleader of the CAR than like the critic of the BOI Report that he was to become in his later testimony before the CIDCFS. It would seem that the facts are presented in a different perspective depending on the degree of pressure applied by an external authority to whom the military is accountable. To repeat: more or less the same information is given, but the light in which it is presented changes its meaning.

Accountability as an Instrument to Foster Compliance

The CF policy on racist conduct (CFAO 19-43) was issued on 25 February 1994. The purpose of the policy was to deter CF members from engaging in racist conduct and holding racist attitudes as well as to determine how the CF members who actually engage in such conduct and hold racist attitudes should be dealt with.

The content of the policy: The policy defines racist conduct as behaviour that promotes, encourages or constitutes discrimination on the basis of race and related concepts (e.g., ethnicity) and, interestingly enough, it includes in the definition of racist conduct: "participation in the activities of, or membership in, a group or organization that a CF member knows, or ought to know, promotes discrimination or harassment on the basis of race, national or ethnic origin, colour or religion" (CFAO 19-43, subsection 3, Definition: Racist conduct).

To include as racist conduct, membership in an organization or group that promotes discrimination is a departure from previous CF positions. The CF had been unwilling to limit the freedom of association protected by the Charter. Whether or not the policy will be effective remains to be seen. Subsection 9 gives a list of examples of racist conduct related to membership or participation in a group or organization that promotes racial discrimination. Example "h" is formulated thus: "attending meetings or rallies for the purpose of supporting the group or organization."

It is actually very difficult to discover whether someone attending a rally simply does it out of curiosity, for fun or because he or she is supporting the group or organization holding the rally. A careful reading of the examples given in the policy raises the possibility that the policy may be reintroducing the distinction between active participation in racist groups or organizations and mere membership in them. The Personnel Policy Review of July 1993 ("Racism and the Canadian Forces") referred to the fact that the U.S. Army prohibited "active participation" in extremist organizations, but did not prohibit "mere membership" or other passive activities such as receiving literature or being present at an event. The list of examples given in subsection 9 of CFAO 19-43 does not mention receiving literature and seems to allow the "passive" attendance at an event (attending without supporting).

CFAO 10-43 describes the role of the military police and of the SIU in reporting and investigating violations of the policy. Different measures are provided for those who do not comply with the policy. These measures are administrative, ranging from informal counselling to a recommendation for release; they also include disciplinary measures and suspension from duty in the case of serious violations that lead to the laying of charges by civilian or military authorities. To monitor the nature and extent of racist conduct in the CF and to assess the effectiveness of counter-racism initiatives, units must report by message to National Defence Headquarters (NDHQ), using a particular kind of format. Annex A to CFAO 19-43 provides two "sample message formats" that should serve as guidelines for the units sending such messages to NDHQ. These samples are interesting. The first is the fictional case of a corporal who has been involved as a member of an organization that promotes racial discrimination. According to the example, he admits attending two meetings but has not actively participated in any discriminatory acts. He claims not to have known that the organization promoted racism. The corporal is counselled that further involvement with the organization will result in more serious administrative or disciplinary action. This is the typical instance to which I referred above of passive involvement with an organization that promotes racial discrimination. The sample case does not say whether the corporal was informally or formally counselled. Since formal counselling is accompanied by probation and there was no probation in this case, we may surmise that the counselling was informal, the most benign administrative measure provided by the policy (the violation is also of the least serious kind). The question is how long passive membership in a racist organization will continue to be sanctioned by informal counselling, before the more severe measures, such as a recorded warning or formal counselling and probation, are imposed.

What makes this last question relevant is the other sample case. It tells of a lieutenant who is convicted of assault on a civilian man belonging to an ethnic minority. Evidence provided by civilian police indicates that the assault appeared to be racially motivated and that the lieutenant was intoxicated at the time. He was sentenced to six months of probation. The administrative measure taken in this case is the following: the lieutenant was placed on report of shortcomings and warned that any further incident of violence, assault or racist conduct of any kind would result in a recommendation for release. There may be a problem with this case. Let us imagine that the lieutenant is not involved in any racist incident for three or four months in Canada. He is sent with his regiment on a peacekeeping mission to an African country, where he is once again involved in an incident of misconduct that appears to be racially motivated. If this incident were reported in the media with the additional information that the lieutenant had already been charged with a racially motivated assault in Canada, whether or not he would be released from the CF following the second incident would matter very little. I believe that the measure would be perceived as too tardy and that the damage to the public reputation of the CF would once again be considerable.

The upshot of these remarks is not that the policy on racism is seriously flawed but that its application must be closely monitored. I shall now present a limited exercise in monitoring.

Monitoring its application: Annex F of the briefing note to the Minister, dated 12 April 1994, consists of a confidential memo entitled "Status of Conclusive Concluded Files." I have classified the offences and also listed the measures taken against the offenders. These measures are, however, not always known.

- *Membership in "non-political" or "non-racist" skinheads:* seven cases (two were released; in other cases the measures taken were unknown or benign);
- *Membership in SHARPSkins (skinheads against racism):* one case (although this member had terminated his association with the SHARPSkins in 1992, he was nevertheless released from the CF);
- *Attendance at cross burnings:* three cases (all three were given the benefit of doubt and were enrolled in the Trades Training Program);
- *Membership in the Ku Klux Klan (KKK):* two cases (the measure taken was to monitor the activities of these members);
- *Membership in an overtly racist group:* one case (measure unknown); and
- *Involvement in a racially motivated violent incident:* two cases (the measures taken were counselling and monitoring, although in one case the offender had been charged with assault before a civilian court).

There are other cases. In one significant case, an individual with a long record of association with U.S. skinheads (he was charged with assault in the United States and received a six month probation sentence) was charged for a "gay bashing" incident in Winnipeg (assault causing bodily harm, assault with dangerous weapons and uttering threats). This individual was removed from his unit and UNPROFOR (UN Protection Forces — to Yugoslavia in 1992) training and subsequently returned to unit training. Pending the result of his trial, he was not suspended from his unit "for his non-racist skinhead activities" (Second Briefing Note for the Minister, 12 April 1994, p. 3).

A briefing note for the CDS and DM prepared in June 1994, gives an update on the current status of administrative action taken in 10 cases of alleged racist conduct. One individual, who was a member of the KKK and skinheads, was released from the CF, because his unit was of the opinion that "it is impossible and no longer relevant to know if he has ceased his activities with the skinheads and the KKK." (Briefing Note for CDS and DM, p.4). In all other cases, "no administrative or disciplinary action has been taken against the member."

Admittedly, all these cases were initiated before the release of CFAO 10-43. Nevertheless, they give an indication of the considerable amount of discretion that is exercised by officers. Of all the cases listed above (28), four members were administratively released from the CF for membership in organizations such as the KKK (one case) and skinheads or SHARPSkins (three cases, two in the same regiment). However, the benefit of doubt was given in cases such as the attendance at cross burnings and the involvement in violent and racially motivated incidents. One might be inclined to believe that the latter incidents were less serious than participation in the activities of skinheads explicitly described as non-political and non-racist.

Another finding is the general leniency toward CF members participating in groups or associations promoting racism or even being themselves involved in violent incidents. Will the attitude change, now that CFAO 10-43 has been released? It is premature to try to answer this question in any definitive way. However, the need for close monitoring of the application of CFAO 10-43 is indisputable.

A framework for accountability? On 2 April 1994, *The Globe and Mail* published an article bearing the headline "Plan to weed out racist peacekeepers not ready." The CF apparently believed that this article was unfair to their efforts or damaging to their reputation, and they decided to answer it by sending a letter to the editor of *The Globe and Mail.*

This letter was apparently difficult to write and there were several drafts, which were commented on in writing by high ranking officials at DND. I have obtained through the CIDCFS, a copy of these annotations and I will quote them, because they reveal the state of the CF policy against racism. Let us remember that this policy was released on 25 February 1994, more than a month before *The Globe and Mail* article.

The notes I quote are handwritten and there are words that I could not decipher. The notes are written on the stationery of the Chief of Defence Staff and of the Deputy Minister. They seem to be addressed to the person assigned to draft this letter.

The first note bears the heading "ADM (Per)" Assistant Deputy Minister (Personnel) and is written on CDS stationery. The ADM (Per) was eventually to sign the letter sent to the editor of *The Globe and Mail.* Here is the handwritten text, as far as I could identify the words:

Further to my note on your proposed letter to the G&M are my (and the DM's) growing concern that we *appear* (emphasis in text) to be vacillating on our handling of the whole racism issue. I would like to meet with you (and whoever you think you need) to discuss this issue this afternoon. [The letter is signed by what seems to be the initials of the Chief of Defence Staff at the time.]

The second note is written on DM stationery and bears the name Robert R. Fowler in printed characters on the stationery. It begins with the headings CDS and Racism (underlined in text). The addressee seems to be the CDS. Here again is the text.

a) *I agree it is* [emphasis in text] too late.
b) You are gentle in your synopsis of the letter. It's worse than that. It is worse than nothing.
c) We might discuss follow-up to the last meeting (sorry I lost my cool. I had intended to listen and avoid becoming enraged. I failed)
d) I remain concerned that :
i) we still don't really know what we're about.
ii) there is no effective accountability/responsibility framework
iii) there is insufficient coordination and cooperation and a lack of common understanding among, PER, JAG and DG Secur.
iv) what has been achieved in 10 months is unacceptable.

Following these comments, the ADM (Per), Lieutenant-General P.G. Addy, wrote another draft of the letter to be sent to *The Globe and Mail.* This new version ended with the sentence: "Any measure designed to further strengthen this policy in the future will be encouraged." The CDS also reacted to this later draft in a handwritten note on his personal stationery. It began, like his previous one, by the underlined heading "ADM (Per)." Here is the text of this note.

1. My impression on reading this draft (especially the last sentence) is:
a. The report that we have not yet come to grips with racism in the Forces is essentially correct;
b. We are on the brink of *doing something*;

c. We *hope* that the public will believe we *really* are serious about doing something (soon) and
d. Any help the public is willing to offer will be welcome.

At the bottom of the same page, there are handwritten notes which are preceded by the heading "DM" and which may have been intended for the Deputy Minister. They are difficult to read: "The note. I think we are reaching the point where we have missed [unreadable word] timing in responding [two or three unreadable words] article."

There is also a fourth note that does not bear on the reply to *The Globe and Mail*. It is handwritten on CDS stationery and comments on the briefing note for the Minister that was prepared on 7 April by a group headed by the ADM (Per). It begins with the heading ADM (Per).

This is comprehensive, although it leaves me with a couple of questions, which the MND may well have too.
a. *Para. 7 b.* What does AETHEP stand for ? Is it a civilian-contracted program?
b. *Para. 11 c.* we note that the SIU determined 17 CF members had been/were involved in racist activities but don't say what, if anything, we are doing about it (unless para. 15c on the new CFAO purport to say that we can now do something). If the latter is correct, it begs the question that we did nothing about racist activity before.
c. *Para. 7d.* States that a screening process for UN deployment has "been stalled to Commands ...for comment", but doesn't say when the process will be implemented. Presumably this lack of forecasted date gives validity to the *Globe and Mail* article saying that we don't have a screening process yet, and may not have one for several years.
2. You must have the answers to these queries ready for the MND (and me) before he meets with the Canadian Jewish Congress on Friday.

The note, dated 11 April, bears what appears to be the initials of General de Chastelain. This note is interesting in several respects. First, the Briefing Note that was prepared for the Minister on 7 April 1994 is an exact repetition of a note that had been prepared for CDS on 23 March 1994. In particular, paragraphs 7b, 7d and 11c are exactly the same in the Briefing Note for CDS dated 23 March 1994. It is, therefore, a bit surprising to read in the CDS' handwritten notes that he objects to the content of certain paragraphs of a briefing note to be submitted to the Minister, when he apparently had none of these objections when the same note was submitted to him earlier.

Second, following the comments of the CDS, the text of the Briefing Note for the Minister was amended along the lines suggested by the CDS. In paragraph 7b the meaning of the acronym AETHRP was provided ("Awareness Education and Training for Harassment and Racism Prevention"). More significantly a paragraph was added at the beginning: "2. (U) In June 1993, ADM(Per) issued guidance to Commanders, based on CDS direction, suggesting they remind commanding officers of the appropriate measures that were available to them under existing regulations to handle cases of CF members associated with racist organizations" (copy in Annex A).

This added paragraph solves the problem raised by the CDS in his handwritten note about paragraph 11c of the original Briefing Note, which could be interpreted to mean that nothing was done by the CF about racist activities before CFAO 10-43 (released on 25 February 1994). The annex referred to in the added paragraph was not transmitted to the CIDCFS. Hence, there is no way to verify the nature of the guidance referred to.

The other troublesome paragraph (7d) was also fixed. It now began in the following way: "As a result of Phase I of the Somalia Board of Inquiry, the CDS directed Commanding Officers of units scheduled for peacekeeping duties to screen their personnel for racist activity. That is being done."

The rest of the text is the same, except that the sentence to which the CDS objected — "These screening procedures have been staffed to Commands and NDHQ Groups for review and comment." — was changed. It began by, "These *enhanced* screening procedures...," the rest of the sentence being the same (we have added the emphasis on "enhanced." This change implies that the screening procedures staffed to Command and NDHQ for comment are only an enhanced version of *screening procedures that already existed*, thus dispelling the impression that no screening procedures were in force, as *The Globe and Mail* article claimed.

This exchange of handwritten notes shows at least three things. First, the CF took reports in the media very seriously and tried hard to maintain the appearance that it had fought against racism in the past and was pursuing the struggle with even more determination. Second, there was no consensus at the highest level, CDS and DM, on whether the CF had reacted against racism in the past and was implementing efficient reforms with enough energy. The DM says that there is no framework of accountability/responsibility (as of April 1994) and that what had been achieved in the last 10 months was unacceptable. Finally, what was good enough for the eyes of the CDS was not good enough for the eyes of the Minister.

Institutions of Accountability: Some Comparisons

Before discussing the structures established to foster accountability for police forces and CSIS, I will identify what I believe to be two major difficulties with respect to the CF.

Although an army is created to protect the national security and hence to defend the frontiers of a country, the CF has not had to defend Canadian territory since 1812.[1] The main operations undertaken by the CF have always occurred in some foreign country. Not only is this a basic difference between Canadian police forces and CSIS, which are domestic agencies in operation on Canadian territory, but the agencies of government that traditionally maintain the staunchest resistance to accountability and external scrutiny are those that operate on foreign territory. Australia provides an interesting illustration of this resistance. Unlike Canada, Australia has two security intelligence agencies. The Australian Security Intelligence Organization (ASIO) is the domestic agency and was made a statutory agency some 16 years ago. However, the Australian Security Intelligence Service (ASIS) was never placed on a statutory footing for fear that a legal charter would limit its operational capacity (Lustgarten, 1995: 165). To borrow Sissela Bok's phrase, the national agencies and organizations that operate on foreign territory are among the deepest "mysteries of the State" (Bok, 1982). The CF has become such an organization, although it is not by any means restricted to operations on foreign territory.

Second, there is an increasingly stronger demand in Canada for agencies to be accountable to Parliament through parliamentary committees, patterned after the different congressional committees created in the United States. The U.S. Armed Forces, the Central Intelligence Agency (CIA) and the Federal Bureau of Investigation (FBI), to name just a few agencies, are accountable to various congressional committees. The report of the Special Committee on the Review of the *Canadian Security Intelligence Service Act* and the *Security Offences Act* is a good example of an attempt to introduce, in Canada, the kind of parliamentary oversight that exists in the United States (Canada, House of Commons, 1990). It failed as most of its recommendations were rejected by the Canadian government (Farson, 1995; for the government response, see Canada, Solicitor General, 1991).[2] I believe that the government's reluctance to follow a course of parliamentary committee oversight on the model of the United States cannot be reduced to an ad hoc resistance to a particular mechanism of accountability, but that it ultimately stems from constitutional differences between the United States and Canada. Like the United Kingdom, Australia and New Zealand, Canada is a Westminster-style democracy (Lustgarten, 1995: 165-168). The original feature of such a democratic structure is that the ministers of the Cabinet are chosen from among the members of Parliament belonging to the party which formed the government (or among the different parties forming a coalition government). All ministers are elected officials holding a seat in the House of Commons. In other word, the executive branch of government is staffed by persons who also act as legislators in the House of Commons. In the United States, the separation between the executive and the legislative powers runs much deeper than in Canada: the President does not select the members of Cabinet from among elected officials who hold a seat in Congress. Furthermore, the President and the appointed Cabinet frequently belong to another party than the one holding the majority in the Senate and that controls it. In this context, there is not only competition between the executive and the legislative branches of government but, at times, open conflict. Bringing government agencies to account before congressional committees is a way for the Congress to assert its power, and it is a practice which is embedded in the nature of the U.S.Constitution, which is based on a system of "checks and balance." By contrast, in a Westminster-style democracy the executive and legislative branches of government cannot be controlled by different political parties, and the government ministers hold a seat in the legislature. The political will of Parliament to hold its own against the executive power has no constitutional basis and, consequently, the establishment of parliamentary committees to oversee the executive agencies of government is not a strong feature of Westminster-style democracies such as Canada. There are indications that the legislators will ask for a bigger role, but it is difficult to predict whether U.S.-style parliamentary oversight through House of Commons committees will become a significant reality in Canada. There are presently numerous standing committees of the House and the Senate (e.g., the House Standing Committees on Justice and the Solicitor General, on Foreign Affairs, and on National Defence and Veterans Affairs; the Senate subcommittee on National Defence); special committees on foreign affairs and national defence were also appointed to examine the participation of Canada in peacekeeping operations. However, these committees are, by their own admission, far from playing as active a role as they should be. *In particular, the reluctance of the executive to provide them with full access to all the information they need in order to operate effectively has proven a major hindrance.* As the Auditor General noted in his last report, the global cost of peacekeeping, both in financial and non-financial terms, does not appear in any public Canadian document (Canada, Auditor General, 1996, Chapter 6, Highlight 6.5).

I believe that, in the context of these difficulties, we should resist the temptation to articulate complex structures that supersede ministerial accountability as we presently know it — an agency is accountable to Parliament through the minister that holds legal authority over it. I also think that the status quo should not be maintained: its limitations have been shown to be too flawed. I would opt for a modest beginning on which something more elaborate could be built as need arose.

Accountability in Policing. I have previously compared the operations of police and the CF (chapter 1) There is no need to compare, once more, policing and soldiering on a descriptive basis, because what we are looking for in this section is not facts but rather norms and models that we could use, at least in part, to make recommendations to enhance the accountability of the CF.

The main point that I made with regard to police is that they perform a service for the community and that they are foremost accountable to the community rather than to the state or some local political equivalent. Hence, the main instrument of accountability, in the case of police forces, is external review boards that process complaints from citizens. Needless to say, national or provincial police forces are also accountable to national or provincial parliaments through their respective ministers.

Except in crises, the CF has few professional contacts with Canadian citizens. Consequently, a complaint committee, whether internal or external, would serve little purpose in Canada. A committee to process complaints from the population was apparently established in Somalia. This is a worthy initiative that should be retained, but it is also very limited. The research on complaint committees has shown that there is little willingness on the part of victims of abuse to complain to these reviewing boards if they have no trust in them, one of the main determinants of trust being whether or not they are familiar with the staff. When the CF operate in a foreign country, the staff of a complaint committee can only appear as strangers to the local population, who would then have little inclination to complain to them. As for the traditional structures of ministerial accountability, they also exist in DND, and there is no need to borrow them from the Ministry of the Solicitor General or one of its provincial equivalents.

It would then seem that there is little to be borrowed from the way in which accountability is exercised in police forces. If, however, there is no overall model that can be imported from police forces into the CF, there are some features which may be incorporated to a model germane to the CF. I have identified four such features.

Public release of data: It is uncertain whether the sanctions imposed by the complaint boards have a substantial effect on the behaviour of individual police. They are usually light and the complaint process is time consuming, especially when complaint boards are legally defined as administrative tribunals where all interested parties are entitled to legal representation. There is, however, a feature of external review boards which has an impact on police organizations, as such. It is the obligation to release, every year, a public report on the operation of the board, i.e., the number of complaints processed, what was at issue, the disciplinary charges that were laid and so forth. Such data present a global picture of the state of discipline and the range of misconduct within a police force. When the picture presented to the public is dramatic because of the amount of police deviance shown, its release acts as a powerful incentive for police forces to remedy the situation.

Proactive review boards: External review boards are not always limited to the reactive processing of individual complaints. In cases such as the RCMP's Public Complaint Commission and its External Review Committee, these institutions are empowered to launch projects of their own. Such projects may be the study of a particular problem, for example, the risks involved in the use of informants and the recommendation of organizational remedies, as opposed to individual sanctions.

External investigators: This is one of the most hotly debated questions in connection to police accountability. When, for instance, a suspect is killed in questionable circumstances, who should investigate whether the use of deadly force was warranted or not? There are three possible answers to this question. (1) A police force should investigate its own members; (2) the investigation should be conducted by another police force; or (3) the investigation should be conducted by an independent unit that does not belong to any police department. All three strategies are presently in use. It is generally recognized that the first one is unsatisfactory and is mistrusted by the public, the police force investigating itself being at risk to pursue its own organizational self-interest rather than the public's. The second course is adopted by many police forces and is presently the one followed in Quebec. It also has serious shortcomings, police solidarity extending beyond the limits of one organizations. When a police force is investigating another one, there is a tendency to conduct the investigation at arm's length for fear that overzealousness would lead to reprisals when the investigated force will in the future become the investigating one. To put matters more concretely, police force A may be investigating police force B in shooting incident X, but some months later it may be the turn of police force B to investigate police force A in shooting incident Y. The present tendency is to adopt the third strategy and to create independent investigative units, as is the case with the Special Investigation Unit created by Ontario.

Non-penal accountability: According to the present *Police Act* of the province of Quebec, the Minister of Public Security who is responsible for policing may appoint an independent inspector or a civilian commissioner endowed with powers similar to the president of a commission of inquiry to conduct an investigation into a police force.[3] The purpose of such an investigation is strictly to learn the facts and recommend remedies for situations that prove problematic. The law explicitly forbids both inspector and special commissioner to lay individual blame and *a fortiori* to recommend criminal prosecution(s). These inquiries have produced useful results in the past. Not being under the threat of individual sanctions, police tend to be more co-operative with the specially appointed inspector or commissioner.

Accountability in CSIS. The Security Service was, until the enactment of the *Canadian Security Intelligence Service Act* (RSC, 1985; henceforth the CSIS Act), a component of the RCMP. It is now a civilian agency created on the recommendations of the McDonald Commission (Canada, 1981a). As we previously saw, the McDonald Commission devised a complex structure designed to ensure that CSIS would be fully accountable and that it would not abuse its powers in the same way as the RCMP Security Service. This structure rested on ministerial accountability, parliamentary oversight, the creation of a position of Inspector General and the appointment of the Security Intelligence Review Committee (SIRC). Not all of the McDonald report recommendations were actually implemented and the creation of CSIS was a thorny process that extended over several years (Whitaker, 1991, 1992).

I shall first review the structure of accountability as it presently exists and then offer some comments.

Accountability to the Minister: The minister responsible for CSIS is the Solicitor General of Canada. However, CSIS is not the only agency involved in the collection, analysis and dissemination of intelligence security, although it is the only one established by statute. As in Australia, none of our agencies engaged in the collection of foreign intelligence is established by statute. The fact that the "intelligence community" comprises several agencies, plus the fact that the McDonald Commission favoured a wider pattern of ministerial accountability than that provided under the single authority of one minister resulted in the creation of a rather intricate web of committees and subcommittees.[4]

Up until the end of the Mulroney administration, the two main threads in this web were the Cabinet Committee on Security and Intelligence (CCSI) and the Interdepartmental Committee on Security and Intelligence (ICSI). Senior ministers (External Affairs, Defence and so forth) sat on the CCSI, which was chaired by the Prime Minister. The CCSI met only occasionally to set goals for the intelligence community and to sketch the policies that were to be developed and implemented to achieve these objectives. This development and implementation of policies were, by and large, performed by the senior bureaucrats of ICSI, assisted by their staff. The CCSI was abolished during the last year of the Tory government, and the newly elected Liberal administration reallocated its responsibilities to the whole Cabinet, the Privy Council Office having been endowed with a special co-ordinating role.

The crucial point is that CSIS and the intelligence community were made accountable to more than one minister, even if the Solicitor General is the officer who exercises legal authority over the Service. The framework for ministerial accountability is collegiate rather than individual. This implies that, *at least at the ministerial level*, this framework is *functional* rather than rigidly *institutional*. Instead of limiting accountability to a narrow match between a minister and an agency, it is the performance of the entire function of intelligence which is made accountable to its various stakeholders. This move from institutional to functional accountability is still very much in its incipient stages at the present time.[5]

Parliamentary Oversight: For reasons I explained previously, there is little to say on this subject, where the recommendations of the McDonald Commission were not applied. There is presently the Standing Committee on Justice and the Solicitor General. However, the Special Committee, appointed to review the CSIS Act, recommended the appointment of a permanent subcommittee of the Standing Committee that would deal with security and intelligence as is the case in the United States (Canada, House of Commons, 1990, rec. 107). This key recommendation was not implemented by the government (for a complete analysis, see Farson, 1995: 209).

The Inspector General: The McDonald, Krever and Keable inquiries revealed that in its fight against terrorism in the 1970s, the RCMP Security Service violated, on several occasions, the rights of Canadian citizens. Several solicitors general of Canada and other high-ranking officials testified during the public hearings of these commissions that they had been kept in the dark by the RCMP on how it was really operating.[6] It was thus felt by the commissions who investigated these scandals that the Solicitor General of Canada needed an officer to be "the eyes of the Minister" into the conduct of the operation of the newly created agency, CSIS, in order to avoid a repetition of the RCMP scandals.

The functions of the Inspector General (IG) are described in section 33 of the CSIS Act. According to this Act, the primary function is to provide the Minister with a certificate stating (1) whether or not the report that the DG of CSIS is required by statute to issue every year is an adequate reflection of the Services's activities and (2) whether CSIS stayed within the parameters given to it by statutes and ministerial instructions and did not abuse its powers. The IG also undertakes in conjunction with the SIRC and officers of the Ministry, research projects which may be linked to a particular problem and would lead to policy recommendations. For instance, in the wake of the scandal that surrounded the infiltration of the right-wing extremist organization known as The Heritage Front, several studies on the use of human sources and related topics had to be conducted. Their results may have served as the basis for the policy recommendations formulated in SIRC's report on this affair (SIRC, 1994).

There are three features of the IG office which are interesting for our purposes. First, the IG has legal access to all CSIS documents, excepting Cabinet confidences, and can also interview members of the service. At least in theory, the IG has the tools necessary to perform his or her duties. Second, the IG belongs to the Ministry of the Solicitor General and he or she is a member of the Minister's staff. Third, and most important, the IG reports exclusively to the Minister, acting in a way as his or her personal auditor of the Service. The IG's reports are not tabled in Parliament and are not made public. The closeness of his relationship to the Minister can be illustrated by drawing a contrast with the duties of the Australian IG of the ASIO. As Lustgarten (1995: 170) indicates, the mandate of the

Australian IG is to assist ministers (notice the plural) in their assessment of the legality and propriety of ASIO activities. The intended audience is the whole executive branch, whereas the Canadian IG is first and foremost "the eyes" of the Solicitor General into CSIS.

The Security Intelligence Review: SIRC was created by the CSIS Act in 1984. Its members are drawn from the body of Privy Councillors, although none of these members can be officially linked to the executive branch or can be holding a seat in Parliament. Before appointing them, the Prime Minister is under a statutory obligation to consult with the leaders of opposition parties with more than 12 seats. Some of the opposition parties are usually represented among the members of SIRC. SIRC has a staff that includes a research and investigative unit. It has the same powers of access to CSIS documents as the IG.

Basically, SIRC has three functions. First, it acts as a complaint board for citizens who believe that their rights have been abused by CSIS. Since CSIS does the investigations leading to a recommendation of the level of security clearance granted to government officials and employees, it hears a certain number of complaints in this field. The complaints are resolved through elaborate *in camera* hearings, where the parties can benefit from legal representation. Second, SIRC oversees and monitors the activities of the Service in a way similar to the IG. For instance, it publicly releases data on the use by CSIS of electronic surveillance. The results of its review of CSIS are released in its annual report, which is tabled in Parliament and is accessible to the public. Third, it may be tasked by the Minister to canvass certain issues — for instance, the state of bilingualism in CSIS — or may proactively task itself to do so. It may also direct the IG, on its own initiative, to conduct an investigation into a particular matter. Finally, it acts as a permanent commission of inquiry into CSIS and may be asked by the Minister to investigate a particular affair that was disclosed to the public. It did so several times and most recently in the Grant Bristow/Heritage Front affair (SIRC, 1994). Although this is not part of its mandate, SIRC organizes consultations and seminars on subjects related to its mandate and the mandate of CSIS.

There are two features of SIRC that deserve mention. The first is that SIRC is, in contrast to the IG, an institution with public accountability. Its annual reports are addressed to the Solicitor General, but the Minister is obligated by section 53 of the CSIS Act to table the reports in Parliament within 15 days. Second, SIRC is independent of the Ministry of the

Solicitor General. It is not part of the Ministry and is not funded by it. Second, even though its annual reports are delivered to the Minister, they must be laid before Parliament and released to the public. Furthermore, the prime object of these reports is not the internal functioning of SIRC but its findings in connection to CSIS. In its annual reports, SIRC reports as much on CSIS — and perhaps much more so — as on itself. Finally, although the Minister may task SIRC to report on a particular issue, the Commission neither controls the process of its inquiry nor the content of its report. In addition, SIRC has the legal power to be proactive and trigger an investigation into an issue which it finds problematic without receiving a specific complaint from the public or being directed to do so by the Minister.

SIRC has been the focus of academic research (Gill, 1989; and Lustgarten, 1995) and of media attention. There is general agreement that it successfully discharged its mandate. On the basis of his comparative analysis of intelligence service accountability in Australia, Canada and the United Kingdom, Lustgarten (1995: 181) identified eight features necessary for the efficiency of an accountability institution in the field of intelligence security. As they might apply in other fields of accountability, I shall quote them:

1) Independence from the executive
2) Proactive capacity
3) Membership representative of the spectrum of political parties acting in a non-partisan manner
4) Access to information about all aspects of the agency's activities
5) The ability to maintain secrecy where necessary
6) Institutional expertise
7) Adequate support staff working full-time
8) The capacity to campaign, that is, to use the media to build up support for its position

The last feature may sound controversial but need not be. When it released its report on the Grant Bristow/Heritage Front affair (SIRC, 1994), SIRC generally exculpated CSIS of any wrongdoing and even commended certain aspects of its performance. One federal political party was highly critical of CSIS and of SIRC's report. SIRC was able to build support for its position in the media. This support put a practical end to the affair and was actually beneficial to CSIS.

Conclusions and Remedies

I will now formulate what appears to me relevant remedies for the defects and problems I have noted. The main conclusions of my analysis could be formulated thus. First, there is a confusion in the CF's notion of accountability between individual penal responsibility and organizational accountability. Second, when accountability is discussed by the CF outside the compliance/sanction frame of reference, this discussion does not go beyond the traditional notions of being accountable to your minister and through him to the elected Parliament. Most assuredly, these are basic requirements of accountability, but the crucial point is that they were not sufficient to avoid the incidents in Somalia and their disastrous aftermath. The major failure of an accountability framework that is exclusively internal is its fostering of complacency. The significance of problems is systematically played down; problems are not addressed before they generate major crises, a state of affairs extremely detrimental to the reputation of the CF.

Last, the margin of discretion commanding officers have in how they make their subordinates accountable and how they insure compliance with regulations and policy is such that it can neutralize implementation.

In proposing remedies, I shall formulate both recommendations and suggestions. My suggestions could become recommendations, if the future investigations of the CIDCFS confirm the analysis that I have made on theoretical grounds and on the basis of documents rather than testimony. My recommendations and suggestions are derived from the preceding analyses.

RECOMMENDATION XV: **I recommend that the results of monitoring the implementation of the policy against racism be made public in annual reports tabled in Parliament and made accessible to the public.**

We previously saw that the CF policy against racist conduct, CFAO 10-43, was released with an annex that provided "sample message formats" that were to be filled out by units and sent to NDHQ. The purpose of this measure was to standardize information on the extent of racist conduct in the CF and the success of the counter-racism policy. This information is of public interest and should reach the public. The state of such projects as AETHRP and SHARP and the results of their implementation should also be part of this annual report. I believe that having to make this information public would act as an incentive to make the implementation of the policy against racist conduct a success and would enhance accountability to the Canadian public in the CF. I see no valid reasons on the grounds of protection of Canada's national security for keeping this information confidential.

RECOMMENDATION XVI: **I recommend that the commanding officers in the CF be provided with guidelines on how to apply the policy against racist conduct and harassment.**

We previously saw that commanding officers exercised considerable discretion in the application of measures against racism and harassment. There are several problems related to the exercise of discretionary powers. One is that they might defeat the purpose of the policy. Another problem is that there may be unwarranted disparity in the different regiments in the imposition of sanctions to CF personnel who violate the policy. One proven remedy against disparity and its potential effects of neutralizing the implementation of a policy is the development and dissemination of guidelines. These guidelines need not be the same for every level of command and could be tailored to meet the particular needs of the various levels of command, if there are different needs. What makes the importance of guidelines crucial is that the policy on racist conduct prohibits membership in groups and associations that promote racial discrimination. Regulations which impose sanctions on persons alleged to be members of deviant groups or organizations are very difficult to apply. There is presently in Canada an intense debate about whether to criminalize membership in organized crime groups and associations, a number of legal scholars and lawyers stressing the potentially harmful character of such legislation for civic rights and freedoms. In some ways, a group of committed skinheads is not very different from a club of delinquent bikers. The arguments showing how difficult it is to apply anti-membership legislation, even in the field of organized crime, suggest the potential problems with the definition of racist conduct proposed by CFAO 10-43.

RECOMMENDATION XVII: **I recommend the creation of the office of Inspector General of the CF.**

This recommendation proposes to amalgamate the office of the IG and an accountability institution, such as SIRC, as they are defined in the CSIS Act. On the one hand, I do not see at present the need for an institution that would be patterned after SIRC, although such an institution may eventually become necessary. One of the main reasons why a review committee does not appear to me to be necessary at the present time is that the CF does not operate in the main domestically, as does the CSIS.

There is, however, a feature of SIRC that should be transferred to the office of the CF Inspector General. As we previously saw, an IG reports exclusively to one or several ministers, whom he or she assists, and his or her reports and findings are not accessible to the public. Furthermore, an IG is part of the staff of a particular ministry of which it is, properly speaking, independent. I believe that the CF Inspector General should be independent of DND and that the role should not be limited to being "the eyes of the Minister." The CF Inspector General should also be a public eye. Consequently, his or her report would be delivered to the Minister of National Defence, who would be under statutory obligation to lay it before Parliament. It should also be accessible to the public.

The primary role of the Inspector General would be to monitor and review the conformity of CF activities with a set of *norms*. I use the word "norm," which means a standard of conduct or an *ethical* value, to specify that the mandate of the Inspector General would not be to oversee the compliance of CF personnel with the great number of regulations in force in the CF. The standards of conduct that would concern the Inspector would be, for example, the Canadian Charter of Rights and Freedoms, the international conventions concerning the conduct of military operations, certain aspects of the ROE in peacekeeping operations and their variations, CF policies that embody norms in the preceding sense (e.g., the policy on racist conduct) and the propriety of conduct with regard to certain ethical values which have not been codified in legal or regulatory form. For example, the hazing rituals that went on in the former Airborne Regiment may not have breached any existing law or regulations but they were clearly in breach of what is generally understood by ethical conduct. These examples are not exhaustive nor do they represent a fully developed concept of the office of CF Inspector General. Such development should be undertaken by a team of experts that would include legal advisors, if my recommendation is perceived as relevant and useful. There are also questions of logistics that must be resolved, such as staffing the office, the duration of a term in office and similar questions.

I shall note that there is now a precedent in the CF for external oversight. The former Chief Judge of Quebec, the Hon. Claude Bisson, will review the activities of the Communications Security Establishment (CSE);

however, most if not all of his reports will remain confidential, as the operations of the CSE are conducted at the highest level of secrecy. I shall finally mention that there is an inspector general in the U.S. Army. A study of that mandate could provide useful information about the nature of such a function for the armed forces.

I do not believe, however, that the creation of the office of an inspector general is, in itself, sufficient to insure CF accountability. As I just said, the primary task of an inspector general would be to monitor and review the compliance of the CF with a set of norms and to report on this compliance. The mandate of the IG would not then consist exclusively of providing information that is both accurate and complete to all authorities which have a legal right to oversee the CF (e.g., a parliamentary committee duly appointed to perform this function). Since accountability, as I defined the concept, is primarily an information process, it appears to us that there is a need to create an office that would be concerned exclusively with the provision of adequate information to whoever is legally entitled to it. The need is all the greater, now that the image of the CF is seriously tarnished in relation to its will to provide an accurate account of its operations and of the conduct of its members.

RECOMMENDATION XVIII: **I recommend the creation of the office of the Registrar General of the CF.**

The Registrar General would not be part of the CF. His basic function would be to ensure that all legally requested information from the CF be complete and accurate and, if need be, to provide this information himself or herself. The Registrar General would have legal access to all information available within the CF, in whatever form it exists — written documents, computer files, videotapes and so forth. Whether the Registrar General should or should not be under the authority of DND is a question that requires careful consideration. The advantage of having such an office as part of DND would be that it would establish an institutional and even a physical proximity with the information — with the documents — to be legally transmitted and disclosed. The advantage of not having the Registrar General under the authority of DND would the assertion of the independence of the office. Both advantages could be secured through a flexible definition of the office that would balance its closeness to all the available information with its independence.

I now formulate three suggestions.

1. In his interviews with officers of the Ministry of the Solicitor General of Canada, Stenning (1995b: 68) found that the most often mentioned strategy for accountability was "public clarifications, through published statements, of the missions, core values, objectives, goals, strategies, standards, plans, policies and procedures of the agencies as a whole and of their constituent parts." Such clarifications provide a yard stick against which the activities of the organizations can be measured. The CF could build on its previous efforts to provide Canadians with such statements. For example, *a general statement of objectives and mission of the CF with regard to international peacekeeping would be very useful.*
2. *The creation of a permanent House of Commons subcommittee on international peacekeeping in its various manifestations might be considered.* There seems to be a growing number of questions to be addressed in relation to peacekeeping. As was remarked by the Auditor General, the cost of peacekeeping operations is spiralling upward; from 1991-1992 to 1995-1996, the costs of the participation of Canada in peacekeeping operations have increased by 410 percent, from $47 million to $240 million (Canada, Auditor General, 1996, Chapter 6, Highlight 6.5; see also Gizewski and Pearson, 1993). The Auditor General has also remarked that the members of Parliament want to be better informed about peacekeeping operations (Canada, Auditor General, 1996, Chapter 7, section 7.22 and ff.) This is only one of the many questions that could be addressed by a subcommittee that would be a permanent appendix to the presently existing Standing Committee on National Defence. This subcommittee would need to have access to all the information relevant for the performance of its mandate. Otherwise, creating it would be useless. *In this respect, a major overhauling of the channel through which DND is disclosing relevant information for external oversight is not only necessary, it is urgent.* The appointment of such a parliamentary subcommittee would enhance the CF's accountability to the elected body.
3. *The possibility of relying on external and independent investigators in certain high profile cases such as the death of Shidane Arone or potential CF disinformation tactics, should also be examined.* This is a complex question, because of the separate status of military justice. For example, is it legally possible to have the RCMP conduct an inquiry into the CF that would lead to charges before a court-martial?

The same question could be asked of a joint investigation task force comprising military police and external investigators. These questions are difficult to answer, but I believe they need to be addressed. According to a press release, the team which has been appointed by the CF to investigate the recent allegations of misconduct on the part of several members of the Canadian peacekeepers in Bosnia will be headed by a retired RCMP officer. I believe that this is a good omen: it shows that my recommendation can be put into practice, at least on an individual case basis. I suggest that this practice be generalized in the case of serious allegations of misconduct, where there is a particular need for transparency in the investigative process. The problems of credibility that plagued police forces when they were investigating themselves may well be raised in relation to the CF, particularly in view of the conspicuously hierarchical nature of the organization.

RELATIONS WITH THE MEDIA

Whitaker (1991) identifies three dimensions of accountability, namely the legislative, the executive and the public. The public dimension includes, but is not limited to, giving accurate accounts to the media. Thus there is a connection between accountability, which was the subject of the previous part of this chapter, and relations with the media, which is the subject of this second and final part.

Besides its connection with the notion of accountability, there are several reasons to examine the subject of the relations of the CF with the media. Since the incidents in Somalia, the CF has been submitted to sustained media coverage. This coverage was so intense, and, in the main, critical of the CF's performance in Somalia and after, that it has taken its toll on CF members and their families. Even the Auditor General has noted that media scrutiny was a cause of increased stress for CF members, particularly ex-members of the CAR, and their families (Canada, Auditor General, 1996, Chapter 7, section 7.67). Ex-members of the CAR are not the only people put under stress by pressure from the media. My previous analysis of the elaborate process the highest authorities at DND — excepting only the Minister — followed in reacting to an article in *The Globe and Mail* reveals that DND is seriously preoccupied with its image in the media. I gave one example of DND's concern with media coverage. I could have given many others. Actually, what developed into the most dramatic part of the CIDCFS inquiry is directly related to DND's responses to queries from the press.

There is no reason to believe that media pressure on the CF and other public agencies will diminish. The safest prediction is that it will increase, because we will be progressively living in what we could call the maximum visibility society. Such diverse phenomena as the impressive rate of growth of the "information highway" and the fascination with devices that provide maximum visibility (cheap cameras, videocams) — to the extent that offenders make film records of their offences — are indications of the emergence of this maximum visibility society.

The last part of this chapter is divided in three. First, I briefly review the literature on the relationship between the armed forces and the media. Second, I focus on the relation between the CF and the Canadian media. I will conclude with recommendations.

The Military and the Media

There is not an extensive body of literature devoted specifically to the relation between the military and the media, although the subject is indirectly discussed in post-Vietnam military literature. There are two important features in what is available. First, it is about the relationship between the media and men *in combat*. The media are no different when they cover military affairs than when they cover other subjects. The prime value of an event for the media is its newsworthiness: combat is much more newsworthy than peacekeeping. Excepting the occasional comment on the role of television — and, in particular, of the Cable News Network (CNN) — in Somalia, there is not much literature dealing specifically with the relationship between the press and peacekeeping. The second feature of this body of research is that it is divided in two, a great divide separating what preceded the Vietnam War and what followed it. The conflict in Vietnam is referred to as the "uncensored war," although it was debatable whether censorship was in fact exercised.(Hallin, 1986; Hammond, 1988). During the Vietnam conflict, censorship was replaced by a set a voluntary guidelines that the press followed in the main.[7] Paradoxically, experienced war correspondents such as American Drew Middleton, lamented the replacement of censorship by media guidelines, which had the effect of making the military far more wary of talking with reporters than during previous conflicts, when there was censorship (Knightley, 1975: 423).

I shall now present what I believe are the main conclusions and themes in this body of literature. Because the relations between the media and the military are not, like accountability, peacekeeping and racism, one of the primary focuses of this report, my review of the literature is not as thorough as it was in those cases. Needless to say, the body of literature on the military and the media is far less extensive than that concerned with my main topics. Here then are the main themes and conclusions.

A Strained Relationship. Lloyds J. Matthews tells us the following story in his preface to a book that he edited on news personnel and national defence. In 1864, during the American Civil War, General William Tecumseh Sherman heard a rumour that three reporters with the Union Army of Tennessee had been captured by the Confederates and executed. His comment was: "Good! Now we'll have news from hell before breakfast." (Matthews, 1991: p. x). The reason for Sherman's wrath against journalists is a familiar one for the military. During his march through Georgia, his intelligence officers had broken the Confederate signal flag code employed from mountain to mountain to transmit details of the Union Army advance. The correspondent of the *New York Herald*, Debow Randolph Keim, learned of the Union's decryption breakthrough and published the story in his paper, thus prompting the Confederates to change their flag code. General Sherman was not pleased.

The relations between the press and the military have been strained since the beginning of war reporting and have remained thus. Hooper (1982: 6) has a list of critical quotations concerning the press collected from generals from the time of the Crimean War to the war in Vietnam. Knightley's history of war correspondents is subtitled "The War Correspondent as Hero, Propagandist, and Myth Maker from the Crimean War to Vietnam" and the jacket of his book portrays a war correspondent writing a story. He is standing with one foot on the corpse of a young woman, who wears a ribbon with the word "Truth" printed in red over her dress. There is actually not a single book on the press and the military that does not discuss the stormy relationship between the two (e.g., Braestrup, 1977; Hallin, 1986; Halloran, 1991a and b; Hammond, 1988; Hooper, 1982, Knightley, 1975; and Matthews, 1991). This relationship is even more inimical than that with the police, although they are the subject of much criticism from the media. Richard V. Ericson, who has written several books on the relationship between the press and the criminal justice system, has declared recently that there were "many respects in which the news media can be conceived as *part of* criminal justice (Ericson, 1995: 138-139, emphasis in text). There are obvious exceptions to this statement that the situation in connection to the media is worse with the military than with the police. During the Oka crisis, for example, the CF enjoyed a much better rapport with the press than did the police forces.

Mutual Distrust. What I have said would suggest that one of the main reasons for the difficult relationship between the press and the military is that war correspondents are always under suspicion of unwittingly providing useful intelligence to the enemy. There are other reasons, one of the most important being the way the media can turn public opinion against a particular war. Although media reporters have been accused since the beginning of press coverage of the military of demoralizing the public and through it the combatants themselves, the turning point for this line of criticism was according, to the research literature, indisputably the war in Vietnam. It is widely believed that this war was not lost on the battlefield but at home, where the press turned public opinion against the war, thus forcing President Johnson not to seek re-election. A typical statement of this opinion was made by Colonel Maurice Tugwell in a lecture delivered on 30 January 1973 at Kingston to the Canadian Land Forces Command and Staff College. The lecture was entitled "Public Opinion in counter-insurgency":

> When historians come to study military events of the late 60s and early 70s they may conclude that the major lesson for soldiers to learn from that period was this — that if you cannot fight an enemy in a way that public opinion at home, and fair-minded opinion overseas, find tolerable, then find another enemy. To endeavour for long to conduct a campaign in the face of really hostile opinion may lose you more than just a battle. (Tugwell, 1973, quoted in Hooper, 1982: 219)

Interestingly enough, Hooper's immediate comment, after quoting this paragraph from Tugwell is "Vietnam has proved the validity of these words." I shall see that this claim is actually controversial.

The war in Vietnam was a turning point in the relations between the media and the military in many respects. In World War I and World War II, there was genuine support from most sides of opinion in the countries at war. Ernie Pyle, perhaps the most famous of all war correspondents, turned GIs into heros in his coverage of the U.S. Army in the World War II. One thing that changed during the Vietnam War was that the distrust between the military and the media became completely reciprocal. The military believed that the press corp was biased against them; the press corp believed no less that the official statements made by the military and the U.S. government about the war were grossly inaccurate. William Hammond wrote an authoritative book on the military and the media during the Vietnam War, which was published by the Center of Military History of the U.S. Army. He is above any suspicion of having written a book that would be biased against the U.S. government and the U.S. Army. Yet, he concluded his book in the following way:

It is undeniable, however, that press reports were still often more accurate than the public statements of the administration in portraying the situation in Vietnam.

In the end, President Johnson and his advisers put too much faith in public relations. Masters of the well-placed leak, adept at manipulating the electorate, they forgot at least two common-sense rules of effective propaganda, that the truth has greater ultimate power than the most pleasing of bromides and that no amount of massaging will heal a broken limb or a fundamentally flawed strategy. (Hammond, 1988: 388)

This criticism is not only directed against U.S. politicians. As the book makes clear, General Westmoreland and the information program of the U.S. Armed Forces in Vietnam also lost a great deal of their credibility. The daily briefing by the military to the press was known as the "five o'clock follies."

The reciprocal distrust between the military and the press in the United States only increased after the invasion of Grenada, where all media were kept in the dark as to what was happening in the field. The media being one of the most internationalized institutions, the attitudes of the most influential media — the U.S. media — are often transmitted to the media of other countries. The media distrust of official statements by the military is as much perceptible in British, French and Canadian published research literature as it is in its U.S. counterpart.

The Erosion of Censorship. The war in Vietnam was in theory uncensored; voluntary guidelines were provided to media personnel who generally complied with them. The Vietnam War began in the early 1960s and ended in the early 1970s. It is fast receding from our collective memory; historical memory in general is surprisingly short at the end of this second millennium.

The release from censorship during the war in Vietnam had the impact of making the press more intolerant of limitations on their access to information and more insistent about their self-proclaimed right to publish all that they perceive as newsworthy. Braestrup (1991: p. XV) recalls that, during the Gulf War, journalists inexperienced in war coverage complained bitterly that they were prohibited from releasing certain kinds of information (e.g., filming particular military installations). Actually, the rules that they operated under were the same as those imposed by the guidelines recommended during the Vietnam War, the latter perceived at the time as reasonable by the press community. Either because of inexperience or a hazy memory, the same guidelines were judged to be unfairly limiting during Operation Desert Shield, a campaign by a coalition against Iraqi forces that outnumbered those of the coalition. The need for prudence was very real and yet some representatives of the media felt that requests for caution were unfounded. The definitive erosion of censorship is now a factor that has to be taken into account in any analysis of the relation between the press and the military.

The Ignorance of the Armed Forces About the Workings of the Press. A fourth strand of the research is the military's lack of knowledge of the working of the press. Halloran (1991a and b) is particularly eloquent in this regard. This ignorance is not limited to U.S. and Canadian journalists. Hooper (1982: 211), who is a British author, also mentions this fact.

There is no doubt that the military are right in complaining about the lack of professionalism of many journalists. However, not only are there persons with knowledge of military matters working in the media — quite a few of them with military experience — but the military's lack of knowledge about the working of the media and their professional constraints are much greater than the reverse. The media are made up of agencies and organizations which are no less complex and self-regulated than armies. Recently, the French army initiated a successful program for acquainting the press with the military and, in particular, with combat life. The CF followed suit and a group of Francophone media reporters underwent a similar crash training course, which took place, at least in part, at Valcartier. Media reciprocation would actually be very beneficial to the CF. A selection of various CF personnel spending at least a day in various newsrooms could be a worthy and useful experience given the CF's ignorance about the media.

De-Mythologizing the Media and Paparazzi Mythology. One of the most frequent assertions made about the media is that they are powerful myth makers. They also are themselves the object of much myth making. Several authors have taken pains to debunk this mythology (Halloran,

1991 a and b; and Hammond, 1988 and 1991). I cannot list all the misunderstandings and the mistaken beliefs, but I will mention some which are among the most current.

As I previously suggested the seminal myth with regard to the power of the media is the belief, exemplified in the quotation from Maurice Tugwell, that the U.S. media turned public opinion against the military during the Vietnam War by depicting it as an irate Goliath using the immense firepower of the most advanced weapon technology against lightly armed Vietnamese fighting to liberate their own country. For historians now enjoying the necessary distance to study these events impartially, this view is increasingly challenged (Mueller, 1973; Hammond 1988 and 1991; and Halloran 1991a and b). As I said, Hammond has the status of a quasi-official historian of the relationship between the U.S. military and the media during the Vietnam War. Here is his overall conclusion: "What alienated the American public in both the Korean and Vietnam Wars, was not news coverage but casualties. Public support for each war dropped inexorably by 15 percentage points whenever total U.S. casualties increased by a factor of ten" (Hammond, 1988: 387; see also Hammond, 1991: 14; and Mueller, 1973).

Actually, the power of the press has been much exaggerated. In this regard, Halloran correctly observes:

> Like "the media", the power of the press is a myth. The press has *influence* not power, and the distinction is important. Military officers have power in that they have the legal and, if necessary, the physical force to have orders obeyed. The press has neither, and cannot enforce anything. (Halloran, 1991a: 41, emphasis in text)

As is implied by the quotation marks around the phrase "the media," Halloran is critical of the belief that the media are some kind of monolith that always follows the same rules. In reality, the media — the written press story, radio, photographs, television, film documentaries, video documentaries and, increasingly, the so-called docu-drama, which does not yet fit any known communication category — are very different from each other.

To realize the diversity of the media is crucial for understanding their functioning. As Braestrup (1991: p. XXIII) stressed, the biases of the press, particularly television, are professional biases in the sense that they are dictated by the medium which it uses. Television, with its small screen, cannot by definition give "the complete picture" and must rely on incidents that have a symbolic value and that may be perceived by the viewers as microcosms. The same applies even more to photography. Generally, both media — photography and television — favour people stories over the depiction of complex situations (perhaps the most famous example of the constraints on this medium is the photograph of the Chief of South Vietnam's National Police, Brig. Gen. Nguyen Ngoc Loan, killing a Viet Cong officer by shooting him through the head. This incident was also filmed by a news cameraman and shown on television).

There is another persistent myth, which I think has not yet received the attention that it deserves in the research literature. It is what I shall call the "myth of the intrusive paparazzi." Paparazzi is the appellation given to press photographers who resort to trickery, who hide themselves, who use extra-powerful camera lenses, etc. to take pictures of people caught unaware. In the maximum visibility society that I previously described, there is no more need for paparazzi. Recording devices of every kind are not only a current fixture in our environment, but people act as self-paparazzi for themselves, their family and friends and the circles in which they move. The pictures they take and the films they make are more compromising than anything that the most intrusive paparazzi might plan to take. These pictures, whether animated or not, always seem to find their way into the hands of the media and then are projected on television networks without any media person spending a single instant in taking them. What probably did the most damage to the CAR and to the CF image was not media coverage per se but the simple showing on television of the gruesome photographs of the distorted face of Shidane Arone next to the smiling faces of his torturers and the retransmission of part of the videotape made during one of the CAR's most debasing hazing rituals. It was not the media but CF members that were the agents of their own fall. To put it plainly, organizations such as the CF, which are the focus of intense media scrutiny, should not be as obsessed as they are with media entrapment but should rather pay more attention to self-entrapment.

The Future. Writing on the military and the media in the early 1980s, more than 15 years ago, Alan Hooper (1982: 209) began the conclusion of his book in the following way:

> This book has been about communication between the military and the media and how that relationship has in turn been affected by the rapid development in the technology of communications, and by the considerable changes which have occurred in society, and in the military, during the last two decades. But the

changes during the last 20 years are as nothing to the likely changes in the next 20. If the development of the transistor changed the technology of communications, then the development of the microchip will change our whole way of life. For instance, we have yet to appreciate the implications of multi-channel and cable television or the full development of ENG (the latter is particularly relevant to military events when one considers the technical feasibility of broadcasting a conflict live via satellite).

What may have seemed speculative some 15 years ago, now appears to us as not having been bold enough in its forecast of the future. When the U.S. Marines disembarked in Mogadishu in the course of Operation Restore Hope the television cameras had preceded them and the operation was broadcast live. Ditto with the air attack on Baghdad during the Gulf War that was broadcast and commented on live by Peter Arnett on CNN. Organizations that want to go on blaming their woes on the media better gear up for a lot of moaning at satellites.

The CF and the Media

I noticed earlier that an incident connected to the relationship between the CF and the media is currently being investigated by the CIDCFS. This incident concerns the public release of policy notes expressed in the format of Response(s) to Query (ies) (RTQ), in response to freedom of access to information legislation. The released document may have been modified before being released; there is also the issue of whether action was taken to avoid having to release more of these completed RTQ forms.

Since the investigation of these issues is not complete, I shall abstain from commenting on them. I will limit my own analysis to what I learned through the examination of documents that I obtained through the Commission.

Before submitting my remarks, there is a general point that I feel is important to make. With customary acuteness, Richard V. Ericson proposes to distinguish between accountability and account ability. Accountability refers to the obligation to provide an account of activities for which a person or an organization is answerable, either because they were performed in the course of the performance of duties or simply because they involve a person's or an organization's responsibility. Account ability is defined by Ericson as the capacity — the skill — to fulfil the previous obligation (Ericson, 1995: 136 and 161). The ability to provide an account is for Ericson a matter of communication and implies the use of a strategy about what will be said, how it is going to be said and also what will be left unsaid (Ericson, 1995: 144). All social and political institutions have, to different degrees, this ability to give an account of themselves. The point of Ericson's argument is that there is no mechanical relation nor pure correspondence between the obligation to provide an account and the ability that will be put to use in fulfilling the obligation. The fulfilment of the obligation is always mediated by a strategy that tries to reconcile what may be divergent elements (the competing needs for secrecy and for public disclosure, the various ways in which a particular piece of information may be given "spin" and so forth). This translation of an obligation through a particular ability occurs in the CF as in all other organizations and is not in itself the symptom of any pathology. Pathology only occurs when the ability subverts the obligation, which is one reason why the process by which an organization accounts for itself must be overseen by a watchful external body.

The elements that come out of my brief analysis of the DND reaction to an article in *The Globe and Mail,* which alleged that the CF had failed as late as April 1994 to implement measures to screen out racists from peacekeeping missions, are the following.

A Reactive Stance. Nearly all that was accomplished in relation to the issue of infiltration of the CF by extreme right-wing elements happened in reaction to revealing press reports. In this regard the CF were always several steps behind the media and never began to learn how to get ahead. It kept its files on the penetration of right wingers at the highest level of secrecy even though there was no perceptible public interest for withholding this information from Canadians.

Involvement of the Top of the Organization. The notes that I quoted expressed an exchange between the CDS, the DM and the ADM (Per) on the content of a reply to *The Globe and Mail*'s article. Other personnel from the Direction of Public Affairs were also involved. This exchange was intense, the DM admitting in one of his notes that he had lost his cool during a meeting on this affair.

This feature of the process raises the issues of the proper delegation of power and the handling of a situation by persons who have the expertise to do so. The article appeared on 2 April 1994. On 8 April 1994, the CDS and the DM were still arguing over the content of the letter and whether the opportunity to send it in a timely way had been missed.

A Lack of Consensus. The civilian and military components of DND did not see eye to eye on this matter. The position of the military was (1) that there had been measures to weed out racists even before the Somalia scandal and (2) that these measures had been strengthened since the adoption of the policy against racist conduct. The position of the DM was (1) that the previous measures had been very slack, (2) that there was no proper accountability on the screening of racist issues, (3) that what had been accomplished over the last 10 months was unacceptable and that (4) there was insufficient co-ordination and co-operation among the DND units concerned with this question. It cannot be denied that the views of the DM and of the highest ranking military staff of DND were antithetical.

Trying to Respond to a Report that was Essentially Correct. Who, it might be asked, was right in this debate? Let us first recall that the DM had prepared a series of "questions and answers" (Q&As) on White supremacist activity at Petawawa. The Q&As are dated 11 May 1993, a time when the BOI had not completed its inquiry. The general line was that, although racism was not tolerable, "CF personnel may hold any political view, as long as their views are not in conflict with carrying out their duties." In line with this principle, Q&As were prepared. For instance:

Q.5 - What regulations, if any, exist concerning CF members attending meetings of this nature?

A.5 - None. Attendance at right wing extremist meetings is not illegal and denying a citizen the right to attend such meetings would be contrary to Charter rights. What Canadian Forces members do on their own time is their own business, as long as it does not contradict the Canadian Forces Code of Service Discipline and/or Canadian law, or bring into question their reliability or loyalty.

To other questions (A.1, A.3), the line was to answer that no comment could be made due to the ongoing military police and BOI investigations regarding some Canadian Forces personnel stationed at Canadian Forces Base (CFB) Petawawa.

The problem with the latter kind of answer is that they assumed that the BOI would recommend measures to curb racism and White supremacists in the CF. Unfortunately, the BOI's report later concluded thus: "The Board has no recommendations to make with regard to selection and screening at this time." (BOI, Annex F, par. 8, p. 3)

Our own research in regard to projects AETHRP and SHARP confirmed that they were, at the time of writing this report, in their incipient stages, to say the best of them. It would therefore follow that in April 1994, it was the DM who was right to claim contra the military that next to nothing had been done to screen racists from Canadian peacekeeping contingents. This means that the article in *The Globe and Mail* was essentially correct, as is recognized several times in CF documents. An attempt to redress a media report that was accurate was an unsuccessful application of the CF's limited ability to account.

From Reacting to the Media to Amending the Briefing to the Minister. We previously saw that there were two versions of the Briefing to the MND, one dated 7 April 1994 and the other 12 April 1996. The first version was thought by the CDS to confirm the content of *The Globe and Mail* article on the lack of screening procedures to avoid recruiting racists among peacekeepers, and so it was amended to correspond to the thinking of the military rather than admitting to what was, in the present case, much closer to the truth. Briefing the Minister by providing disinformation is a clear instance of organizational pathology — the ability to distort an account undermines accountability as a legal obligation.

Remedies

As in the previous section I will make recommendations and suggestions.

RECOMMENDATION XIX: **I recommend that the academic curriculum of the CF provide better education for the military on the workings of the media and relations with them.**

This recommendation flows from my analyses and from the research literature (Hooper, 1982: 212-213; and Halloran, 1991a and b). It is crucial to understand that such education should also include courses on the ethics of the relationship with the media. If these courses were only to teach the military how to be spin doctors, it would then be better not to give them at all.

RECOMMENDATION XX: **There is an urgent need for a CF policy on the use by its personnel of all recording devices (cameras, videocams, film cameras, tape recorders, etc.).**

Our idea here is not that soldiers should be trained to avoid recording examples of their own turpitude and pathologies. No amount of training could prevent twisted minds of whatever venue from photographing the people they are in the process of beating to death nor would it convince them not to film their own unsavory rituals. In what I called the maximum visibility society, there is, however, a vital need to educate the military on the use of visibility enhancing devices (when, where and how to use them, and what for).

Our suggestion is that there should be much better co-ordination in dealing with the media. Members of the top echelon should limit their role to final approval of a public release, and this only when there is a real need for their authorization.

CHAPTER SIX

Conclusion

This report focussed on three topics: the concept of peacekeeping and its derivatives, racism and extremist right-wing penetration into the Canadian Forces (CF), and accountability. I briefly discussed other subjects, such as the relations between the military and the media. For all the topics that I just mentioned, I provided conclusions and recommendations and suggestions at the end of the relevant chapters. Hence, there is little to add at the end of a report which is twice as long as I originally envisaged. Nevertheless, to avoid ending my work too abruptly, I will provide a few general concluding comments on each of the main topics on which I focussed my analyses.

PEACEKEEPING AND RELATED CONCEPTS

Peacekeeping, up until the last few years when the number of missions increased and their cost was multiplied by four, was a relatively piecemeal affair. The general wisdom was that forces with the normal soldiering skills were up to the job and that when they were not, then it was up to the military to improvise in the field.

This kind of thinking is no longer adequate. As the missions get more complex in terms of their mandate and the deployment of personnel and equipment, improvisation will no longer be enough to fill the gaps in training and military doctrine. There is, however, another reason, running much deeper, why this perspective on peacekeeping is too narrow. If I were to provide a structural or a systemic answer to what went wrong in Somalia, including all three phases of the operation, I would answer that now that the Cold War is over, we have yet to come to terms with the true proportions of and the considerable difficulties involved in maintaining an army in a state of readiness in times of peace. In his groundbreaking study on the psychological costs of learning how to kill, U.S. Lieutenant

Colonel Dave Grossman quotes studies undertaken from the U.S. Civil War up to the last two world conflicts that show that, contrary to what is generally believed, doing "the killing job" does not come naturally nor easily to soldiers. Even in times of war more than 50 percent of the troops do not fire at all or fire in a way that makes sure they will not hit an enemy.[1] Teaching human beings how to kill and, even more difficult, imparting in them the will to do so, is a tall endeavour. The lesson is not easily learned. What gets to be a nearly impossible task to perform is to nurture these killing skills and the will to use them in a time when the occasions to go into combat are getting fewer and when even the possibility of facing the situations, for which one has been undergoing a gruelling kind of training, are dwindling. If the psychological cost of learning to kill in war is considerable, the resulting anxieties and role conflicts of not being able to apply what was learned at great cost must also be significant. It may well be that the best description of the "hell of peace" for professional soldiers is to be found in novels such as *Buffalo Soldiers* by Robert O'Connor (1992). This novel is about the plight of American soldiers who have very little to do at their stations in Germany. On the first page, O'Connor quotes Nietzsche: "When there is peace, the warlike man attacks himself."

There is now some urgency to review the ethos according to which the "killing job" is the raison d'être of the CF. This review could examine whether it is possible to integrate peacekeeping in its diverse forms into a revised military ethos. I are not, at this point, advocating such an integration. What I am saying is that there may be a problem with the "killing job" ethos in times when there may not be a pressing need for professionals at this job. This then is the problem. One way of solving it is to find new missions, such as peacekeeping, for the CF and to draw the consequences of *formally* integrating this kind of mission into a redefined CF mandate. There are many groups, such as Canada 21, which advocate this solution. There may be better ones. What is certain is that training the Canadian military for the "killing job" when there is less and less killing to be done is self-defeating.

I have one last comment to make about getting serious about peacekeeping. There was not one discussion of peacekeeping among those I read that did not advocate the need for reforming the United Nations. In his latest report, the Auditor General takes a very firm stand on this question (Canada, Auditor General, 1996, Chapter 6, section 6.56 and ff.). I believe that he is fundamentally right.

RACISM AND EXTREME RIGHT-WING PENETRATION

One of the sound concepts which is currently used in the CF is that of the Total Force. This concept refers to the integration of the Regular Force and the Primary Reserve.

However, in contrast with this notion of integration I found that the thinking in the CF was often dichotomous. In addition to the current dichotomy in the military between officers and non-commissioned members (NCMs), I also noted dichotomies between mind and body, between time spent on duty and off duty and between words and deeds. These dichotomies allowed what seemed to us a fairly high level of tolerance for racism and membership in White supremacist and extremist right-wing groups. The official position, before the adoption of the policy against racist conduct, was that although the CF disapproved of racism, its members were entitled to their own political opinions and their right to free association guaranteed that they could belong to any group they wanted and attend whatever meetings they wished, as long as it did not break any law or interfere with the performance of their duties. In a similar way, the use of racial slurs was viewed as a fairly benign practice that could be found in most armies, although it was "most inappropriate." The Board of Inquiry (BOI) concluded that the use of racial slurs (euphemized in its report as "nicknames") appeared to be more prevalent at the soldier level and did not seem to indicate racism (BOI, Report, Phase I, Volume XI, Annex I, par. 16-18). Not being true behaviour, verbal behaviour could not, in principle, be conceived as racist behaviour per se.

Although the position that what CF members do with their time when they are not on duty is their own business may be formally correct, it falls short of what might be expected of professionals. Professions have a symbolic dimension in that they represent values in the eyes of their members and of the public. This fact has been recognized by many professions, including the police, which have voluntarily adopted codes of professional ethics. These codes actually limit the formal rights and freedoms possessed by Canadian citizens, but these limitations are viewed as reasonable and consistent with the values embedded in the profession. Doctors usually do not militate in favour of the rights of tobacco smokers, and police do not consort with skinheads. Whatever the CF might be otherwise, they surely belong to what David Bayley (1994) appropriately called "forces of order." And whatever they also may be, skinheads, right-wing extremists and other groups of the same ilk are forces of disorder. Although it is theoretically and also perhaps legally possible to belong professionally to the forces of order and, when off duty, associate with the forces of disorder, there is a serious inconsistency in this double affiliation, both at the level of practice and at the level of public image. How is a CF member, working in a recruiting centre, to respond when one of his skinhead buddies wants to join up?

The CF has taken important and commendable steps against racist conduct within its ranks, and it will take some more. Screening mechanisms, training courses, exercises in cultural sensitization are all necessary instruments for eradicating racism. But they are no more than technical measures which have to be superseded by the development of an *ethos*, if they are to succeed at all.

A good illustration of this point would be the development of a set of technical measures to monitor language. For example, progressive sanctions could be applied against persons who use racial slurs. Such disciplinarian attempts to police language have generally failed: racial epithets crop back into the vocabulary as soon as those taking pleasure in using them think they are escaping notice and will not be caught. Yet there are professions, organizations and corporations where the use of such racist language has been successfully rooted out. In these institutions people do not abstain from using racial epithets because they fear individual sanctions. Rather they abstain mainly because they are wary of losing the respect of persons whom they themselves respect, by breaching the ethos that unites them.

I firmly believe that the development of a binding professional ethos with which racist conduct is felt to be absolutely incompatible is the only way that all the other measures taken by the CF will gain significance and be effective.

ACCOUNTABILITY

There is little to be said about accountability that I have not said before and that has not actually been acknowledged within the CF. I am in full agreement with the CF Deputy Minister who stated in a handwritten note that there was no framework of accountability in connection with screening out racist applicants from the CF and particularly from being involved in peacekeeping missions. I tried to argue that the gap in accountability was greater than simply not having good screening instruments. Accountability is confused with personal responsibility toward the Minister and the disciplinary process. Both have been shown to fall seriously short of providing an accountability framework, as it is now conceived.

There are two major challenges which the CF is facing with regard to accountability. Armed forces in general, and not only the CF, are among the few remaining institutions which can be properly described as closed and largely impervious to outside influence. Academics and lawyers are often derided for their use of jargon. I belong to the academic community and, working in the field of criminal law, I have frequently co-operated with lawyers and legal scholars. Never have I experienced jargon as impenetrable as that which is used in the CF. I say this partly in jest, but also because it is an apt indication of the separation between armed forces and civilian institutions. As I repeatedly emphasized, accountability is a process which is directed to the outside — one truly accounts to another who is not merely a clone of oneself. Recognizing this external dimension of accountability is one of the challenges facing the CF.

The other challenge is to confront the facts as they are instead of defensively playing down the problems. Clinging to the "can do" attitude in a working environment that has broken down is courting disaster. There are many illustrations of this in the events that surrounded the CAR's deployment in Somalia. Illustrations are also to be found in relations with the media. But perhaps none is as convincing as the number of warnings that the officer in command of 2 Commando was unfit for this mission. These warnings were sounded at all levels throughout the pre-deployment phase and particularly during training; they continued to be issued in-theatre. Yet they were all ignored until the crisis exploded.

FAIRNESS

My last word is addressed to the Commission of Inquiry into the Deployment of the Canadian Forces to Somalia (CIDCFS) and to the CF. As I was at pains to show, not only the Canadian Airborne Regiment Battle Group (CARBG) blundered in Somalia; so did most of the other participants in the First United Nations Operation in Somalia (UNOSOM I), the United Nations Task Force (UNITAF) and the Second United Nations Operation in Somalia (UNOSOM II). Fairness requires that the behaviour of the Canadian troops be assessed in the context of what can only be described as a failed UN mission. This is how it is perceived, and I believe that this perception is essentially correct.

What is required in assessing the CARBG's performance is a sense of urgency which will negate any tendency to complacency *and* a sense of fairness which will take into account the difficulties posed by the context of the operation. A sense of urgency is precisely what was most wanting in the Hewson and BOI reports.

But what is finally at issue is the success or failure of the recommendations that will be made by the CIDCFS. If the final report is perceived to be unfair by those who are primarily concerned with it, only lip service will be paid to its recommendations, and the whole painful exercise will have been in vain. If, on the other hand, the report is unduly complacent, nothing will be done to prevent other crises like the Somali affair occurring in the future.

THE SCOPE AND DEPTH OF THE PROBLEMS

My admonition to be fair to the CF in taking into account the context in which Operation Deliverance was conducted rests in part on an assumption that misbehaviour such as that which occurred in Somalia is not current in the CF and that the explanation should therefore reach out for external factors. Although I believe that this assumption is still, on the whole, valid, the allegations about serious misconduct on the part of Canadian peacekeepers in Bosnia make it obvious that the whole explanation for the misconduct of Canadian peacekeepers cannot rest exclusively nor principally on the consideration of external factors and on the specific difficulties of a particular mission. Having restricted its investigations to the Somalian case, the CIDCFS cannot fully appreciate the scope and the depth of the problems with which the CF are confronted. It thus falls on the CF itself to make this assessment and to apply the necessary remedies. It is to be hoped that its credibility has not been so damaged that the public will not be able to entrust it with this grave responsibility.

Recommendations

RECOMMENDATION I: I recommend that the CF collect data on its racial and ethnic composition in a much more systematic way.

RECOMMENDATION II: I recommend that the CF follow the example of some police forces and government agencies and use socio-psychological tests for screening purposes before personnel embark on a peacekeeping mission.

RECOMMENDATION III: I recommend that the disciplinary record of persons selected by their unit for participation in a peacekeeping mission should be thoroughly reviewed, particularly in connection with convictions under the *National Defence Act* (NDA) and the Criminal Code.

RECOMMENDATION IV: I recommend that the CF take the necessary steps to grant equality of access to all Canadian ethnic groups to the units that might be sent on peacekeeping missions.

RECOMMENDATION V: I recommend that the CF conduct a study on the presence and role of women in the military in the context of peacekeeping missions.

RECOMMENDATION VI: I recommend that a basic course on peacekeeping and related notions such as peacemaking, peace enforcing, peace and nation building be integrated in the general training of all members of the CF.

RECOMMENDATION VII: I recommend that cross-cultural training and contextual sensitization be much more elaborate than it was for Operation Cordon (later Deliverance).

RECOMMENDATION VIII: The CF should look beyond its own personnel resources and tap into the experience and knowledge of other agencies and of civilians.

RECOMMENDATION IX: A special unit should be established either in the CF or at DND whose mandate would be to extract systematically from every Canadian experience in peacekeeping those elements — lessons learned, best and worst practices — that could be used to develop a knowledge base for the preparation for future missions.

RECOMMENDATION X: All ranks should receive a refresher course on all legal and human rights matters pertaining to their peacekeeping mission.

RECOMMENDATION XI: When taking part in a mission that requires the formulation of rules of engagement, the CF should make sure that these rules are formulated early enough for their meaning to be carefully explained to all persons who will have to apply them.

RECOMMENDATION XII: All rituals of initiation allowed in the CF should be supervised by officers who would be accountable for any behaviour incompatible with military law and professional ethics.

RECOMMENDATION XIII: To the extent that CF authorities believe that initiation rites and similar activities are beneficial to the military esprit de corps and should therefore be allowed, they should write policy guidelines governing the conduct of such ceremonies.

RECOMMENDATION XIV: If the CF is to continue being involved in peacekeeping operations, it should develop an intelligence capacity which it completely lacked at the time of the intervention in Somalia. Let us remember in this regard that Captain Hope, the CAR' s intelligence officer, rushed to the *Encyclopedia Britannica* when he learned that Canadians were to be deployed in Somalia under Chapter VII of the UN Charter. Such shortcomings must be remedied.

RECOMMENDATION XV: I recommend that the results of monitoring the implementation of the policy against racism be made public in annual reports tabled in Parliament and made accessible to the public.

RECOMMENDATION XVI: I recommend that the commanding officers in the CF be provided with guidelines on how to apply the policy against racist conduct and harassment.

RECOMMENDATION XVII: I recommend the creation of the office of Inspector General of the CF.

RECOMMENDATION XVIII: I recommend the creation of the office of the Registrar General of the CF.

RECOMMENDATION XIX: I recommend that the academic curriculum of the CF provide better education for the military on the workings of the media and relations with them.

RECOMMENDATION XX: There is an urgent need for a CF policy on the use by its personnel of all recording devices (cameras, videocams, film cameras, tape recorders, etc.).

A Postscriptum on Accountability

After submitting my report to the Commission of Inquiry into the Deployment of Canadian Forces in Somalia, I got stimulating feedback from one member on the conception of accountability that I developed in it. This concept rested in part on the broad distinction that accountability is a matter that has to do with organizations and institutions while responsibility has to do with people.

The member of the Commission of Inquiry who commented on my view of accountability objected to this distinction and stated:

> In my view, accountability is a process which applies as much to individuals as to organizations and institutions. One can speak as much in terms of individual accountability as in terms of collective or corporate accountability. Conversely, responsibility is not only a matter to do with people as one can speak in terms of corporate or collective responsibility. An institution may develop its own culture which becomes [a] source of responsibility for the institution as such. In the same vein, the military chain of command may act collectively and engage the responsibility of the institution itself (personal communication).

In an attempt to respond to this objection, I believe I have clarified my conception of accountability. This postscriptum on accountability is a slightly expanded version of my initial response to the Commission of Inquiry. It is addressed to the member of the Commission who commented on my view and takes, in part, the form of a personal letter.

In saying that I make a broad distinction between accountability (a matter that has to do with organizations and institutions) and responsibility (a matter of people), you have not misread me. Although I tried to avoid dichotomizing the two notions, I did devote some effort to see if and how they were distinct from each other. Again, broadly speaking, I came up with the idea that accountability relates *more* to institutions and organizations than to people and that the reverse was true in the case of responsibility. I will now try to argue for this position.

Let me be clear about one thing. I do not disagree with anything in the paragraph I quoted above. It is obvious that individuals as such, and not only institutions and organizations, can be said to be accountable. Indeed, institutions and organizations do not exist separately from the people involved in them. In the end, it is always one or several persons who give an account; things and moral *personae* don't speak and cannot, in consequence, account for themselves. It is no less evident that responsibility can be extended to collective entities as much as to individuals (at least in civil law, the criminal prosecution of corporate organizations raises serious problems that are not yet solved).

However, in stressing what is common between the concepts of accountability and responsibility, we make little headway in setting out what *distinguishes* them. Indeed, they are often used as synonyms. There is also a tendency, exemplified by a recent article by Norman Spector, who was chief of staff for Brian Mulroney, to use the phrase "accountability *and* responsibility," without attempting to distinguish between the two concepts which are used in conjunction (see Norman Spector, *The Globe and Mail*, Saturday, February 1, 1997).

There is another trend which attempts to distinguish between the meaning of responsibility and that of accountability. This trend is all the more interesting because it shows how difficult it is to differentiate these concepts. I provide two examples, both from *The Globe and Mail*.

In an editorial criticizing the Minister of Justice and the RCMP Commissioner for the half-hearted apologies they offered ex-Prime Minister Mulroney in relation to alleged kickbacks in the government purchase of planes from the Airbus corporation, the editorialist writes: "They *admit responsibility and deny accountability*. They feign to make amends as they shilly-shally around the core of the case..." ("O, what a tangled web they weave," *The Globe and Mail*, January 7, 1997, p. A24). In this quote, responsibility refers to the *fact* that persons actually did something (they wronged Mr. Mulroney), whereas accountability represents the element of *liability* (being faced with the consequences of one's behaviour). What the editorialist says is that they admit their guilt but refuse to face the consequences of their behaviour (that is, having to make true amends).

In the second example, interestingly enough, Jeffrey Simpson raised the question "How can there be accountability without responsibility?" the following day. In his article, he draws a connection between accountability and responsibility that is exactly the opposite of the previous one:

In the Mulroney affair, apparently everyone is fully accountable — from middle-ranking civil servants and police officers right up to ministers and RCMP commissioners — but no one is responsible. People are accountable in the sense of making decisions or answering questions, but they are not responsible in terms of whatever consequences might flow from errors of judgment unless some form of public outcry occurs (*The Globe and Mail*, January 8, 1997, p. A20).

In this quote, it is accountability that represents the factual element (they have taken decisions and they do *answer questions* on their behaviour), but they refuse to face the consequences flowing from their errors of judgment unless there is public outcry. As you will notice, the meaning attached to the words are in precise contradiction with the previous case, where it was responsibility that designated the factual element ("they admit responsibility") and accountability that referred to the element of liability ("they deny accountability").

This is not to say that the members of the editorial staff of *The Globe and Mail* don't know what they are talking about. They actually make all the right distinctions:

- the actual provision of answers (of an account), that is giving factual information on what truly occurred;
- the factual guilt, that is, whether persons actually performed a misdeed and who these persons are; and
- the moral or legal liability, that is, having to face and eventually to suffer the consequences of the misdeed.

The problem is that the vocabulary has not yet settled, particularly in relation to the element of liability, which can be referred to by using either "accountability," as in the first quote from *The Globe and Mail* or "responsibility," as in the second.

I have briefly reviewed some law dictionaries and found that responsibility, accountability and liability were indeed used as synonyms:

- "*Accountability*: adj., liable, responsible" (Dukelow and Nuse, 1991).
- In *The Dictionary of Canadian Law* (Dukelow, 1995), accountability is also defined in terms of responsibility and liability. As for the latter, here is what one finds in this dictionary: *responsibility* "...is meant to signify a legal duty or obligation or the situation in which one is potentially or actually subject to some obligation" (p. 1075), whereas *liability* signifies "the situation in which one is potentially or actually subject to some obligation" (p. 677), which is the same meaning as that for responsibility. In another dictionary, I found that liability "is a general

[word] and has no judicially fixed meaning.... Generally it means an obligation or responsibility."

The only conclusion that can be drawn from this examination, however brief it may be, is that there is a great degree of semantic laxity in respect to words such as responsibility, accountability and liability; they can, in fact, be used almost interchangeably, and all are defined through the broad notion of being under an obligation or duty. Questions remain as to who precisely is meant to be under such an obligation and what the content of the obligation itself is.

So, what are we to do? I believe that we should try to clarify the presently inconsistent use of the terminology of responsibility/accountability, while remaining conscious of the fact that we are not lexicographers and that our goal should be modest (clearing up the ground a bit).

The simplest and most straightforward way of drawing a distinction between responsibility and accountability is to note that only the latter explicitly refers to the provision of an account or of a true description of who did what. That much cannot be put in doubt, that is, *being accountable implies providing an account*. I dwell at length on this aspect of accountability in my report, which stresses the fact that accountability is originally an information process. According to this view, accountability would then be defined as an information process (the obligation to provide a true account of what occurred), while the element of fact (actual guilt or authorship of the deed) and the element of liability (the obligation to face up to the consequences of one's behaviour) would be covered by responsibility. However, whatever its neat character, I don't believe that the simple distinction between accountability considered as an information process and responsibility as both the attribution or recognition of guilt and the acceptance of its consequences is satisfactory nor that it solves all problems. It is, nevertheless, a fundamental aspect of accountability. In one of the best discussions of the concept of accountability that I have read, the Canadian Study Team on Public Service Values and Ethics (ST-PSVE), a Task Force appointed by the Clerk of the Privy Council Office, used the word "answerability" to refer to this aspect of accountability:

Answerability is also often used as a synonym for "accountability", especially in relation to minister's answerability to Parliament. The Study Team uses "answerability" as a term to describe a key aspect of accountability, the duty to inform and explain (Canada, Privy Council Office, 1996, p. 10).

The reason I think this distinction (accountability as an information process and responsibility as attribution/recognition of guilt) is not fully satisfactory is not that I believe it to be wrong. Rather, I think it narrows the meaning of accountability too much. The same point is made in the report of the Task Force (Canada, Privy Council, 1996, p. 10). When people use the word accountability, it seems that they mean more than merely giving an adequate report of one's action. They generally include an element of justification: one who is accountable is not only under the obligation to supply a description of his or her conduct but also to justify the action pursued and, if no justification can be given, to show readiness to accept blame. In addition to this element of justification, there is also one of facing up to the consequences which appears to be saying that a person or an organization is, or should be, accountable. For instance, I recently watched an exposé on television which revealed that more than 100 Nazi war criminals fled to Canada after the war. They were never bothered by the police and most of them, now being quite elderly, will die peacefully "without having had to account for their past crimes." Clearly, by these words, the reporter meant more than just compelling these war criminals to provide a descriptive account of their past behaviour and then requiring them to explain or justify why they acted in such an abhorrent way. There is an obvious element of retribution involved in such a use of the notion of accountability. However, if we reintroduce the element of liability to retribution into the meaning of accountability, we are immediately at risk of blurring the distinction between accountability and responsibility. The earlier examples from *The Globe and Mail* show how very easy it is to confuse the meaning of these words. The real question, then, can be formulated thus: *How can we extend the meaning of accountability beyond the mere provision of accounts and explanations without confusing it with the meaning of responsibility?*

An interesting attempt to draw a distinction between the two concepts was made by the Task Force. In its report, the ST-PSVE elucidated the meaning of accountability:

> In most circumstances, accountability can be thought of as enforcing or explaining responsibility. It is often used as a synonym for "responsibility" because both are defined by the office holder's authority; they cover the same ground. Accountability involves rendering an account to someone, such as Parliament or a superior, on how and how well one's responsibilities are being met, on actions taken to correct problems and to ensure they do not recur. It also involves accepting

personal consequences, such as discipline, for problems that could have been avoided had the individual acted appropriately. All public office holders are accountable to the courts because of the requirements of the rule of law. Ministers are also accountable to Parliament, while deputy ministers are accountable to ministers, not Parliament (Canada, Privy Council, 1996, p. 10).

This characterization of accountability as the enforcement of responsibility includes both the aspect of answerability (what we previously called "the factual element") and the element of liability (facing up to the consequences of one's behaviour). It might then be asked whether this characterization actually conflates the respective meaning of accountability and responsibility, thus denying the need to distinguish between the two concepts. The Task Force's report actually draws a neat distinction between responsibility and accountability by providing a definition of responsibility that makes it wholly distinct from accountability. This definition or characterization of the meaning of responsibility is the following:

The term "responsibility" does not of course apply only to ministers. Within the public sector, all office holders have responsibilities that are defined by their authority. They are responsible for carrying out their authority well, within the law and with respect for ethical values, and, should problems arise, they are responsible for correcting them and doing whatever is reasonable to ensure that they do not reoccur. *The term is most often used in respect to the authority of ministers under a system of parliamentary government and to the duties and obligations that come with this authority: ministerial responsibility* (Canada, Privy Council Office, 1996, p. 9, my emphasis).

This quote is very telling of the difficulty in drawing a distinction between responsibility and accountability. The italicized sentence is the most successful in distinguishing responsibility from accountability. However, it is not really "responsibility" which is made distinct from accountability, but the plural form, responsibilities, which is a notion different from responsibility and which means a set of duties and obligations. This paragraph of the Task Force's report eventually leads to a clear distinction between the concepts of accountability and of responsibilit*ies* (duties and obligations). However, it is much less successful in providing a distinction between responsibility and accountability. It is actually the first part of the quote that discusses the meaning of responsibility as such and provides a characterization of it — "carrying out...authority well, within the law and with respect for ethical values" — which is very similar to the definition of accountability previously quoted (submitting to the requirements of the rule of law, among other things). It is only when the dynamic concept of responsibility (being liable for the way authority is actually carried out) is replaced by the static notion of responsibilities (the duties and obligations that come with authority) that it becomes possible to distinguish clearly between responsibilit*ies* and accountability. The distinction between responsibility as such and accountability itself is much less clear.

Drawing a clear distinction between responsibility and accountability is not an easy task, and it is doubtful that it can be wholly solved without any artificiality, that is without the establishment of some linguistic convention on the meaning of the terminology to be used (for example, in simply saying "by accountability, I mean M and by responsibility, I mean N"). The convention which I tried to establish in my report was that accountability is *more* a matter which has to do with organizations and institutions while responsibility is *more* a matter that has to do with individual people. I now try to show that this convention is far from being arbitrary and that, furthermore, it does not have the undesirable consequence of making individuals unaccountable.

To begin, accountability is not a new idea,[1] but its emphasis, as measured by increasing occurrences of the word "accountability" in research and government literature, is fairly recent (late 1950s). My research into the introduction and growing proliferation of this word shows that it was first used in connection with performance assessment in the fields of education and health. It was also used in relation to audits and financial accountability. In all these cases, accountability refers to an obligation that applies more to an organization considered as a whole than to individuals considered separately.

Accountability does not exclude individuals, but does put an emphasis on organizational processes. I quote in this regard from a recent essay on accountability. In this paragraph, Andrew Goldsmith (1995) comments on the views of Herman Goldstein, both authors being known experts on police accountability:

This enlarged sense of accountability has previously been described as follows:

> Accountability, in its broadest sense, includes much more than responsibility for determining policies in discretionary areas. It covers every aspect of administration of an agency, including, for example, its operating efficiency, its hiring and promotion practices, and its financial management. Accountability

encompasses as well responsibility for the conduct of individual employees — for the use which they make of their authority and for their integrity (Goldstein, 1997, p. 131).

It is clear from this approach that accountability presupposes an interest in system-wide issues, as well as matters arising from specific acts by individual officers. While this distinction is largely artificial, it does illustrate the possibility of pursuing accountability at different levels of organizational generality or abstraction, as well as seeing police work as the result of a number of interrelated organizational processes. Thus, a legalistic preoccupation with the conformity or otherwise of individual officers to substantive or procedural standards is only a very partial and incomplete approach to the issue of accountability. The strongly punitive treatment of offenders in individual complaints ignores a number of considerations, including the most fundamental of all: "Both the police and the public lose sight of the primary objective of control, which is to achieve maximum conformity with legal requirements, establishing policies, and prevailing standards of propriety" (Goldstein, 1997, p. 160). Thus, the multifaceted nature of the task of ensuring conformity to established standards of policing should not be lost sight of in any balanced and effective mechanism of accountability (Goldsmith, 1995, pp. 112-113).

Admittedly, both Goldstein and Goldsmith acknowledge that accountability encompasses both the individual and the organizational. But the point of their remarks is to stress that the distinctive feature of accountability — as opposed to the traditional legalistic approach of punishing individual misconduct — is the recognition of how crucial the need is for controlling organizational processes. There are no other grounds for writing this paragraph but to make this very point.

There are several good reasons to emphasize the organizational and institutional dimensions of accountability. The first one can be derived from the professional culture of the armed forces; it reminds me of the so-called "police mentality" that prevailed some 50 years ago in large police departments and still does in small forces. The Canadian Forces shows a marked inclination to explain away any misconduct — from the Hewson Report until today — by reducing it to the mischief of a few rotten apples and rabble-rousers who can be expelled from the organization. This rigid refusal to even consider that problems may extend beyond a limited number of maladjusted individuals is what I call the zero degree of accountability. In this state of non-accountability, problems are neatly packaged into small cans that can be swiftly disposed of internally to silence any call for external scrutiny. Insisting on the fact that accountability is at least as much, and perhaps even more, a challenge to the organization as a whole than to its individual members might be helpful in trying to remedy this unfortunate military tendency to look for convenient scapegoats among the "other ranks," as the British say.

Another reason to stress the organizational dimension of accountability is its actual organizational or institutional foundation. By this, I mean the following. In an important sense, bearing responsibility for a particular act is not a matter of institutional or organizational convention but a matter of fact: one either performed or did not perform a particular deed. In this sense the true answer to the question of who is responsible for this can be given in the overwhelming majority of cases without having to refer to institutional and organizational rules. I am aware that to be *held* responsible (to bear the consequences of one's action) is somewhat different than to be identified as the author of a particular deed. In other words, there is a difference between factual responsibility (which points to strict liability) and legal responsibility. However, the ground that separates legal responsibility from factual responsibility has been thoroughly covered for ages and, although law is a matter of institution, there is a general consensus on the content of the rules that allow us to disassociate factual agency from legal responsibility (for example, in criminal law, youth, consciousness of one's action, the capacity to control one's behaviour, duress, self-defence and the other recognized instruments of legal defence).

In contrast, the scope of institutional or organizational discretion is much broader in the definition of accountability and in the determination of its practical implications. There is really no equivalent of the distinction between factual and legal responsibility in relation to the notion of accountability. With accountability, one must begin with institutions, there being no factual or natural accountability preceding instituted accountability and providing grounds for it. Hence it can be decreed that an agency such as, for instance, the Communication Security Establishment (CSE) will not be accountable to Parliament to preserve its secrecy (until the recent appointment of the former Chief Justice of Quebec, Claude Bisson, as CSE's civilian inspector, this agency was for all practical purposes unaccountable). An organization can also use several means to preclude its staff from being accountable to an external body (an oath of secrecy, the obligatory shredding of documents on a regular basis, barring individuals from an access to their own documents to preclude them from giving an adequate account of their behaviour and so forth). Such examples showing that accountability is not a matter of fact but almost entirely a matter of institutional convention could be multiplied. In complete contrast, responsibility in the sense of being the true author of a deed *cannot be erased by an institutional fiat*: the order, "Although you did A, you are forbidden to *be* responsible for A" is at best incomprehensible and at worst self-contradictory (admitting that you did A is in a sense the same thing as saying that you are responsible *of* it — although you may not be found nor held responsible *for* it). However, it makes perfect sense for an organization to say to one of its individual members, "Although you did A, you are absolutely forbidden to give an external account for A." Such orders are actually current in military and police organizations and in agencies such as security services and, more generally, in all organizations strictly bound by confidentiality. These considerations, which could be further developed, lead me to conclude that accountability is indeed a concept that applies more specifically to organizations than to individuals, notwithstanding that the latter should also be held to accountability. Organizations and institutions actually shape the meaning of accountability and they generally do it with noted differences among themselves.

This conception of accountability, as I just stated, does not preclude the assertion that individuals are accountable. However, saying that accountability has more to do with institutions and organizations than with individuals has an interesting consequence with respect to individual accountability. It implies that the more weight an individual carries within an organization, the more this person should be held accountable. The higher one is in the structure of an organization, the more one's person and organization tend to blend. For example, no army has been identified with any individual soldier, whereas great generals have provided an identification for armies (the U.S. Third Army in World War II was as much General Patton's army as the French Grande Armée was Napoleon's army). The same point can be made with regard to great ministers or great entrepreneurs (ITT was Geneen's corporation). I believe this feature of the conception of accountability that I tried to articulate — establishing a positive ratio between rank and accountability — is a desirable one in the context of the Commission of Inquiry and that it may help to remedy the objectionable ethos (or lack of ethos) displayed during the testimony of many high-ranking officers, as they appeared to believe that one's accountability decreased as one got nearer to the top.

I hope that I have made myself clearer with these remarks and that they will be somewhat useful.

March/April 1997

Notes

CHAPTER ONE — CONCEPTUAL ANALYSIS

1 For a description of this process, see Board of Inquiry, Canadian Airborne Battle Group (1993), Report, Phase I, Volume XI, Ottawa: Department of National Defence (henceforth, BOI, Vol. XI) and United Nations, Security Council (1994), *Report of the Commission of Inquiry Established Pursuant to Security Council Resolution 885 (1993) to Investigate Armed Attacks On UNOSOM II Personnel Which Led to Casualties Among Them*, New York, 24 February 1994 (henceforth, COI, REP). Also see Coulon (1993: chapters 6-7) and Durch (1993: 441-474).

2 The original German version of the *Civilizing Process* was published in 1939 (*Uber den Prozess der Zivilisation*, Basel: Haus zum Falken). By a tragic irony, editions and translations of this work published after the war are dedicated by Norbert Elias to the memory of his parents, his mother Sophie Elias having died at the Auschwitz concentration camp, probably in 1941.

3 Canada 21 Council (1994), *Canada 21: Canada and Common Security in the Twenty-First Century*, Toronto: University of Toronto, Centre for International Studies. This report is published both in English and French.

4 See Carnegie Endowment for International Peace (1992), *Changing Our Ways: America and the New World/Carnegie Endowment National Commission on America and the New World*, Washington, D.C.: The Brookings Institution, p. 64.

5 This assertion is based on extensive analysis of policy statements made by Canadian officials, including Prime Minister Mulroney, External Affairs Minister Barbara McDougall and UN Ambassador Louise Fréchette (see Cox, 1993: 50-55).

6 David Cox gives a personal interview with G.A.H. Pearson in November 1993 as his source for this story (see Cox, 1993, note 8).

7 For an expanded definition of these terms, see Canada 21 Council (1994: 77-78). "Actions to Restore International Peace and Security," added by the Council for the purposes of information, is not to be found in *An Agenda for Peace*. "Peace enforcement," which appears in chapter 4 of the Secretary-General's statement under the heading of "peacemaking," is not defined by the Canada 21 Council (see Boutros-Ghali, 1992: Chapter IV).

8 The primary source for the original principles of peacekeeping is to be found in the *Public Papers of the Secretaries-General of the United Nations* (Cordier and Foote [eds.], 1973, Volume III of Hammarskjold's papers [1956-57] and Cordier and Foote [eds.], 1974, Volume IV of Hammarskjold's papers [1958-1960]). See particularly Hammarskjold's *Second and Final Report on the Plan for an Emergency International United Nations Force* (dated 6 November 1956, Cordier and Foote, 1973: 344-351) and his *Summary Study of the Experience Derived from the Establishment and Operation of the United Nations Emergency Force* (dated 9 October 1958, Cordier and Foote, 1974: 230-292). The Hammarskjold public papers do not record the contribution of L.B. Pearson, but they remain the invaluable source for a record of the birth of peacekeeping. The original principles of peacekeeping are discussed by Fabian (1971), *Soldiers Without Enemies: Preparing the United Nations for Peacekeeping,* Washington, D.C. : The Brookings Institution. Urquhart (1987: 131-139), James (1990: 1-16) and Fetherston (1994: 13), among many others, provide enlightening discussion of these principles.

9 In 1984, the only instance of an operation carried out under the umbrella of Chapter VII was the Korean War, which was understandably not seen as a model for peacekeeping. However, since then the Gulf War and the operations in Somalia have occurred.

10 Needless to say, this observation was prophetic, as the "peacekeeping force" sent to the former Yugoslavia, under the terms of the Dayton (Ohio) agreement, is operating under NATO rather than the UN.

11 According to BOI, Volume XI: Annex A, par. 7, the Canadian Joint Force Somalia was deployed until the end of the first rainy season (June). It was withdrawn before the UNOSOM II mission developed into a combat situation.

12 I will quote the testimonies given at the court-martial proceedings conducted against different members of the CAR according to the pagination given in their reproduction in Folio Views.

13 See the memorandum of Ouafaa Douab to David Pomerant, dated June 20, 1995, p. 3 (internal CIDCFS document).

14 As an example, I might quote the following fact. Using "racism" as a keyword, there are no less than 485 titles to be found in the libraries of McGill University. Reviewing all titles, I found only two that dealt with racism in a military context (Dower, 1986 and Tinker, 1987). Needless to say, this issue is addressed tangentially in many books, but it is never confronted per se in these books.

15 Some of these incidents are very well known, such as the beating of Rodney King in Los Angeles or unjustified police shootings of Blacks in Canada (the Griffin/Gossett and the Marcellus François affairs in Montréal and similar incidents that occurred in Toronto). In Nova Scotia, the wrongful prosecution and condemnation of Donald Marshall were perceived as motivated in part by racial prejudice against Aboriginal Canadians.

16 Following Abercrombie (1984: 173), Armitrage and Kennedy(1989) eventually specify that the actions referred to are oppressive in nature. This is precisely my point: in order not to be misleading, any worthwhile definition of racism has to elaborate on a series of points.

17 *The Nature of Prejudice* was first published in 1954. I quote the 25th anniversary edition, published in 1979 by the Addison-Wesley Publishing Company).

18 "Like the soldiers who confessed in 1945 that their goal had become killing rather than simply winning, even after the war ended and the Japanese turned their energies to the tasks of peaceful reconstruction, a surprising number of Americans expressed regrets that Japan surrendered so soon after the atomic bombs were dropped. A poll conducted by *Fortune* in December 1945 found that 22.7 percent of respondents wished the United States had had the opportunity to use 'many more of them [atomic bombs] before Japan had a chance to surrender'" (Dower, 1986: 54).

19 "Once the war was over and the Americans were installed on the home islands as occupiers, the Japanese capacity for subtlety and delicacy could again be recognized. Then, Admiral Nimitz, doubtless with Halsey and such in mind, reminded his subordinates that "the use of insulting epithets in connection with the Japanese as a race or individuals does not **now** become the officers of the United States Navy" (Fussell, 1989: 120, emphasis added).

20 Allport (1979) does not make the later distinction between racism based on exclusion (which comprehends segregation) and racism based on

domination (which uses discrimination). Consequently, he conflates these two forms of racism, segregation being for him a manifestation of discrimination.

21 See Gunnar Myrdal, with the assistance of Richard Sterner and Arnold Rose (1944), *An American Dilemma: The Negro Problem and Modern Democracy*, New York: Harper and Row, 20th Anniversary Edition, 1964. Henceforth to be quoted as Myrdal (1964).

22 In this regard compare the praise of Beaumont for the German SS in his book entitled *Military Elites* (Beaumont, 1974: 22-24, 152-156 and 187-189) to Sydnor's much more thorough and scholarly study of the same corps (*Soldiers of Destruction: The SS Death's Head Division, 1933-1945*, Princeton: Princeton University Press, 1977). Also compare the official history of the French Foreign Legion by Gandy (1988) to the account of the dismal experiences of a Jewish soldier within the Legion (Szajkowski, 1975).

23 Racism is ubiquitous in colonial wars. For studies of the Spanish conquest of Mexico, Central and South America, see Todorov (1989) and Delacampagne (1983). This literature, being by necessity of a historical character, I have chosen not to review it extensively.

24 It is all the more remarkable that in their war propaganda the Japanese did not describe their Western enemies as subhumans and animals, although they were surely not depicted as honourable men. However, the Nazis, as is well known, depicted Slavs, Gypsies and Jews as subhumans [untermenschen].

25 The recent events in Rwanda are no exception to this rule. Although the colour of their skin is for us the same, in their genocidal radio propaganda, Hutus described their Tutsi enemies as separated from them by physical differences.

26 It was, for instance, believed that their skin did not reflect moonlight, thus they were used for night attacks when there was a full moon (White, 1990: 79).

27 "In seeking Indian aid, the whites were requiring the Indian to be precisely what they most detested in him - savage, bloodthirsty, relentless, skulking, terrorizing and tearing the scalp from his opponents" (Johnson, 1977: 650).

28 According to January 1995 CF statistics, there were seven Aboriginal soldiers in the CAR.

29 See both the *Oxford English Dictionary* and the *Concise Oxford Dictionary*. See also the different meanings provided for the verb "to account" by *Webster's Third New International Dictionary*: "to furnish a

justifying analysis or a detailed explanation of one's financial credits and debits or of the discharge of any of one's responsibilities." This latter formulation captures the whole meaning of the verb "to account."

30 Even granting that the Constitution of Canada is to be dated back to the *Constitutional Act of 1791*, and not merely to the *British North America Act* of 1867, which created the Dominion of Canada, it must be recognized that the Canadian constitution does not itself date back much more than 200 years and that, consequently, the "recent vintage" of collective ministerial responsibility has been served, so to speak, for at least half of the time this country has been in existence. The recent character of this vintage is thus very relative.

31 Thus, the *Police Accountability Act of 1991* (H.R. 2972) was introduced into the U.S. House of Representatives, in July 1991. The content of this legislation was the establishment of criminal and civil penalties for misconduct by police officers; the Act would also allow private citizens to sue. I will eventually undertake to show that the conception of accountability which is presented in the *Brief for the Commission of Inquiry Into the Deployment of Canadian Forces to Somalia: Chain of Command and Accountability*.

32 A striking illustration of the ambiguity in this doctrine was recently provided in France. Mr. Laurent Fabius was Prime Minister of France during the period when blood contaminated by the HIV virus was given to unsuspecting patients by doctors, who could be said to be ultimately under the supervision of the French Ministry of Health and of the French government. The fury of public opinion when this scandal was made public was such that Mr. Fabius was finally criminally charged before the highest court of the country for having failed in his capacity as French Prime Minister to prevent the transfusion of contaminated blood. This is an example of the doctrine of political accountability as personal responsibility pushed to its extreme and, dare I say, absurd limits.

33 Part III of the *Third Report of the Commission of Inquiry Concerning Certain Activities of the Royal Canadian Mounted Police: Certain RCMP Activities and the Question of Governmental Knowledge* (third report of the McDonald Commission, Canada, 1981b) is a very thorough description of the doctrine of plausible deniability in its practical application.

34 For the sake of clarity, Annex A of this report includes Light's own comments on this table.

35 For example, when "military accountability" was used in a key word search of a social science data bank, no corresponding titles of books or articles turned up.

36 For this conflict, see Hornung (1991), and York and Pindera (1991).
37 Historically, the British police was created as an attempt to put an end to the social warfare that was developing between mass rioters and the army in British cities. See Radzinowicz (1956), vol. 3.
38 The provision of a list of references to sustain this assertion would be rather perfunctory. The second section of Matthews' bibliography on accountability in the administration of justice is specifically devoted to police accountability. It is entitled "Police Complaints and Civilian Review of Police" and lists 41 annotated references (Matthews, 1993: 45-57).

CHAPTER TWO — DATA

1 Mobile Command Study (1985, September), *A Report on Disciplinary Infractions and Antisocial Behaviour Within FMC with Particular Reference to the Special Service Force and the Canadian Airborne Regiment* (this is generally referred to as the Hewson Report), Ottawa: D, par. 66.
2 All my sources are taken from official Canadian Forces history: "Airborne/Aéroporté Regiment" Canadian Armed Forces, 1979 *Tenth Anniversary Journal 1968/1978*, *Ducimus: The Regiments of the Canadian Infantry* Canadian Armed Forces, 1990 Grimshaw (1981), *The Badges and Insignia of the Canadian Airborne Forces*, Ottawa: Department of National Defence, 1990.
3 See the Information Briefing to the Commission entitled "OP Deliverance: The CARBG in Somalia," p. 3.
4 See the review of the literature in Donna Winslow's report to this Commission.
5 The questionnaires were distributed in January 1995 and were for the most part completed by March 22 of the same year.
6 "Recruitable" means here persons who conform to basic requirements of the CF: Canadian citizens, between the ages of 17 to 52, with no disqualifying physical disability.
7 The fact that the CF drastically changed the ethnic make-up of the CAR after the incidents in Somalia can not be ignored as a possibility. This would imply some machiavellian design on the part of the CF, which it is not my mandate to unearth, and which, furthermore, I find improbable.
8 The percentage of Black officers in the *regular CF* cannot be tabulated for comparison purposes on the basis of the Chouinard and Chiasson (1995) survey, because the numbers and percentages which they give refer both to the regular force and the primary reserve.

9 In particular, compare Annex C (discipline), D (leadership) and K (values and attitudes) of Volume XI (Phase I) of the BOI Report to Part 2 (recruiting, selection, indoctrination, special training and elitism) and Part 3 (leadership, taskings and discipline, junior leadership levels and the effects of taskings) of the Hewson Report. Their content is eerily similar.

10 See in this regard the description of the content of *The Turner Diaries* by Kinsella (1994). This book, which is not on the market, was published by William L. Pierce, under the pseudonym of Andrew Macdonald and is a cult book for the extreme right. It tells how extreme right wingers forge an alliance with the U.S. military and use nuclear weapons to conquer the world.

11 Actually, the investigator makes a mistake about the meaning of one of Kinsella's allegations. On page 407, Kinsella writes that "One unidentified member of McKay's regiment testified, under oath at a military Board of Inquiry, that a colleague openly boasted that he was going to Somalia to 'shoot me a nigger'." This allegation is judged by the investigator to be false, because "it was not possible to conclusively prove or disprove this statement was ever spoken." As is quite clear from the book's text (p. 407), Kinsella's point is *not* that this racist statement was made but rather that a member of the CAR actually *testified* before the BOI to the effect that such a statement was made (Kinsella's whole paragraph is devoted to the *testimony* of a soldier and not to an examination of its truth). Kinsella's point about a soldier testifying at the BOI's inquiry that someone of the rank of sergeant said "I came to Somalia to shoot me a nigger" is absolutely correct. This soldier is no longer unidentified. His name is Private Grant: Q. On what basis would you classify them, as racist? A. If I hear someone say, "I came to Somalia to shoot me a nigger," I'd kind of find that kind of racist. Q. What rank level is this individual at? Q. OC, Sergeant-Major, sergeants." (BOI transcriptions, Phase 1, Volume 4, 63-24).

12 This information was given to me by Auditeur Général Fobe, who is the Belgian equivalent of our own Judge Advocate General.

CHAPTER THREE — PRE-DEPLOYMENT

1 I am using the electronic transcripts of the witnesses' answers to the questions of the BOI, the courts-martial and of this Commission. These transcripts were processed through the computer software Folio Views. However, there is a major difference in the way in which these testimonies were input in the Commission's computers. A print-out of the

testimonies made before the Commission gives the date on which the witness testified and the volume and page of the transcripts, as they can be found on paper support in the Commission's archives (e.g., Vol. 23, Wednesday 29 November 1995, PG. 4277, Master Warrant Officer Mills). However, a print-out of the BOI or the courts-martial testimonies just gives the name of the witness. Since the transcripts that I am working with are only available in electronic form, I will have to refer to the witnesses' answers in two different ways. For the Commission's own transcripts, I shall indicate the witness's name, and the volume and page where his or her testimony can be found in printed sources (e.g., CIDCFS, MWO Mills, 23: 4277). For the BOI and the courts-martial transcripts, for which I have no such source, I will write down the witness's name and the number of the *record* where the testimony can be found in the *inquiry.nfo* (BOI) and in the *newgcm.nfo* (courts-martial) files in Folio Views (e.g., BOI, MWO Mills, 7507 and CM, Mills, 6154).

2 There is the very frustrating problem of having to complete my study while the Commission's investigation still has a long way to go. This problem can be solved at least partially with respect to topics which are less dependent on factual findings. It admits of no solution with regard to factual issues. In the latter case, the problem is not only that conclusions are speculative, but that they may be unfair and may prejudice views about persons and institutions.

3 One scholarly journal, *The International Journal of Intercultural Relations,* has been published since 1977 (with four issues each year). A significant number of articles on the screening of international advisors and their cross-cultural training has appeared in this journal.

4 This was the position finally taken by Lieutenant-Colonel Turner, who was first surprised by the CAR's selection: "In the final analysis, particularly when the mission changed from Cordon to Deliverance, I was convinced with the benefit of hindsight obviously that the Canadian Airborne Regiment was absolutely the right unit to deploy in Somalia." (CIDCFS, L.Col. Turner, 20: 3634).

5 The morale problem was not solved but merely transferred to 1 Royal Canadian Regiment "who were incredibly disappointed they weren't picked and had to turn their vehicles to another unit that wasn't mech infantry." (BOI, Major Turner, 236)

6 Since writing this section, I learned from General De Chastelain's testimony before the Commission that morale played a role in the decision to send the CAR on its mission to Somalia. General De Chastelain made it clear that morale was not the most important

consideration but that "taken in conjunction with everything else...I think it makes a great deal of sense."*(The Globe and Mail*, 21 February 1996, p. A7). Based on my examination of the evidence, I would tend to believe that bolstering morale was more than a secondary consideration in selecting the CAR.

7 The issue of the integration of 24 Militia (in accord with the concept of the Total Force which was avidly supported by B.Gen. Beno) was a bone of contention between L.Col. Morneault and B.Gen. Beno himself. There is one letter from L.Col. Morneault to B.Gen. Beno that bears witness to the tension between them on this issue (Reference: 3350-OC [COMD] 6 October 1992).

8 The meeting occurred after May 1993. The document has no precise date.

9 In cases of spousal abuse or shoplifting, it will generally be judged that the link of confidence has not been severed and that the officer can still be employed by a police force. In the case of outright theft or of a serious crime of violence, the officer will be dismissed. These rules are, in view of my experience, approximate given the range of discretion left to police management. The crucial test cases are those involving officers convicted of abusing their powers and of brutality. There are not yet any hard and fast rules in this connection. Different police departments pursue very different courses.

10 Section 25 of the CF policy on Racist Conduct (CFAO 19-43) provides for the "elimination of racist attitudes and the prevention of racist conduct" through a program of training and education that would reach all members of the CF.

11 The name is spelled differently in various transcripts of the testimonies (e.g., Hosain, Hossan, etc.). In spite of these spelling variants, they undoubtedly refer to the same individual.

12 For a discussion of the recommendations on peacekeeping training formulated by government bodies in Canada, see Fetherston, 1994, chapters 7-8.

13 The titles of these three manuals are respectively, *Raising your Effectiveness in Today's Neighbourhoods through Community Policing; The Recruitment, Selection and Retention of Visible Minorities; and Training within a Diverse Society.*

14 Interestingly enough, CWO Raymond was critical of what the CAR had become. He believed, for instance, that its members were too young and that dressing them in paratrooper garb was an unnecessary way of bolstering their identity ("...the only thing that should be different is the

fact that they wear a maroon beret and a different cap badge," BOI, CWO Raymond, 1013) .

15 Recruits were submitted to electric shocks only during the 1994 hazing ritual, according to the content of the videotapes.

16 This nausea-provoking ritual was performed both years. In the 1994 version, the breadball goes from mouth to mouth without having been urinated on.

17 The commanding officer of one of the CAR's Commandos has the rank of a major.

18 The "Rebel flag" was a reproduction of the Southern Confederate flag during the American civil war (1860-1865). It was used by 2 Commando as its unofficial and, apparently, forbidden emblem.

19 Sergeant Gravelle repeats the same description in another part of his testimony: "Je ne suis...comme je vous ai dit, c'est une argument personnel. Je crois que les ... c'est une façon que les jeunes a se valoriser. Qui qui est le plus tough, qui qui joue le plus dur. C'est ce que je vois moi. J'ai pas besoin de vous dire d'autres choses." (Translation: I am not ... as I said, it is a personal view. I believe that the ... is a way that youngsters have to uplift themselves. Who is the toughest, who plays the hardest. I, this is what I see. I don't need to tell you more.) (CIDCFS, Sergeant Gravelle, 6: record 1112).

20 "Elite" comes from the participle of a verb in Old French, which originally meant "the chosen ones," "the elected ones" and evolved into meaning "the best ones."

21 During the video made in 1994, a drunken soldier hollers to the video camera, "Your brother, he shits through his mouth, that's really (untranslatable swear word) Airborne!" (in the original French: "Ton frère, y (il) chie de gueule, Airborne en tabarnak!"). This is a typical outcry in a context from which only twisted rallying cries can spring.

22 Corporal Robin spells his first name either in English (Christopher) or French (Christophe). He probably is more comfortable in French, since this is the language he testified in.

23 Because the person undergoing this treatment is lying on the ground, it is difficult to identify for certain if he is Corporal Robin. What is absolutely clear is that soldiers were urinating on each other during this hazing ritual.

24 The witness alleged lack of memory when asked questions probing whether he was submitted to maltreatment due to racial discrimination. He answered "no idea" [aucune idée] when asked how he could reconcile contradictory statements made to the military police in January 1995 and to the Commission when he testified. Corporal Robin, along with several

other military, was interrogated on the hazing ritual of July 1992, after parts of the videotape recording it were shown on TV.

25 Corporal Robin's selective memory so incensed one of his interrogators that he asked Robin whether he knew what day it was when he testified (Corporal Robin, 6: 1032).

26 The Commission's Counsel lost his cool, after a barrage of "Je m'en souviens pas (plus), Monsieur" from Corporal Robin. [translation: I don't remember].When the Counsel then asked him whether he knew what day of the week it was and insisted he identify it ("Jeudi, Monsieur"), the Counsel showed himself as unnecessarily sarcastic and unaware of the culture of Corporal Robin's profession (CIDCFS, Robin, 6: 1032).

27 This strategy is adapted from the Commission's own future proceedings. One of its counsels — Mr. Noel — has stated that witnesses such as Commander Jenkins who are involved with this file will be recalled (CFDCFS, Noel, 6: 1159). I believe that it is now premature to delve into the issue of right-wing extremism.

CHAPTER FOUR — IN THEATRE

1 I will not mention as a difficulty that I cannot rely in this chapter on the hearings of the Commission, which are still to come. Hence all the ensuing analyses have to be considered "inside brackets." They will be reviewed according to the findings of the Commission.

2 When he attempted suicide shortly after Arone's death, Master Corporal Matchee inflicted on himself what appears to be permanent brain damage. He never testified in any court-martial proceedings. The military police inquiry and all the testimonies elicited in this affair leave very little doubt that he played the major role in Arone's death.

3 Corporal Purnelle has recently published a book on his experiences in the CAR and now faces disciplinary charges. He has not yet testified before the CIDCFS. If he eventually does, it is unlikely that he will contradict what he has written in his book, unless faced with testimonial or documentary evidence showing that he was wrong. Even in this latter case, he may refuse to recant and be justified, witnesses and documents not being above suspicion. Furthermore, it is reasonable to assume that his book contains most of what he believes to be significant in his experience. He may not be the best judge of that, as his eventual testimony before the CIDCFS may reveal. In this report, it is assumed that Corporal Purnelle's book is a significant addition to our knowledge of events that occurred in Somalia.

4 Captain Rainville was the object of a news report in the Montréal daily *Le Journal de Montréal*, where he was shown in armed regalia (the weapons he displayed in these photographs exceeded service weapons, rather resembling those a Hollywood-style commando would carry). It was later reported that he led a simulated terrorist assault on the so-called Quebec Citadel, where his team brutally overwhelmed the lethargic members of the garrison. Captain Rainville's taking of the Quebec Citadel is a paradigm example of the extreme frustration that aggressive soldiers suffer when they have no enemy to fight. That such a man was sent to Somalia as a peacekeeper was a grievous mistake.

5 LeBlond describes 35 separate incidents, including the serious incidents that I previously mentioned. Some of LeBlond's entries (e.g., item No. 2) list several incidents under the same heading. In a few cases, LeBlond describes incidents consisting of hostile behaviour displayed by Somalis toward Canadian troops (mischief and theft).

6 Many of these abuses took the form of humiliating public displays of prisoners. Photographs included in a Canadian Broadcasting Corporation (CBC) (French) television program portrayed young Somalis (young adolescents or children) tied up and posted with placards announcing they were thieves (*Le Point*, 2 July 1996; see also *Le Devoir*, 3 July 1996. pp.1 and 8, to which I referred previously).

7 Some of these MVAs involved injuries to Somalis.

8 Needless to say, persons who are part of a minority in their own unit may be confronted with people of their own racial-ethnic group, who are the majority in a foreign country (e.g., a Canadian Black policing Somalis in Somalia).

9 I am on the steering committee of a community policing project in a Montréal neighbourhood populated by minority groups. They are in part policed by officers from minorities, although there are few of these on the Montréal Urban Community Police Force. I have also been involved as the main researcher in a research-action project in the policing of the Crees of James Bay (Quebec). The project was subsidized both by the Crees and the federal ministry of the Solicitor General.

10 There were times when people using derogatory language toward ethnic groups were not even aware that there were alternatives. There is little probability that any significant number of Canadians still believe that words now identified as racial slurs are only descriptive words without any pejorative connotations. Nevertheless, some of the troopers that testified before the BOI apparently still held that kind of belief.

11 Unless one wants to make a gesture of provocation or rebellion, one does not use drugs against someone (one's parents, for instance). This is why the issue of responding in kind to the use of drugs does not really arise in non-pathological circumstances (it is a rare parent that will begin using cocaine to deter his child from consuming soft drugs). There is of course a social and a legal response to the use of drugs, but it has no relation whatsoever to a deliberate escalation in drug abuse.

12 Needless to say, these figures were probably inflated. Nonetheless, thousands of Somalis died in the fighting according to more reliable sources such as non-governmental organizations (NGOs).

13 Technicals were 4X4 vehicles similar to pick-up trucks with a heavy machine gun or a recoilless cannon mounted on the rear platform. They were a familiar sights in the CNN TV broadcasts. According to Marchal (1993a and b), the U.S. troops succeeded in clearing the streets of Mogadishu and other Somali cities of these "technicals."

14 The French words which Smith uses to describe them ("un melange de resistant, de sans-culotte et de brigand" are difficult to translate (Smith, 1993: 181). A "resistant" is an armed opponent to a political regime; what a "brigand" is is transparent enough. "Sans culotte" refers to the poorest French revolutionaries who literally did not even own a pair of pants and who were notorious rioters that could not be controlled.

15 Corporal Matchee's attempted suicide in the days that followed Shidane Arone's death makes it unlikely that he fully intended to kill him, although he was willing to inflict serious harm.

CHAPTER FIVE — POST-DEPLOYMENT

1 The exception to this assertion is the protection of Canadian territorial waters, particularly against intrusions by submarines.

2 "...the government has chosen to ignore every single recommendation for statutory change made by the special committee." (Farson, 1995: 209).

3 The Minister usually appoints an inspector when the case to be investigated is less serious and a special commissioner in the more serious cases. See the Quebec *Police Act*, section 180 and ff.

4 What is referred to as the "intelligence community" in Canada is a rather complex network of agencies and policy-making committees. In 1987, SIRC issued a paper that outlined the workings of this network and is now only partially valid *(The Security and Intelligence Network in the Government of Canada: A Description*). The Senate also issued a report

(Canada, Senate, 1989, Terrorisme, Rapport du deuxième Comité spécial du Sénat sur le terrorisme et la sécurité publique). For a description of the various components of the Canadian intelligence community, also see Brodeur, 1990 and 1992.)

5 On this distinction between functional and institutional oversight, see Lustgarten (1995: 182).

6 Whether the RCMP keeps its "political masters" abreast of how it was circumventing the law is still a very controversial question. A former DG of the Security Service, M. John Starnes, testified repeatedly before the McDonald Commission to the effect that he asked the Cabinet for directives and guidelines on the issue of the legality of the Security Service operations, particularly in the field of counter-terrorism (see Canada, 1981a and b). He also clearly stated during the hearings of the Quebec Keable Commission that he intentionally refrained from advising the Minister of the theft of the list of all members of the Parti Québécois in 1973, the purpose of this omission being to protect the Minister's plausible deniability of any knowledge of this illegal operation (it led to the criminal prosecution of several members of the RCMP).

7 The greatest press leak of the Vietnam War did not happen in Vietnam but in the United States. It was the publication of the Pentagon Papers by *The New York Times*.

CHAPTER SIX — CONCLUSION

1 Relying on Marshall's studies of U.S. soldiers in the World War II, Grossman (1995: 25) puts this figure much higher at 80 to 85 percent, all factors for not shooting or aiming wildly being taken into account.

A POSTSCRIPTUM ON ACCOUNTABILITY

1 The phrase *logon didonai* (to provide a reason) was used in ancient Greek philosophy in much the same sense as the present notion of accountability.

Bibliography

Abdallah, W.M. (1984). *Internal Accountability: An International Emphasis*, Ann Arbor, MI: UMI Research Press.

Abercrombie, N. (1984). *The Penguin Dictionary of Sociology*, London: Allen Lane.

Africa Watch (1990). *Somalia: A Government at War With Its Own People. Report*, London: Africa Watch.

African Rights (1993). *Somalia: Human Rights Abuses by the United Nations Forces*, African Rights, London: July 1993.

Alex, N. (1969). *Black in Blue: A Study of the Negro Policeman,* New York: Appleton-Century Crofts.

Allport, G.W. (1954/1979). *The Nature of Prejudice*, Reading, MA: Addison-Wesley Pub. Co., 25th Anniversary Edition.

Anderson, D. (1995). "The New Peacekeeping Partnership," *Joint Forces Quarterly,* Vol. 5.

Aristotle (1988). *The Politics*, Stephen Everson (ed.), Cambridge, U.K.: Cambridge University Press.

Armitrage, P. and J.C. Kennedy (1989). "Redbaiting and Racism on Our Frontier: Military Expansion in Labrador and Québec," *The Canadian Review of Sociology and Anthropology*, Vol. 26, No. 5, 798-817.

Armstrong, J.G. (1990). "The Unwelcome Sacrifice: A Black Unit in the Canadian Expeditionary Force 1917-1919," in N.F. Dreisziger (ed.), *Ethnic Armies*, Waterloo: Wilfrid Laurier, 178-197.

Bailey, S.D. (1988). *The Procedure of the UN Security Council*, Oxford: Clarendon Press, Second Edition.

Balibar, É. and I. Wallerstein (1988). *Race, Nation, Classe : les identités ambiguës*, Paris: Éditions de la découverte.

Ball, D.Y. (1994). "Ethnic Conflict, Unit Performance, and the Soviet Armed Forces," *Armed Forces and Society,* Vol. 20, No. 2, 239-258.

Banton, M. (1987). *Racial Theories*, Cambridge: Cambridge University Press.

Bard, M. and J. Zacker (1976). *The Police and Interpersonal Conflict: Third Party Approaches*, Washington, DC: The Police Foundation.

Barrett, S.R. (1987). *Is God a Racist?* Toronto: University of Toronto Press.

Barua, P.B. (1992). "Ethnic Conflict in the Military of Developing Nations," *Armed Forces and Society*, Vol. 19, No. 1, 123-138.

Bauman, Z. (1989). *Modernity and the Holocaust,* Ithaca, NY: Cornell University Press.

Bayley, D. (1994). *Police for the Future*, New York: Oxford University Press.

Beaumont, R. A. (1974). *Military Elites*, Indianapolis, NY: Bobbs-Merrill Company, Inc.

Belgique, Auditorat Général près la Cour militaire (1993). *Commission d'enquête SOMALIE, Rapport,* Bruxelles: Palais de Justice.

Berdal, M.R. (1993). *Whither UN Peacekeeping?* London: International Institute for Strategic Studies, Adelphi Paper 281.

Bettelheim, B. and M. Janowitz (1964). *Social Change and Prejudice, Including the Dynamics of Prejudice*, New York: The Free Press.

Binkin, M. and M.J. Eitelberg, with A.J. Schexniner and M.M. Smith (1982). *Blacks and the Military*, Washington, DC: The Brookings Institution.

Bittner, E. (1990). *Aspects of Police Work*, Boston: NorthEastern University Press.

Bland, D.L. and J.D. Young (1988). "Trends in Canadian Security Policy and Commitments," *Armed Forces and Society*, Vol. 15, No. 1, 113-130.

Board of Inquiry, Canadian Airborne Regiment Battle Group (1993). Report, Phase I, Vol. XI, Ottawa: Department of National Defence.

Boivin, J.-R., J. Bernier and L. Bessette (1995). *Groupe de travail sur l'arbitrage de différend chez les policiers et pompiers municipaux. Rapport,* Québec: ministère de l'Emploi et ministère des Affaires municipales.

Bok, S. (1982). *Secrets: On the Ethics of Concealment and Revelation*, New York: Basic Books.

Borich, G.D. (1974). *Evaluating Educational Programs and Products*, Englewood Cliffs, NJ: Educational Technology Publications.

Bourke, J.G. (1891). *Scatalogic Rites of All Nations*, Washington, DC: W.H. Lowdermilk & Co. (First reprinting, 1968, Johnson Reprint Corporation).

Boutros-Ghali, B. (1992). *An Agenda for Peace*, New York: United Nations.

Boyd, B. and R. Bell (1985). *Crisis Intervention and Conflict Management Training in the RCMP: A Prescriptive Package*, Ottawa: Minister of the Solicitor General.

Braestrup, P. (1991). "Introduction," in L.J. Matthews (ed.), *Newsmen & National Defense: Is Conflict Inevitable?* Washington: Brassey's (U.S.), XV-XXIV.

Brauman, R. (1993). *Le crime humanitaire*, Paris: Arléa.

Brodeur, J.-P. (1994). *Inspection du Service de police de la Communauté urbaine de Montréal,* Volume II, Québec: Ministère de la Sécurité publique.

– (1992). "Undercover Policing in Canada: Wanting What is Wrong," *Crime, Law and Social Change*, Vol. 18, 105-136.

– (1990). "Security Intelligence and Policing in Canada," in A.G. Gagnon and J.P. Bickerton (eds.), *Canadian Politics: an Introduction to the Discipline*, Peterborough, Ont.: Broadview Press.

Bryant, C.D. (1979). *Khaki-Collar Crime: Deviant Behavior in the Military Context*, London: Collier Macmillan, New York: The Free Press.

Butcher, T. (1995). *Delivering Welfare: The Governance of the Social Services in the 1990s*, Philadelphia: Open University Press.

Buttney, R. (1993). *Social Accountability in Communication*, London, Newbury Park, CA: Sage.

Byers, R.B. and M. Slack (eds.) (1983). *Canada and Peacekeeping: Prospects for the Future*, Downsview, Ont.: York University, Research Program in Strategic Studies.

Caforia, G. (1988). "The Military Professional: Theories of Change," *Armed Forces and Society*, Vol. 15, No. 1, 55-70.

Cambridge, A.X. and Stephan Feuchtwang (1990). *Anti-Racist Strategies*, Aldershot, U.K.: Avebury.

Campbell, V. (1980). "Double Marginality of Black Policemen," *Criminology*, Vol. 17, No. 4, 477-484.

Canada (1994). *1994 Defence White Paper*, Ottawa: Minister of Supply and Services Canada.

– (1981a). *Second Report of the Commission of Inquiry Concerning Certain Activities of the Royal Canadian Mounted Police: Freedom and Security Under the Law* (the 2nd McDonald Report), 2 vols., Ottawa: Minister of Supply and Services.

– (1981b). *Third Report of the Commission of Inquiry Concerning Certain Activities of the Royal Canadian Mounted Police: Certain RCMP Activities and the Question of Governmental Knowledge* (the 3rd McDonald Report), Ottawa: Minister of Supply and Services.

– (1969). *Abridged Report of the Royal Commission on Security* (the Mackenzie Report), Ottawa: Queen's Printer.

– (1966). *Report of the Commission of Inquiry into Complaints Formulated by George Vistor Spencer* (the D.C. Wells Report), Ottawa: Queen's Printer.

Canada, Auditor General (1996). *Report of the Auditor General of Canada to the House of Commons*, Ottawa: Minister of Supply and Services, chapters 6 and 7.

Canada, Canadian Armed Forces (1990). "The Canadian Airborne Regiment," in *Ducimus: The Regiments of The Canadian Infantry,* Ottawa: Department of National Defence, 33-37.

– (1979). "Airborne Regiment/ Régiment Aéroporté," *Journal du Dixième Anniversaire 1968-1978,* Ottawa: Department of National Defence.

Canada, Commission of Inquiry into the Deployment of Canadian Forces to Somalia (1995). *Commissioners' Statement about the Terms of Reference of the Commission, 3 August 1995*. Ottawa.

– Commission of Inquiry on Equality in Employment (1984). *Report of the Commission on Equality in Employment*, (the Abella Report), Ottawa: Supply and Services Canada.

Canada, Department of National Defence (1993). *Racism and the Canadian Forces*. Director General Personnel Policy, Personnel Policy Review, July 1993, Ottawa: Department of National Defence.

– (1992). *Somalia Handbook*. Ottawa: Department of National Defence, North York, Ontario: G2 Land Force Central Area Headquarters.

– (1991). *Demographic Profile of Department of National Defence Personnel*. Ottawa: Department of National Defence, Demographic Profile of Department of Defence National Defence Personnel Information System (DPIS).

– (1987). *Challenge and Commitment: A Defence Policy for Canada* (White Paper). Ottawa: Department of National Defence.

Canada, House of Commons (1995). *Minutes of Proceedings and Evidence of the Standing Committee on National Defence and Veterans Affairs,* Ottawa, Issue No. 13.

– (1993). *The Dilemmas of Committed Peacekeepers: Canada and the Renewal of Peacekeeping*, Ottawa: Standing Committee on National Defence and Veterans Affairs.

– (1990). *In Flux but Not in Crisis*, Report of the Special Committee on the Review of the *Canadian Security Intelligence Service Act* and the *Security Offences Act*, Ottawa: Queen's Printer.

– (1989). *Standing Orders*, Ottawa: Canadian Government Publishing Centre.

Canada (1992). National Defence Consultative Committee on Social Change: *First Report*, Ottawa: Department of National Defence.

Canada, Privy Council Office (1996). *A Strong Foundation*. Report of the Study Team on Public Service Values and Ethics (John C. Tait, Chairperson). Ottawa: Privy Council Office.

– (1993). *Responsibility in the Constitution*, Ottawa: Minister of Supply and Services Canada.

Canada, Security Intelligence Review Committee (1994). *The Heritage Front Affair. Report to the Solicitor General of Canada*. Ottawa: Security Intelligence Review Committee.

– (1987). *The Security and Intelligence Network in the Government of Canada: A Description*, Ottawa: Security Intelligence Review Committee.

Canada, Senate (1990-1995). *Proceedings of the Subcommittee on National Defence*, Ottawa.

– (1989). *Terrorisme : Rapport du deuxième Comité spécial du Sénat sur le terrorisme et la sécurité publique*, (Chairman, W.M. Kelly), Ottawa: Minister of Supply and Services Canada.

– (1984-1989). *Report of the Special Committee on National Defence,* Ottawa.

Canada, Solicitor General (1991). *On Course: National Security for the 1990s.* Ottawa: Minister of Supply and Services Canada.

– (1990). *A Vision of the Future of Policing in Canada: Police-Challenge 2000*, (also known as the Leighton/Normandeau Report) Ottawa: Minister of Supply and Services Canada.

Canada 21 Council (1994). *Canada 21: Canada and Common Security in the Twenty-First Century*, Toronto: University of Toronto, Centre for International Studies.

Canadian Airborne Regiment (1992). "Operation Cordon: Training Plan," Petawawa: Canadian Airborne Regiment Headquarters.

Canadian Association of University Teachers, Independent Study Group on University Governance (1993). *Governance and Accountability: The Report of the Independent Study Group on University Governance*, Ottawa: Canadian Association of University Teachers.

Canadian Sentencing Commission (1987). *Sentencing Reform: A Canadian Approach*, Ottawa: Minister of Supply and Services.

Cannizzo, C.A. (guest ed.) (1989). "Canadian Defense Policy: Challenges and Continuity: A Symposium," *Armed Forces and Society*, Vol. 16, No. 1, Fall.

Cantrill, H. (1941). *The Psychology of Social Movements,* New York: John Wiley and Sons Inc.

Carbine, M.E. (1990). *Health Care and the Law III: Reports on Three Roundtable Discussions,* Washington DC: Roscoe Pound Foundation.

Carnegie Endowment for International Peace (1992). *Changing Our Ways: America and the New World/Carnegie Endowment National Commission on America and the New World*, Washington DC: The Brookings Institution.

Carter, Cardinal G.E. (1979). *Report to the Civic Authorities of Metropolitan Toronto and its Citizens* (The Carter Report), Toronto: Council of Metropolitan Toronto.

Cerky, J. (1985). *Perceptions of Cultural Differences Among Selected Black and White Police Officers: A Sociological Analysis,* Dissertation, Fordham University, Ann Arbor MI: University Microfilms International.

Chouinard, A. and D. Chiasson (1995). *Analysis of the Canadian Forces Diversity Survey,* Project Report ORA-PR-9523, Ottawa: Department of National Defence (October, 1995)

Conteh-Morgan, E. (1994). "The Military and Human Rights in a Post-Cold War Africa," *Armed Forces and Society*, Vol. 21, No. 1, 69-88.

Cordier, A.W. and W. Foote (eds.) (1974). *Public Papers of the Secretaries-General of the United Nations,* Volume IV: *Dag Hammarskjold 1958-1960*, New York and London: Columbia University Press.

– (1973). *Public Papers of the Secretaries-General of the United Nations,* Volume III: *Dag Hammarskjold 1956-1957*, New York and London: Columbia University Press.

Cot, J.-P. and A. Pellet (eds.) (1991). *La Charte des Nations Unies : Commentaire article par article*, 2nd edition, revised and expanded, Paris: Economica.

Coulon, J. (1993). *Les casques bleus*, Montréal: Fides.

Cox, D. (1994). "Canada and the United Nations: Pursuing Common Security Through Global Institutions," *Canadian Foreign Policy/La politique Étrangère du Canada,* Spring/Printemps, 63-78 (abridged version of the following report).

– (1993). *Canada and the United Nations: Pursuing Common Security Through Global Institutions*, a study undertaken for the Canada 21 Council, Toronto: University of Toronto, Centre for International Studies, 75 pages.

– (1983). "The International Commission of Control and Supervision in Vietnam," in H. Wiseman (ed.), *Peacekeeping: Appraisals and Proposals*, New York and Willowdale, Ont.: Pergamon Press, 298-333.

– (1968a). "Peacekeeping in Canadian Foreign Policy," in S. Clarkson (ed.), *An Independent Foreign Policy for Canada?* Toronto: McClelland and Stewart, 187-197.

– (1968b). "Peacekeeping: The Canadian Experience," in *Peacekeeping: International Challenge and Canadian Response*, Lindsay, Ont.: John Dyell Ltd, 41-62.

– (1968c). "Issues and Options: A Report on the Queen's/CIIA Conference on Peacekeeping February 24-26, 1967," in *Peacekeeping: International Challenge and Canadian Response,* Lindsay, Ont.: John Dyell Ltd, 63-92.

Critchley, T.A. (1967). *A History of Police in England and Wales*: *900-1966*, London: Constable and Co.

Cronbach, L.J. (1982). *Designing Evaluations of Educational and Social Programs*, San Francisco: Jossey-Bass.

Cryderman, B.K. and A. Fleras (1992). *Police, Race and Ethnicity: a Guide for Police Services*, Toronto, Vancouver: Butterworths, 2nd Edition.

Cunningham, G. (1982). "Who Shall Control the Police?" *Police*, April 1982.
Dandeker, C. (1994). "National Security and Democracy: The UK Experience," *Armed Forces and Society*, Vol. 20, No. 3, 353-374.
de Chastelain, General A.J.G. (1992). "Changes in the Canadian Armed Forces," in D.E. Code and C. Ursulak (eds.), *Leaner and Meaner: Armed Forces in the Post Gulf War,* Ottawa: The Conference of Defence Institute, 7-30.
Decraene, P. (1977). *L'expérience socialiste somalienne*, Paris: Berger Levrault.
Delacampagne C. (1983). *L'invention du racisme*, Paris: Fayard.
Diehl, P.F. (1993). *International Peacekeeping*, Baltimore and London: The Johns Hopkins University Press.
Doern, B. (1993). *Political Accountability and Efficiency*, Kingston, Ont.: Government and Competitiveness, School of Policy Studies, Queen's University.
Dorn, E. (1989). *Who Defends America? Race, Sex and Class in the Armed Forces,* Washington, DC: Joint Center for Political Studies Press, Lanham. MD, distributed by arrangement with University Press of America.
Dominguez, J.I. (1976). "Racial and Ethnic Relations in the Cuban Armed Forces: a Non-Topic," *Armed Forces and Society,* Vol. 2, No. 2, 273-290.
Doubnov, S. (1988). *Histoire d'un soldat juif : 1880-1915*, Paris: Éditions du Cerf, coll. Judaïsmes.
Dower, J.W. (1986). *War Without Mercy: Race and Power in the Pacific War*, New York: Pantheon Books.
Dreisziger, N.F. (ed.) (1990). *Ethnic Armies: Polyethnic Armed Forces From the Time of the Habsburgs to the Age of the Superpowers*, Waterloo: Wilfrid Laurier Press.
Dreisziger, N.F. with R.A. Preston (1990). "Polytechnicity and Armed Forces: An Introduction," in N.F. Dreizinger (ed.), *Ethnic Armies*, Waterloo: Wilfrid Laurier University Press, 1-20.
Drysdale, J. (1994). *Whatever Happened to Somalia?* London: Haan Associates.
Dukelow, D.A. (1995). *The Dictionary of Canadian Law,* 2nd edition, Scarborough, Ont.: Carswell.
Dukelow, D.A. and J. Nuse (1991). *The Pocket Dictionary of Canadian Law,* Scarborough, Ont.: Carswell.
Dunphy, B. (1994). "Is Hatred Catching? A New Book on Organized Racism Tries to Isolate the Virus," *The Toronto Sun*, Sunday, April 3, 1994, p. M 21.
Durch, W.J. (ed.) (1993). *The Evolution of Peacekeeping: Case Studies and Comparative Analysis,* New York: St. Martin's Press.
Echenberg, M. (1990). "Race, Ethnicity and Social Class in the French Colonial Army: The Black African *Tirailleurs* 1857-1958," in N.F. Dreisziger (ed.), *Ethnic Armies,* Waterloo: Wilfrid Laurier University Press, 50-68.

Ehrlich, H.J. (1971). *The Social Psychology of Prejudice*, New York: John Wiley and Sons.
Eliade, M. (1965). *Rites and Symbols of Initiation. The Mysteries of Birth and Rebirth*, New York: Harper and Row.
Elias, N. (1994). *The Civilizing Process*, Cambridge, MA: Basil Blackwell (New Edition).
– (1983). *The Court Society*, Oxford: Basil Blackwell.
– (1982a). *The History of Manners* (Vol. I of *Uber den Progress der Zivilisation*, 1939), New York: Pantheon Books.
– (1982b). *The Civilizing Process: State Formation and Civilization* (Vol. II of *Uber den Progress der Zivilisation*, 1939), Oxford: Basil Blackwell.
– (1982c). *Power and Civility,* New York: Pantheon Books.
Enloe, C. (1980). *Ethnic Soldiers: State Security in Divided Societies*, Athens, GA: The University of Georgia Press.
Ericson, R.V. (1995). "The News Media and Account Ability in Criminal Justice," in P.C. Stenning (ed.), *Accountability for Criminal Justice*, Toronto: University of Toronto Press, 135-161.
Fabian, L. (1971). *Soldiers Without Enemies: Preparing the United Nations for Peacekeeping*, Washington, DC: The Brookings Institution.
Faes, G. (1993). "Vivre et laisser mourir," *Jeune Afrique*, No. 1671, Jan., 14-20.
Farson, S. (1995). "The Noble Lie Revisited: Parliament's Five-Year Review of the CSIS Act — Instrument of Change or Weak Link in the Chain of Accountability," in P.C. Stenning (ed.), *Accountability for Criminal Justice*, Toronto: University of Toronto Press, 185-212.
Fesler, J.W. and D.J. Kettl (1991). *The Politics of the Administrative Process,* Chatham, NJ: Chatham House.
Fetherston, A.B. (1994). *Towards a Theory of United Nations Peacekeeping*, New York: St. Martin's Press.
Foner, J.D. (1974). *Blacks and the Military in American History*, New York: Preager.
Friedland, M.L. (1995). *A Place Apart: Judicial Independence and Accountability in Canada*, Ottawa: Canada Communication Group.
Friedman, R. (1975). "Institutional Racism: How to Discriminate Without Really Trying," in T. Pettigrew (ed.), *Racial Discrimination in the United States*, New York: Harper and Row, 384-424.
Frisch, H. (1993). "The Druze Minority in the Israeli Military: Traditionalizing an Ethnic Policing Role," *Armed Forces and Society*, Vol. 20, No. 1, 51-68.
Frymier, J.R. (1996). *Accountability in Education: Still an Evolving Concept*, Bloomington, IN: Phi, Delta, Kappa, Educational Foundation.

Fussell, P. (1989). *Wartime: Understanding and Behavior in the Second World War*, New York, Oxford: Oxford University Press.

Fyfe, J. (1981). "Who Shoots? A Look at Officer Race and Police Shootings," *Journal of Police Science and Administration*, Vol. 9, No. 4, 367-382.

Gabrenya, W.K. (1988). *Effective Change at the Defense Equal Opportunity Management Institute: Studies of Assessment and Symbolic Racism*, Patrick Air Force Base, FL: Defense Equal Opportunity Management Institute, U.S. Army.

Galtung, J. (1976). "Three Approaches to Peace: Peacekeeping, Peacemaking and Peace Building," in J. Galtung (ed.), *Peace, War and Defence: Essays in Peace Research,* Vol. 3, Copenhagen: Christian Ejlers.

Galtung, J. and H. Hveem (1976). "Participants in Peacekeeping Forces," in J. Galtung (ed.), *Peace, War and Defence: Essays in Peace Research,* Vol. 3, Copenhagen: Christian Ejlers.

Gandy, A. (1988). *La légion en Indochine, 1885-1955*, Paris: Presses de la Cité.

Gardam, Col. J. (1992). *The Canadian Peacekeeper*, Burnstown, Ont.: General Store Publishing House Inc.

Geller, W.A. and K.G. Karales (1981). *Split-second Decisions: Shootings of and by Chicago Police,* Chicago: Chicago Law Enforcement Group.

Giddings, P. (ed.) (1995). *Parliamentary Accountability: A Study of Parliament and Executive Agencies*, London, Houndmills, Basingstoke, Hampshire: Macmillan Press.

Gill, P. (1989). "Symbolic or Real? The Impact of the Canadian Security Intelligence Review Committee," *Intelligence and National Security*, Vol. 4, 550-575.

Gimenez-Salinas, A. (1995). *La Formation Policière de Base au Québec et en Catalogne : Étude Comparative et Évaluation*, Mémoide de Maitrîse: École de Criminologie, Université de Montréal.

Gizewski, P. (1993). *The Burgeoning Cost of UN Peacekeeping: Who Pays and Who Benefits?* Ottawa: Canadian Centre for Global Security.

Goldring, A.J. and R. Wettenhall (1980). "Three Perspectives on the Responsibility of Statutory Authorities," in P. Weller and D. Jaensch (eds.) *Responsible Government in Australia,* Richmond, Vic.: Drummond, 136-150.

Goldsmith, A. (1995). "Public Complaints Procedures in Police Accountability," in P. Stenning (ed.), *Accountability for Criminal Justice,* Toronto: University of Toronto Press, p. 110-134.

Goldstein, H. (1997). *Policing a Free Society,* Cambridge, MA: Ballinger.

– (1990). *Problem-Oriented Policing,* Philadelphia: Temple University Press.

– (1987). "Toward Community-Oriented Policing: Potential, Basic Requirement and Threshold Questions," *Crime and Delinquency*, Vol. 33, No. 1, 6-30.

– (1979). "Improving Policing: A Problem-Oriented Approach," *Crime and Delinquency*, Vol. 25, No. 2. 236-258.
– (1967). "Administrative Problems in Controlling the Exercise of Police Authority," *The Journal of Criminal Law, Criminology and Police Science*, Vol. 58, No. 2, 160-172.
Goodrich, L.M., E. Hambro and A.P. Simons (1969). *Charter of the United Nations, Commentary and Documents*, New York: Columbia University Press, 3rd and rev. ed.
Gould, S.J. (1981). *The Mismeasurement of Man*, New York: W.W. Norton.
Goulding, M. (1993). "The Evolution of United Nations Peacekeeping," *International Affairs*, Vol. 69, No. 3, 451-464.
Greater Toronto Region Working Group on Policing in Multicultural, Multiracial Urban Communities (1993). *Guidelines for Crisis Management and Conflict Mediation Involving Police Services and Diverse Racial and Ethnocultural Groups*, Toronto, Dr. Monica Armour, ed.
– (1990). *Guidelines for Intercultural/Race Relations Training for Police Personnel,* report, Toronto: Council of Metropolitan Toronto.
Greenberg, M.A. (1990). "The Control of Police Conduct: A Key Issue for Security Executives," *Journal of Security Administration*, Vol. 13, 63-72.
Greenspan, J. (1980). *Accountability and Quality Insurance in Health Care,* Bowie, MD and Etobicoke, Ont.: Charles Press.
Grimshaw, L.E. (1981). *The Badges and Insignia of the Canadian Airborne Forces: A History of Canadian Airborne Badges and the Men Who Wore Them*, Edmonton: Lone Pine Publishing.
Groll-Ya'Ari, Y. (1994). "Toward a Normative Code for the Military," *Armed Forces and Society*, Vol. 20, No. 3, 457-472.
Grossman, Lt.Col. D. (1995). *On Killing*, Toronto: Little, Brown and Co.
Grundy, R.W. (1976). "Symposium on Racial and Ethnic Relations in the Armed Forces: An Introductory Note," *Armed Forces and Society,* Vol. 2, No. 2, 227-232.
Guyot, D. (1991). *Policing as Though People Matter*, Philadelphia: Temple University Press.
– (1985). "Police Interference versus Political Accountability in Municipal Policing," in E.S. Fairchild and V.J. Webb (eds.), *Politics of Crime and Criminal Justice*, Newbury Park, CA: Sage.
Guyot, J.F. (1976). "Efficiency, Responsibility and Equality in Military Staffing: the Ethnic Dimension in Comparative Perspective," *Armed Forces and Society*, Vol. 2, No. 2, 291-304.
Haas, E.B. (1990). *When Knowledge is Power*, Berkeley and Los Angeles: University of California Press.

Hallin, D.C. (1986). *The "Uncensored War": The Media and Vietnam*, New York: Oxford University Press.

Halloran, R. (1991a). "Soldiers and Scribblers: A Common Mission," in L.J. Matthews (ed.), *Newsmen and National Defense: Is Conflict Inevitable?* Washington: Brassey's (U.S.), 39-59.

– (1991b). "Soldiers and Scribblers Revisited: Working With the Media," in L.J. Matthews (ed.), *Newsmen and National Defense: Is Conflict Inevitable?* Washington: Brassey's (U.S.), 131-141.

Hammond, W.M. (1991). "The Army and Public Affairs: A Glance Back," in L.J. Matthews (ed.), *Newsmen and National Defense: Is Conflict Inevitable?* Washington: Brassey's (U.S.), 1-18.

– (1988). *United States in Vietnam: Public Affairs: The Military and the Media, 1962-1968,* Washington, DC: Center of Military History, United States Army.

Hannigan, T.P. (1990). "Traits, Attitudes, and Skills that are related to Intercultural Effectiveness and their Implications for Cross-Cultural Training: A Review of the Literature," *International Journal of Intercultural Relations*, Vol. 14, No. 1, 89-111.

Hawes, F. and D. Kealey (1981). "An Empirical Study of Canadian Technical Assistance: Adaptation and Effectiveness on Overseas Assignment," *International Journal of Intercultural Relations,* Vol. 5, No. 3, 239-258.

– (1980). *Canadians in Development*, Ottawa: Canadian International Development Agency.

Heron Lepper, J. (translator) (1936). *Les sociétés secrètes, de l'Antiquité à nos jours,* Paris: Payot.

Hewson Report (1985). *A Report on Disciplinary Infractions and Antisocial Behaviour within FMC with Particular Reference to the Special Service Force and the Canadian Airborne Regiment.* Ottawa.

Hilberg, Raul (1983). *The Destruction of the Jews of Europe*, New York: Holmes and Meier, 3 vols.

Hirsch J.L. and R.B. Oakley (1995). *Somalia and Operation Restore Hope*, Washington, DC: United States Institute of Peace Press.

Holdoway, S. (1995). "Culture, Race and Policy: Some Themes in the Sociology of the Police," *Policing and Society*, Vol. 5, No. 2, 109-120.

– (1991). *Recruiting a Multi-Racial Police Force: A Research Study*, London: HMSO.

– (1988). *Crime and Deviance*, Walton-on-Thames: T. Wilson.

– (1983). *Inside the British Police: A Force at Work*, Oxford: Basil Blackwell.

Holdoway, S. and M.R. Chatterton (eds.) (1979). *The British Police*, Beverly Hills, CA: Sage.

Holm, T. (1986). "Culture, Ceremonialism and Stress: American Indian Veterans and the Vietnam War," *Armed Forces and Society*, Vol. 12, No. 2, 237-252.
Hooper, A. (1982). *The Military and the Media*, Aldershot, U.K.: Gower.
Hornung, R. (1991). *One Nation Under the Gun*, Toronto: Stoddart, Pub. Co.
Horton, T.R. (1992). *The CEO Paradox: The Privacy and Accountability of Leadership*, New York: American Management Association.
Hughes, E.C. and MacG.H. Hughes (1952). *Where People Meet: Racial and Ethnic Frontiers*, Glencoe, IL: The Free Press.
Huntingdon, S.P. (1957). *The Soldier and the State*, New York: Vintage Books.
International Peace Academy (1984). *Peacekeeper's Handbook*, New York and Toronto: Pergamon Press.
Jacobs, J.B. and J. Cohen (1978). "The Impact of Racial Integration on the Police," *Journal of Police Science and Administration*, Vol. 6, No. 2, 168-183.
James, A. (1990). *Peacekeeping in International Politics*, Houndmills, Basingstoke, Hampshire and London: Macmillan in association with the International Institute for Strategic Studies.
Jamin, Jaen (1977). *Les lois du silence*, Paris: Maspéro.
Janowitz, M. (1982). "Consequences of Social Science Research on the U.S. Military," *Armed Forces and Society*, Vol. 8, No. 4, 507-524.
– (1977). "From Institution to Occupation: the Need for Conceptual Continuity," *Armed Forces and Society*, Vol. 4, No. 1, 51-54.
– (1971). *The Professional Soldier*, New York: The Free Press.
Janowitz, M. and C.C. Moskos (1976). "Comment," *Armed Forces and Society*, Vol. 2, No. 3, 433-434.
– (1975). "Radical Composition in the All-Volunteer Force: Policy Alternatives," *Armed Forces and Society*, Vol. 1, No. 1, 109-124.
Jockel, J.T. (1994). *Canada and International Peacekeeping*, Toronto: Canadian Institute of Strategic Studies and Washington, DC: Center for Strategic and International Studies.
Johnson, R.R. (1977). "The Search for a Usable Indian: An Aspect of the Defense of Colonial New England," *Journal of American History*, Vol. 64, 649-650.
Jones, P. (1989). *Peacekeeping: An Annotated Bibliography*, Kingston, Ont.: Ronald P. Frye & Company.
Kalshoven, F.F. (1991). "State Responsibility for Warlike Acts of the Armed Forces," *International and Comparative Law Quarterly*, Vol. 40, 827-851.
– (1989a). *Assisting the Victims of Armed Conflict and Other Disasters*, Boston: Kluwer Academic Publishers.
– (1989b). *Implementation of International Humanitarian Law*, Dordrecht, Boston: Nijhoff.

– (1973). *The Laws of Warfare: A Summary of its Recent History and Trends in Development*, Leiden: Sijtnoff.
Kealey, D.J. (1990). *Cross-Cultural Effectiveness: A Study of Canadian Technical Advisors Overseas,* Hull, Canada: Canadian International Development Agency.
– (1989). "A Study of Cross-Cultural Effectiveness: Theoretical Issues, Practical Applications," *International Journal of Intercultural Relations*, Vol. 13, No. 3, 387-427.
– (1988). *Explaining and Predicting Cross-Cultural Adjustment and Effectiveness: A Study of Canadian Technical Advisors Overseas,* Hull, Canada: Canadian International Development Agency.
Kealey, D.J. and B.D. Ruben (1983). "Cross-Cultural Personnel Selection: Criteria, Issues and Methods," in D. Landis and R. Brislin (eds.), *A Handbook of Intercultural Training*, New York: Pergamon Press, Vol. 1, pp. 155-175.
Kemp, R.W. and C. Hudlini (1992). "Civil Supremacy over the Military: Its Nature and Limits," *Armed Forces and Society*, Vol. 19, No. 1, 7-26.
Kerstetter, W.A. (1985). "Who Disciplines the Police? Who Should?" in W.A. Geller (ed.), *Police Leadership in America*, New York: Praeger, 149-182.
– (1979). "Peer Accountability as a Primary Control Mechanism in Police Agencies," *Criminal Justice Review*, Vol. 4, No. 2, 113-120.
Ki-Zerbo, J. (1972). *Histoire de l'Afrique noire, D'Hier à Demain*, Paris: Hatier.
Kinder, D.R. and D.O. Sears (1981). "Prejudice and Politics: Symbolic Racism versus Threats to the Good Life," *Journal of Personality and Social Psychology*, Vol. 40, 414-431.
Kinsella, W. (1994). *Web of Hate: Inside Canada's Far Right Network*, Toronto: Harper Collins, Pub.
– (1992). *Unholy Alliances: Terrorists, Extremists, Front Companies and the Libyan Connection in North America,* Toronto: Lester Publishing Co.
Kirk-Greene, A. (1980). "'Damnosa Hereditas'; Ethnic Ranking and the Martial Races Imperative in Africa," *Ethnic and Racial Studies,* Vol. 3, October, 393-414.
Knightley, P. (1975). *The First Casualty*, New York: Harcourt Brace Jovanovich.
Knouse, S.B. (1991). "Introduction to Racial, Ethnic, and Gender Issues in the Military: the Decade of the 1990s and Beyond," *International Journal of Intercultural Relations* (Special Issue: Racial, Ethnic, and Gender Issues in the Military), Vol. 15, No. 4, 385-388.
Kroeker, H.V. (1981). *Sovereign People or Sovereign Governments: Proceedings of a Conference*, Montreal: Institute for Research on Public Policy.
– (1978). *Accountability and Control: the Government Expenditure Process*, Montreal: C.D. Howe Research Institute.

Kuykendall, J.L. and D.E. Burns (1980). "The Black Police Officer: An Historical Perspective," *Journal of Contemporary Criminal Justice*, Vol. 1, No. 4, 4-12.
Lafontaine, J.S. (1985). *Initiation,* Manchester: Manchester University Press.
Lane, Christel (1981). *The Rites of Rulers*, Cambridge: Cambridge University Press.
Langbein, H. (1975). *Hommes et femmes à Auschwitz,* Paris: Fayard.
Larson, R.C. (ed.) (1978). *Police Accountability: Performance Measures and Unionism*, Lexington, MA: Lexington Books.
Last, Maj. D.M. (1992). "Negotiating Skills for Peacekeepers," *Peacekeeping and International Relations*, Vol. 21, No. 3 and 4: 5-6.
Laurance, E.J. (1980). "The Congressional Role in Defense Policy-Making: The Evolution of the Literature," *Armed Forces and Society*, Vol. 6, No. 3, 431-454.
Law Reform Commission of Victoria (1991). *Access to the Law: Accountability of the Legal Profession*, Melbourne (Australia): The Commission.
Leemon, T.A. (1972). *The Rites of Passage in a Student Culture: a Study of the Dynamics of Transition*, New York: Teachers College Press.
Leinen, S. (1984). *Black Police, White Society,* New York; University Press.
Leman-Langlois, S. and J.-P. Brodeur (1995). *Le droit national et international et les événements du 16 mars 1968 à My Lai, Viêt-nam*, Research report presented to CIDCFS.
Lepper, J.H. (1932). *Famous Secret Societies,* London: S. Low Marston and Co.
Lévi-Strauss, C. (1970). *Race et Histoire*, Paris: Gonthier.
Lewis, I.M. (1994). *Blood and Bone,* Lawrenceville, NJ: The Red Sea Press.
– (1993). *Understanding Somalia: Guide to Culture, History and Social Institutions,* London: Haan Associates.
Light, P.C. (1995). *Thickening Government: Federal Hierarchy and the Diffusion of Accountability*, Washington, DC: the Governance Institute.
– (1993). *Monitoring Government: Inspectors General and the Search for Accountability*, Washington, DC: The Brookings Institution and the Governance Institute.
Lindsay, J.M. (1990). "Congressional Oversight of the Department of Defense: Reconsidering the Conventional Wisdom," *Armed Forces and Society*, Vol. 17, No. 1, 7-26.
Liu, F.T. (1992). *United Nations Peacekeeping and the Non-Use of Force*, Boulder and London: Lynne Rienner Publishers.
Lustgarten, L. (1995). "Security Services, Constitutional Structure, and Varieties of Accountability in Canada and Australia," in P.C. Stenning (ed.), *Accountability for Criminal Justice*, Toronto: University of Toronto Press, 162-184.

MacGregor, M.J. (1981). *Integration of the Armed Forces 1940-1965*, Washington, DC: Center of Military History, U.S. Army. (For sale by the Superintendant of Documents, U.S. Government Printing Office.)

MacKenzie, L. (1994). *Peacekeepers: The Road to Sarajevo*, Toronto: Harper & Collins.

Maguire, M., J. Vagg and R. Morgan (eds.) (1985). *Accountability in Prison: Opening Up a Closed World,* London and New York: Tavistock Publications.

Mahdi, L.C., S. Foster and M. Little (1987). *Betwixt and Between: Patterns of Masculine and Feminine Initiation*, La Salle, IL: Pen Court.

Malek, S.J. (1977). *Accountability: Its Meaning and Relevance to the Health Care Field*, Hyattsville: U.S. Department of Health.

Malouf, A. and J.-P. Brodeur (1993). *Rapport de l'enquête spéciale tenue sur les désordres qui ont fait suite à la conquête de la Coupe Stanley par le club Canadien*, Québec: Ministère de la Sécurité publique.

Marchal, R. (1993a). "Les *mooryaan* de Mogadiscio. Formes de la violence dans un espace urbain en guerre," *Cahiers d'études africaines*, Vol. 33, No. 2, 295-320.

– (1993b). "Somalie : Autopsie d'une intervention," *Politique Internationale,* No. 61, 191-208.

– (1993c). "L'invention d'un nouvel ordre régional," *Politique Africaine,* No. 50, 2-8.

– (1992). "La guerre à Mogadiscio," *Politique Africaine*, No. 46, 120-125.

Mariel, P. (1974). *Rituels et initiations des sociétés secrètes*, Paris: Maison Mame.

Marshall, G. (ed.) (1989). *Ministerial Responsibility*, New York and Oxford: Oxford University Press.

Marshall, S.L.A. (1978). *Men Against Fire*, Gloucester, MA: Peter Smith.

Marshall, The Honourable Justice T.D. (1995). *Judicial Independence and Accountability*, Scarborough, Ont.: Carswell.

Martin, L. (1996). *Measuring the Performance of Human Service Programs*, Thousand Oaks, CA: Sage.

Martin, M.L. (1994). "National Security and Democracy: The Dilemma from a French Perspective," *Armed Forces and Society,* Vol. 20, No. 3, 395-422.

Matthews, C.J. (1993). *Accountability in the Administration of Criminal Justice: A Selective Annotated Bibliography*, Toronto: Centre of Criminology, University of Toronto.

Matthews, L.J. (ed.) (1991). *Newsmen and National Defense: Is Conflict Inevitable?* Washington: Brassey's.

McConahay, J.B. and J.C. Hough (1976). "Symbolic Racism," *Journal of Social Issues*, Vol. 32, No. 4, 23-45.

McCubbins, M.D. and T. Schwartz (1984). "Congressional Oversight Overlooked: Police Patrols versus Fire Alarms," *American Journal of Political Science*, Vol. 28, 165-177.

Miller, L.L. and C. Moskos (1995). "Humanitarians or Warriors? Race, Gender, and Combat Status in Operation Restore Hope," *Armed Forces and Society*, 615-637.

Morrison, A. (forthcoming). "Canada and Peacekeeping," in D. Dewitt and D. Leyton-Brown (eds.), *Canada's International Security Policy.*

Morrison, A., R. Tychonick and S. McNish (1993). *The Changing Face of Peacekeeping*, Toronto: The Canadian Institute of Strategic Studies.

Mosher, F.C. (1982). "Comment," in L.R. Bruce and J.D. Carroll (eds.), *Improving the Accountability and Performance of Government*, Washington, DC: The Brookings Institution, 70-76.

Moskos, C.C. (1986). "Institution/Occupation Trends in Armed Forces: An Update," *Armed Forces and Society*, Vol. 12, No. 3, 377-382.

– (1977). "From Institution to Occupation: Trends in Military Organizations," *Armed Forces and Society*, Vol. 4, No. 1, 41-50.

– (1976). *Peace Soldiers: The Sociology of a United Nations Military Force*, Chicago and London: The University of Chicago Press.

– (1975). "UN Peacemakers: The Constabulary Ethic and Military Professionalism," *Armed Forces and Society*, Vol. 1, No. 4, 388-401.

– (1972). "Minority Groups in Military Organizations," in S.E. Ambrose and J.A. Barber (eds.), *The Military and American Society: Essays and Readings*, New York: The Free Press, 192-201.

Moskos, C.C., J.H. Butler, A.N. Sabrosky and A.J. Schexnider (1980). "Symposium: Race and the United States Military," *Armed Forces and Society*, Vol. 6, No. 4, 586-613.

Mueller, J. (1973). *Presidents and Public Opinion*, New York: John Wiley.

Müller, J.W. (1992). *The Reform of the United Nations*, New York, London and Rome: Oceana Publications Inc., 2 vols.

Myrdal, G. (1964). *An American Dilemma: The Negro Problem and Modern Democracy* (with the assistance of Richerd Sterner and Arnold Rose, originally published in 1944), New York: Harper and Row, 20th anniversary ed.

Nirls, D. (1991). "Congress and the Politics of Military Reform," *Armed Forces and Society*, Vol. 17, No. 1, 487-512.

Noone, M.F. (1989). "Military Social Science Research and the Law," *Armed Forces and Society*, Vol. 15, No. 2, 193-206.

Nova Scotia (1989). *Report of the Royal Commission on the Donald Marshall Jr. Prosecution* (the Marshall Inquiry Report), Halifax: published on the authority of the Lieutenant Governor in Council by the Royal Commission on the Donald Marshall Jr. Prosecution.

O'Connor, R. (1992). *Buffalo Soldiers*, New York: Vintage.

Ontario (1995). *Report of the Commission on Systemic Racism in the Ontario Criminal Justice System*, Toronto: Queen's Printer for Ontario.

– (1989). *The Report of the Race Relations and Policing Task Force* (The Lewis Task Force Report), Toronto: Ministry of the Solicitor General.

– (1980). *Policing in Ontario for the Eighties: Perceptions and Reflections,* Report of the Task Force on the Racial and Ethnic Implications of Police Hiring, Training, Promotion and Career Development (the Gerstein Report), Ontario: Minister of the Solicitor General.

– (1974). *The Police Are the Public and the Public Are the Police,* Report of the Task Force on Policing in Ontario, Toronto: Ministry of the Solicitor General.

Palmer, S. (1977). *Public Accountability and Peer Review in Health Care Delivery in the United States and the United Kingdom*, Bethesda, MD: U.S. Department of Health.

Pelletier, C. (1990). *L'apprentissage de la diversité au Service de Police de la Communauté Urbaine de Montréal*, Montréal: CIDIHCA.

Philpot, R. (1991). *Oka : dernier alibi du Canada anglais*, Montréal: VLB Éditeur.

Pistone J. (1992). *The Ceremony: The Mafia Initiation Tapes* (Introduction by Joseph Pistone). New York: Dell Publishing.

Pitman, W. (1977). *Now is Not Too Late,* Report submitted to the Council of Metropolitan Toronto by the Task Force on Human Relations (the Pitman Report), Toronto: Council of Metropolitan Toronto.

Poliakov, L. (1955-1968). *Histoire de l'antisémitisme.* Paris: Calman Lévy, 3 vols.

Preston, R.A. (1990). "Bilingualism and Multiculturalism in the Canadian Armed Forces," in N.F. Dreisziger (ed.), *Ethnic Armies*, Waterloo: Wilfrid Laurier University Press, 158-177.

Pruegger, V.J. and T.B. Rogers (1994). "Cross-Cultural Sensitivity Training: Methods and Assessment," *International Journal of Intercultural Relations,* Vol. 18, No. 4, 369-387.

Prunier, G. (1990-1991). "A Candid View of the Somali National Movement," *Horn of Africa*, Vol. XIII, No. 3-4, 239-241 and Vol. XIV, No. 1-2: 113-115.

Purnelle, M. (1996). *Une armée en déroute*, Montréal: Liber.

Québec (1988). *Rapport final du Comité d'enquête à la Commission des Droits de la personne du Canada* (Rapport Bellemare), Comité d'enquête sur les relations entre les corps policiers et les minorités visibles et ethniques, Québec: Ministère des Communications: Commission des droits de la personne.

Radzinowicz, L. (1956). *A History of English Criminal Law and its Administration from 1750,* Vol. 3: *Cross Currents in the Movement for the Reform of the Police*, London: Stevens and Sons Limited.

Raper, A.F. (1933). *The Tragedy of Lynching*, Chapel Hill: The University of North Carolina Press.

Rehr, H. (ed.) (1979). *Professional Accountability for Social Work Practice: A Search for Concepts and Guidelines*, New York: Prodist.

Reiss, A.J. (1971). *The Police and the Public*, New Haven: Yale University Press.

– (1968). "Police Brutality: Answers to Key Questions," *Trans-Action*, No. 5, July-August 1968, 5-18, reprinted in R. Lundman (ed.), 1980, *Police Behaviour*, New York: Oxford University Press.

– (1967). *Studies in Crime and Law Enforcement in Major Metropolitan Areas,* in *The President's Commission on Law Enforcement and the Administration of Justice*, Field Surveys III, Washington, DC: U.S. Government Printing Office, 2 vol.

Reith, C. (1975). *The Blind Eye of History: A Study of the Origins of the Present Police Era*, Montclair, NJ: Patterson Smith.

Rhine, E.E. (ed.) (1993). *Reclaiming Offender Accountability: Intermediate Sanctions for Probation and Parole Violators: A Project of the American Correctional Association 1990-1991*, Laurel, MD: American Correctional Association.

Rikhye I.J., M. Harbottle and Bjorn Egge (1974). *The Thin Blue Line: International Peacekeeping and Its Future*, New Haven and London: Yale University Press.

Roulier, R.P. (1995). *Bank Governance Contracts: Establishing Goals and Accountability in Bank Restructuring*, Washington, DC: World Bank.

Rourke, F.E. (1979). "Bureaucratic Autonomy and the Public Interest," in C.H. Weiss, and A.H. Barton (eds.), *Making Bureaucracies Work*, Beverly Hills, CA: Sage, 99-117.

Ruben, B.D. and D.J. Kealey (1979). "Behavioral Assessment of Communication Competency and the Prediction of Cross Cultural," *International Journal of Intercultural Relations*, Vol. 3, No. 1, 15-47.

Sahnoun, M. (1994). *Somalia: The Missed Opportunities*, Washington, DC: United States Institute of Peace Press.

St. Pierre, M. (1991). "Accession and Retention of Minorities: Implications for the Future," *International Journal of Intercultural Relations*, Vol. 15, No. 4, 469-489.

Sallis, J. (1988). *Schools, Parents and Governors: A New Approach to Accountability*, London: Routledge.

Schexnider, A.J. and J.S. Butler (1976). "Race and the All-Volunteer System: A Reply to Janowitz and Moskos," *Armed Forces and Society,* Vol. 2, No. 3, 421-432.

Schraeder, Peter J. (1992). "The Horn of Africa: US Foreign Policy in an Altered Cold War Environment," *Middle East Journal*, Vol. 46, No. 3.

Sears, D.O. and J.B. McConahay (1973). *The Politics of Violence: the New Urban Blacks and the Watts Riot*, Boston: Houghton Mifflin.

Segal, D. (1994). "National Security and Democracy in the U.S.," *Armed Forces and Society*, Vol. 20, No. 3, 375-394.

– (1986). "Measuring the Institution/Occupation Change Thesis," *Armed Forces and Society*, Vol. 12, No. 3, 351-382.

Segal, D.R., J.J. Harris, J.M. Rothberg and D.H. Marlowe (1984). "Paratroopers as Peacekeepers," *Armed Forces and Society*, Vol. 10, No. 4, 487-506.

Shearing, C.D. and P.C. Stenning (1980). *Police Training in Ontario: An Evaluation of Recruit and Supervisory Courses*, Toronto: Centre of Criminology, University of Toronto.

Shields, P.M. (1993). "A New Paradigm for Military Policy: Socioeconomics," *Armed Forces and Society*, Vol. 19, No. 4, 479-510.

Sibony, Daniel (1988). *Écrits sur le Racisme*, Paris: Christian Bourgeois.

Skogan, W.G. (1995). "Community Policing in the United States," in J.-P. Brodeur (ed.), *Comparisons in Policing: An International Perspective*, Aldershot, U.K. and Brookfield, U.S.: Avebury, 86-111.

– (1994). "The Impact of Community Policing on Neighborhood Residents: A Cross Site Analysis," in D.R. Rosenbaum (ed.), *The Challenge of Community Policing: Testing the Promises*, Thousand Oaks, CA: Sage, 167-181.

– (1990). *Disorder and Decline*, New York: The Free Press.

Smith, Maj. D.P. (1995). *The Employment Equity Profile of the CF Recruitable Population*, Technical Note 04/95, Willowdale, Ontario: Canadian Forces Personnel Applied Research Unit.

Smith, S. (1993). *La guerre perdue de l'humanitaire,* Paris: Calmann Lévy.

Smith, A.D. (1981). "War and ethnicity: the role of warfare in the formation, self-images and cohesion of ethnic communities," *Ethnic and Racial Studies*, Vol. 4, No. 4, 375-397.

Snyder, L.L. (1962). *The Idea of Racialism: Its Meaning and History*, Princeton, NJ: D. Van Nostrand Company, Inc.

Solomon, P.H. (1985) "The Law Reform Commission of Canada's Proposals for Reform of Police Powers: An Assessment," *Criminal Law Quarterly*, Vol. 27, No. 3, 321-351.

Stahl, M.J., C.W. Nichols and T.R. Manley (1980). "An Empirical Examination of the Institution/Occupation Model," *Armed Forces and Society*, Vol. 6, No. 2, 257-269.

Stanley, J. (1994). "National Security and Democracy: A Symposium: Introduction," *Armed Forces and Society*, Vol. 20, No. 3, 351-352.
Statistics Canada (1991). *Adults with Disabilities: Their Employment and Education Characteristics, 1991 Health and Activity Limitation Survey*, Ottawa.
Steinberg, S. (1981). *The Ethnic Myth: Race, Ethnicity and Class in America*, New York: Atheneum.
Stenning, P.C. (ed.) (1995a). *Accountability for Criminal Justice*, Toronto: University of Toronto Press.
– (1995b). "Accountability in the Ministry of the Solicitor General of Canada," in P.C. Stenning (ed.), *Accountability for Criminal Justice*, Toronto: University of Toronto Press, 44-73.
Stretch, R.S. (1990). "Effects of Service in Vietnam on Canadian Armed Forces Military Personnel," *Armed Forces and Society*, Vol. 16, No. 1, 113-130.
Sydnor, C.W. (1977). *Soldiers of Destruction: the SS Death's Head Division, 1933-1945*, Princeton, NJ: Princeton University Press.
Szajkowski, Z. (1975). *Jews and the French Foreign Legion*, New York: Ktav Pub. House.
Taguieff, P-A. (1992). "L'antiracisme en crise : Éléments d'une critique réformiste," in M. Wieviorka (ed.), *Racisme et modernité*, Paris: Éditions de la découverte, 357-392.
– (1988). *La force du préjugé*, Paris: Éditions de la découverte.
Taylor, E.K. (1953). *Public Accountability of Foundations and Charitable Trusts*, New York: Russell Sage Foundation.
Tinker, H. (1977). *Race, Conflict and the International Order: from Empire to United Nations*, New York: St. Martin's Press.
Todorov, T. (1989). *Nous et les autres : la réflexion française sur la diversité humaine*, Paris: Seuil.
– (1984). *The Conquest of America: the Question of the Other*, trans. from the French by Richard Howard, New York: Harper and Row.
Tonnies, F. (1957). *Community and Society*, trans. and ed. by Charles P. Loomis, East Lansing: Michigan State University Press, originally published in 1887.
Torbiorn, I. (1982). *Living Abroad: Personal Adjustment and Personnel Policy in the Overseas Setting*, Chichester: Wiley.
Torrenzano, A. (1995). *L'imbroglio somalien*, Paris: L'Harmattan.
Tucker, M. (1974). *Screening and Selection for Overseas Assignment: Assessments and Recommendations to U.S. Navy*, Denver, CO: Center for Research and Education.
Uglow, S. (1988). *Policing Liberal Society*, Oxford and Toronto: Oxford University Press.

United Nations (1978). *Déclaration et programme d'action contre le racisme et la discrimination raciale*, New York: United Nations.

United Nations, Security Council (1994). *Report of the Commission of Inquiry Established Pursuant to Security Council Resolution 885 (1993) to Investigate Armed Attacks On UNOSOM II Personnel Which Led to Casualties Among Them,* New York, 24 February 1994.

Urquhart, B. (1987). *A Life in Peace and War,* New York: Harper & Row.

Urquhart, B. and E. Childers (1990). *A World in Need of Leadership: Tomorrow's United Nations*, Uppsala, Sweden: Dag Hammarskjold Foundation.

Vagg, J. (1994). *Prison Systems: A Comparative Study of Accountability in England, France, Germany and the Netherlands*, Oxford: Clarendon Press, New York: Oxford University Press.

Vagts, A, (1959). *A History of Militarism, Civilian and Military,* New York: The Free Press.

Van Dijk, T.A. (1991). *Racism and the Press*, London, New York: Routledge.

– (1987). *Communicating Racism: Ethnic Prejudice in Thought and Talk*, Newbury Park, CA: Sage.

Vandevort, L. (1979). *Political Control of Independent Administrative Agencies,* Study Paper, Ottawa: the Law Reform Commission of Canada.

Wagner, R.B. (1989). *Accountability in Education: A Philosophical Inquiry*, New York: Routledge.

Wakin, M.M. (ed.) (1986). *War, Morality and the Military Profession*, 2nd ed. rev. and updated, Boulder, CO: Westview Press.

Walker, D.B. (1983). "Black Police Values and the Black Community," *Police Studies*, Vol. 5, No. 4, 20-28.

Walker, N. (1971). *Sentencing in a Rational Society,* New York: Basic Books (first published in the United Kingdom in 1969).

Walzer, M. (1977). *Just and Unjust Wars: A Moral Argument with Historical Illustrations*, New York: Basic Books.

Ward, J. (1996). "Boyle warned Collenette of Somalia Affair 3 years ago," *The Montreal Gazette*, Tuesday, August 13, A7.

Weber, M. (1946). "Bureaucracy," in H.H. Gerth and C. Wright Mills (eds.), *Max Weber: Essays in Sociology*, Oxford: Oxford University Press.

Weigel, R.H. and P.W. Howes (1985). "Conceptions of Racial Prejudice: Symbolic Racism Reconsidered," *Journal of Social Issues*, Vol. 41, No. 3: 117-138.

Wheeler, D.L. (1976). "African Elements in Portugal's Armies in Africa (1961-1971)," *Armed Forces and Society*, Vol. 2, No. 2, 223-250.

Whitaker, R. (1992). "The Politics of Security Intelligence Policy-Making in Canada: II," *Intelligence and National Security*, Vol. 7, 53-76.

– (1991). "The Politics of Security Intelligence Policy-Making in Canada: I," *Intelligence and National Security*, Vol. 6, 649-668.

White, B. (1990). "The American Army and the Indian," in N.F. Dreisziger (ed.), *Ethnic Armies*, Waterloo: Wilfrid Laurier Press, 69-88.

Wieviorka, M. (1994). *Racisme et xénophobie en Europe*, Paris: Éditions de la découverte.

– (ed.) (1993). *Racisme et Modernité*, Paris: Éditions de la découverte.

– (1992). *La France Raciste*, Paris: Seuil.

– (1991). *L'espace du racisme*, Paris: Seuil.

Wilbanks, W. (1987). *The Myth of a Racist Criminal Justice System*, Monterey, CA: Brooks/Cole Publishing Company.

Willame, J.-C. (1994). *Gouvernance et pouvoir, Essai sur trois trajectoires africaines. Madagascar, Somalie, Zaïre*, Bruxelles: Institut Africain-CEDAF.

Wilson, D., S. Holdoway and C. Spencer (1984). "Black Police in the United Kingdom," *Policing*, Vol. 1, No. 1, 20-30.

Winslow, D. (1997). *The Canadian Airborne Regiment in Somalia: A Socio-cultural Inquiry,* Ottawa, Commission of Inquiry into the Deployment of Canadian Forces to Somalia.

Wiseman, H. (1983). "The Dynamics of Future Developments," in H. Wiseman (ed.), *Peacekeeping: Appraisals and Proposals*, New York and Willowdale, Ont.: Pergamon Press, 341-370.

Woodhouse, D. (1994). *Ministers and Parliament: Accountability in Theory and Practice*, Oxford: Clarendon Press.

Wright, B. (1987). *Black Robes, White Justice*, New York: Lyle Stuart.

York, G. and L. Pindera (1991). *People of the Pines*, Boston, London, Toronto: Little, Brown and Co.

Young, F.W. (1965). *Initiation Ceremonies: A Cross-Cultural Study of Status Dramatization.* Indianapolis: Bobbs-Merrill.

Young, W.L. (ed.) (1982). *Minorities and the Military*, Westport, CT and London, U.K.: Greenwood Press.

APPENDIX A

The following, reprinted with permission, is from a chapter entitled "The Newest Monitors," in *Monitoring Government: Inspectors General and the Search for Accountability* by Paul C. Light (Washington, D.C.: The Brookings Institution and the Governance Institute, 1993).

The fall of 1978 was a busy season in the search for government accountability. During the final month of the legislative session, Congress passed the Civil Service Reform Act, the Ethics in Government Act, and a government-wide Inspector General Act, which together were arguably the most important package of administrative reforms in fifty years.

The statutes, however, were not part of a unified reform strategy. The bills came from different committees, involved different advocates, and never crossed paths in the legislative process. Nor were they the only efforts to strengthen accountability. First came the congressional reforms of the 1960s and early 1970s — an expanded committee system, a stronger General Accounting Office, a new Office of Technology Assessment, a new Congressional Budget Office, the War Powers Resolution, and the Budget and Impoundment Control Act. Then came the "sunshine" laws and other limits on bureaucratic discretion of the early to mid1970s: the Federal Advisory Committee Act, the Freedom of Information Act Amendments, the Privacy Act, the Federal Election Campaign Act, the Government in the Sunshine Act, as well as assorted legislative vetoes, appropriations riders, personnel floors and ceilings, deadlines, and triggers.

Then, late in this reform history came the two IG acts — the first at HEW, the second covering the whole of government. Each made up a small portion of what was a wide swath of reform measures, the most important of which was the dramatic increase in the number of reviewers of and in government — from congressional staff to the media, GAO investigators to interest groups. The IGs merely were the newest monitors of the era.

The 1978 IG Act was the only reform statute to combine all three contemporary strategies of accountability — ruled-based compliance, performance incentives, and improvements in basic governmental capacity. While the Civil Service Reform Act, for example, operated primarily through performance incentives and capacity building, and the Ethics in Government Act through compliance, the IG Act invited questions of design, implementation, organization, and effectiveness. The IGs' eventual reliance almost exclusively on compliance in their search for accountability reveals a great deal about the politics of administrative reform.

PATHS TO ACCOUNTABILITY

Despite experiments with performance incentives, such as merit pay, and occasional investments in civil service reform, the definition of accountability in government has remained relatively constant over the past fifty years: limit bureaucratic discretion through compliance with tightly drawn rules and regulations. Public administration scholar Francis Rourke wrote in the late 1970s that,

> The reformers of the 1960s and the 1970s seem bent not on extending but on curtailing the independence of bureaucratic organizations. They argue that bureaucracies represent formidable concentrations of power in contemporary society, and that executive agencies should be brought back within the political system and made more accountable. If traditional efforts at reform could be described as an attempt to depoliticize the administrative process in the United States, the reform movement in our day seems rather aimed at repoliticizing administration — at least in the sense of restoring public control over previously independent agencies.

This definition is well supported to this day in public administration scholarship. Even a cursory review of contemporary public administration textbooks suggests that the dominant definition is one of command-and-control. Accordingly, accountability is seen as the product of limits on bureaucratic discretion — limits that flow from clear rules (commands), and the formal procedures, monitoring, and enforcement that make them stick (controls). Public administration scholars Dennis Palumbo and Steven Maynard-Moody, for example, summarized the two methods of bureaucratic accountability as follows:

> One is *external*; this involves controls by legislatures through such mechanisms as legislative oversight, by the courts through review of decisions and administrative

law, or by citizen participation. The second method is through *internal* controls; this includes development of professional standards and ethics, the use of rules, whistle-blowers, representative bureaucracy, and opening up administrative proceedings.

The problem with the control definition, however, is that it creates an artificial trade-off between accountability and values such as creativity and innovation. Addressing the potential cost of accountability, for example, Frederick Mosher wrote:

> I begin with the premise that accountability, anymore than any other single value, is not an absolute. If everyone were held accountable for everything he did or tried or thought or imagined, this would be a pretty sterile, dull, and static world. Accountability is not commonly associated with invention or novelty or serendipity, but rather with carrying out assignments, which are more or less specifically defined, honestly, efficiently, effectively, and at minimal cost. Thus, at the very outset, there is a conflict between the value associated with accountability and the values of originality, experimentation, inventiveness, and risk-taking.

Defined in such command-and-control terms, accountability breeds just such conflicts. However, other definitions exist in the field. Public administration scholars James Fesler and Donald Kettl, for example, divided accountability into two dimensions: "One is accountability; faithful obedience to the law, to higher officials' directions, and to standards of efficiency and economy. The other is ethical behavior: adherence to moral standards and avoidance even of the *appearance* of unethical actions." After conducting an inventory of the familiar methods for achieving faithful obedience and ethical behavior, they rejected the conventional compliance path to accountability: "In the end, we come back to the recruitment and retention of individuals dedicated to public service, respectful of its call for bureaucratic accountability and ethical behavior, and both knowledgeable about and committed to the constitutional, democratic system."

This capacity-based view of accountability is particularly appropriate for understanding the enactment of the three 1978 reform statutes. With a Democrat in the White House, the Watergate scandal receding from memory, and the taking of American hostages in Iran still two years away, the accepted definition of accountability widened briefly. The Civil Service Reform Act, for example, contained a clear commitment — albeit never fully funded — to performance pay, merit bonuses, and greater stewardship of the government's human capital. The Ethics in Government Act created a financial disclosure process that, while heavily laden with compliance measures, still envisioned presidential appointees coming into office with a greater incentive to perform in the public's interest. And the 1978 IG Act involved a mix of all three definitions of accountability.

The choice of one strategy does not necessarily exclude the need for a second or third. Compliance accountability, for example, is not inconsistent with performance incentives or capacity building and may be essential for assuring the fairness and equity that sometimes undermine employee confidence in performance systems. Nevertheless, considering each definition as separate is useful. In part because of budget pressures, in part because of political incentives, Congress and the president may have created an artificial trade-off among the three, substituting compliance accountability for performance systems and capacity building. Moreover, the three involve different targets, mechanisms, and timetables (see Table 1.1).

Table 1.1.
Aspects of Accountability

	Definition of accountability		
Characteristic	*Compliance*	*Performance*	*Capacity building*
Point of intervention	Post-activity	Mixed pre- and post-activity	Pre-activity
Primary targets	Individuals and accounts	Individuals and programs	Agencies and government
Primary mechanism	Rules	Incentives	Technologies
Role of sanctions	Negative	Positive	Positive
Role of management	Supervision and discipline	Goal setting and reinforcement	Advocacy and stewardship
Role of oversight	Detection and enforcement	Evaluation and bench marking	Analysis and design
Complexity of strategy	Simple	More complex	Most complext
Durability of effects	Short-term	Intermediate	Long-term

To compare the methods, start with the point of intervention. Under a compliance model, intervention awaits the activity. Although compliance also assumes some pre-activity deterrence effect, it primarily is a tool for catching mistakes after they have occurred. In contrast, both performance and capacity-based accountability rely on having an impact before something happens.

The definitions also aim at different primary targets, ranging from individuals and accounts at one end of a continuum, to individuals and programs toward the middle, and agencies and government as a whole at the other end. Accepting the imprecision of such a continuum, compliance can reasonably be placed at the individual and accounts end, performance in the middle, and capacity building at the agencies and government side. Investigations, for example, focus on individual violations, whether perpetrated by beneficiaries, contractors, or federal employees.

The primary mechanisms for achieving change are more precise. The common currency of compliance accountability is rules and regulations, whether dealing with procurement, travel, personnel, paperwork, or a specific policy. The rules are reinforced through cumbersome signature and approval systems. In contrast, performance accountability emerges from the establishment of incentives, the most familiar of which are the pay for performance provisions of the Civil Service Reform Act. Capacity building focuses on technologies, broadly defined to include people and the tools of management. In addition, capacity building includes both program and organizational redesign; that is, the development of workable programs and responsive structures.

Consider next the role of sanctions. Compliance accountability places its emphasis on negative sanctions, whether formal or informal. Medicare providers who violate the rules can be debarred, or removed from the program; contractors who cheat can be fined; employees who steal can be prosecuted. Performance accountability puts its focus instead on positive sanctions, notably bonuses and awards, although members of the Senior Executive Service (SES) created under the Civil Service Reform Act also can receive sabbaticals. The problem with positive sanctions, particularly during the 1980s, is inadequate funding. Without funding, pay for performance has a hollow ring. Capacity-based accountability does not involve formal sanctions per se but can be seen as a similarly positive approach to performance incentives. Increases in training budgets, purchase of state-of-the-art technology, and so forth can be powerful inducements for organizational effectiveness.

Managers and overseers play different roles in the three types of accountability. Capacity building views the role of management as one of advocacy and stewardship. Managers are responsible for both securing and maintaining the tools and resources to achieve effectiveness. By comparison, performance accountability requires goal setting and reinforcement, while compliance accountability requires tight supervision and needed discipline. The role of oversight is equally distinct. Compliance accountability demands detection of violations and enforcement of sanctions, performance, evaluation of effectiveness and bench marking, which organizational scholar Janet Weiss defines as "the results achieved by the very best organizations doing the same work", and capacity building, analysis and design.

The complexity of the strategy for achieving change also varies. Compliance and performance accountability are both relatively simple concepts to implement, while capacity building is much more complex. Holding individuals accountable to clearly stated rules and regulations is one thing, finding and training a new generation of public managers is another.

The durability of each approach flows from the other characteristics. When accountability resides in rules, not the individual or organization, it must be constantly reinforced. When it resides in incentives or organizations, it is more easily remembered. Individuals do the right thing not because they fear detection, but because the incentives and organization lead them, usually over a longer period of time. Compliance, however, is not less valid than capacity building, instead, each method has a different staying power.

The differences do not indicate an inherent trade-off among the three approaches. If Congress saw such a trade-off, as some scholars do, it is not apparent in the statute. The IGs were to play a role in all three, working to assure conformity with internal rules and regulations (compliance), economy and efficiency across a wide range of programs and activities (performance), and more general effectiveness in the management and operation of their departments and agencies (capacity).

As much as Congress hoped the three goals could coexist — compliance reinforcing performance, capacity reinforcing compliance — many IGs were forced to choose. The IGs became instruments of retrospective, or backward-looking, compliance rather than catalysts for either performance incentives or capacity building.

MONITORING TOWARD ACCOUNTABILITY

Whatever the type of accountability, whether tracking down cheaters or assessing the prospects for efficiency, the IGs have but one tool at hand: *monitoring*. They are to look, not act; recommend, not implement. The IGs were neither created as line, or operating, officers of their departments and agencies nor given any powers to suspend, or otherwise interfere with, program activities. Monitoring can be exercised through traditional financial compliance audits, highly individualized criminal investigations, program evaluation, or policy analysis, but it relies on others for action.

Instead, IGs were given complete access to information — information that Congress wanted, too. The IGs were free to audit, investigate, review, assess, analyze, evaluate, oversee, and appraise every problem, abuse, deficiency, and weakness relating to the programs and operations of their establishments, but they were specifically prohibited from accepting any program operating responsibilities. Ironically, even as Congress began its hearings on ways to reduce paperwork load imposed on the private sector by government, it signaled its willingness to impose an ever-increasing level of regulatory and reporting requirements on executive agencies and their employees. So, too, did presidents Carter and Reagan, and their Offices of Management and Budget.

At the same time Congress and the president increased the regulation of the federal government's employees, the private sector began to embrace the management philosophy of W. Edwards Deming, which focused on designing quality into a product at the front end of the process, instead of inspecting it in at the back end. Although all of Deming's fourteen points toward quality may be relevant to federal management improvement, the third point is most relevant to the IGs: cease dependence on mass inspection. As biographer Mary Walton explained, "American firms typically inspect a product as it comes off the line or at major stages. Defective products are either thrown out or reworked, both are unnecessarily expensive. In effect, a company is paying workers to make defects and then to correct them. Quality comes not from inspection but from improvement of the process."

Quality, however, does not eschew measurement. Implementation of Deming management is highly dependent on statistical process control and careful performance monitoring that allows managers, not inspectors, to track and tune the process. What makes monitoring different under Deming is its role in the management process. Unlike the American public sector, where monitoring is sometimes an end in itself and is almost always part of individual performance review, the Deming philosophy views monitoring as valid only for checking the progress of an overall management plan. Monitoring is never to be used for assessing employee blame. Evaluation by individual performance, merit rating, or annual review is one of Deming's seven deadly diseases blocking quality in manufacturing.

APPENDIX B

The following, reprinted with permission, is from *Somalia and Operation Restore Hope: Reflections on Peacemaking and Peacekeeping* by John L. Hirsch and Robert B. Oakley (Washington, D.C.: United States Institute of Peace Press, 1995).

A. UN SECURITY COUNCIL RESOLUTION 794
DECEMBER 3, 1992

The Security Council,

Reaffirming its resolutions 733 (1992) of 23 January 1992, 746 (1992) of 17 March 1992, 751 (1992) of 24 April 1992, 767 (1992) of 27 July 1992 and 775 (1992) of 28 August 1992,

Recognizing the unique character of the present situation in Somalia and mindful of its deteriorating, complex and extraordinary nature, requiring an immediate and exceptional response,

Determining that the magnitude of the human tragedy caused by the conflict in Somalia, further exacerbated by the obstacles being created to the distribution of humanitarian assistance, constitutes a threat to international peace and security,

Gravely alarmed by the deterioration of the humanitarian situation in Somalia and underlining the urgent need for the quick delivery of humanitarian assistance in the whole country,

Noting the efforts of the League of Arab States, the Organization of African Unity, and in particular the proposal made by its Chairman at the forty-seventh regular session of the General Assembly for the organization of an international conference on Somalia, and the Organization of the Islamic Conference and other regional agencies and arrangements to promote reconciliation and political settlement in Somalia and to address the humanitarian needs of the people of that country,

Commending the ongoing efforts of the United Nations, its specialized agencies and humanitarian organizations and of non-governmental organizations and of States to ensure delivery of humanitarian assistance in Somalia,

Responding to the urgent calls from Somalia for the international community to take measures to ensure the delivery of humanitarian assistance in Somalia,

Expressing grave alarm at continuing reports of widespread violations of international humanitarian law occurring in Somalia, including reports of violence and threats of violence against personnel participating lawfully in impartial humanitarian relief activities; deliberate attacks on non-combatants, relief consignments and vehicles, and medical and relief facilities; and impeding the delivery of food and medical supplies essential for the survival of the civilian population,

Dismayed by the continuation of conditions that impede the delivery of humanitarian supplies to destinations within Somalia, and in particular reports of looting of relief supplies destined for starving people, attacks on aircraft and ships bringing in humanitarian relief supplies, and attacks on the Pakistani UNOSOM contingent in Mogadishu,

Taking note with appreciation of the letters of the Secretary-General of 24 November 1992 (S/24859) and of 29 November 1992 (S/24868),

Sharing the Secretary-General's assessment that the situation in Somalia is intolerable and that it has become necessary to review the basic premises and principles of the United Nations effort in Somalia, and that UNOSOM's existing course would not in present circumstances be an adequate response to the tragedy in Somalia,

Determined to establish as soon as possible the necessary conditions for the delivery of humanitarian assistance wherever needed in Somalia, in conformity with resolutions 751 (1992) and 767 (1992),

Noting the offer by Member States aimed at establishing a secure environment for humanitarian relief operations in Somalia as soon as possible,

Determined further to restore peace, stability and law and order with a view to facilitating the process of a political settlement under the auspices of the United Nations, aimed at national reconciliation in Somalia, and *encouraging* the SecretaryGeneral and his Special Representative to continue and intensify their work at the national and regional levels to promote these objectives,

Recognizing that the people of Somalia bear ultimate responsibility for national reconciliation and the reconstruction of their own country,

1. *Reaffirms* its demand that all parties, movements and factions in Somalia immediately cease hostilities, maintain a cease-fire throughout the country, and cooperate with the Special Representative of the Secretary-General as well as with the military forces to be established pursuant to the authorization given in paragraph 10 below in order to promote the process of relief distribution, reconciliation and political settlement in Somalia;

2. *Demands* that all parties, movements and factions in Somalia take all measures necessary to facilitate the efforts of the United Nations, its specialized agencies and humanitarian organizations to provide urgent humanitarian assistance to the affected population in Somalia;

3. *Also demands* that all parties, movements and factions in Somalia take all measures necessary to ensure the safety of United Nations and all other personnel engaged in the delivery of humanitarian assistance, including the military forces to be established pursuant to the authorization given in paragraph 10 below;

4. *Further demands* that all parties, movements and factions in Somalia immediately cease and desist from all breaches of international humanitarian law including from actions such as those described above;

5. *Strongly condemns* all violations of international humanitarian law occurring in Somalia, including in particular the deliberate impeding of the delivery of food and medical supplies essential for the survival

of the civilian population, and *affirms* that those who commit or order the commission of such acts will be held individually responsible in respect of such acts;

6. *Decides* that the operations and the further deployment of the 3,500 personnel of the United Nations Operation in Somalia (UNOSOM) authorized by paragraph 3 of resolution 775 (1992) should proceed at the discretion of the Secretary-General in the light of his assessment of conditions on the ground; and *requests* him to keep the Council informed and to make such recommendations as may be appropriate for the fulfillment of its mandate where conditions permit;

7. *Endorses* the recommendation by the Secretary-General in his letter of 29 November 1992 (S/24868) that action under Chapter VII of the Charter of the United Nations should be taken in order to establish a secure environment for humanitarian relief operations in Somalia as soon as possible;

8. *Welcomes* the offer by a Member State described in the Secretary-General's letter to the Council of 29 November 1992 (S/24868) concerning the establishment of an operation to create such a secure environment;

9. *Welcomes also* offers by other Member States to participate in that operation;

10. *Acting* under Chapter VII of the Charter of the United Nations, *authorizes* the Secretary-General and Member States cooperating to implement the offer referred to in paragraph 8 above to use all necessary means to establish as soon as possible a secure environment for humanitarian relief operations in Somalia;

11. *Calls* on all Member States which are in a position to do so to provide military forces and to make additional contributions, in cash or in kind, in accordance with paragraph 10 above and *requests* the Secretary-General to establish a fund through which the contributions, where appropriate, could be channelled to the States or operations concerned;

12. *Authorizes* the Secretary-General and the Member States concerned to make the necessary arrangements for the unified command and control of the forces involved, which will reflect the offer referred to in paragraph 8 above;

13. *Requests* the Secretary-General and the Member States acting under paragraph 10 above to establish appropriate mechanisms for coordination between the United Nations and their military forces;

14. Decides to appoint an ad hoc commission composed of members of the Security Council to report to the Council on the implementation of this resolution;

15. *Invites* the Secretary-General to attach a small UNOSOM liaison staff to the Field Headquarters of the unified command;

16. Acting under Chapters VII and VIII of the Charter, *calls upon* States, nationally or through regional agencies or arrangements, to use such measures as may be necessary to ensure strict implementation of paragraph 5 of resolution 733 (1992);

17. *Requests* all States, in particular those in the region, to provide appropriate support for the actions undertaken by States, nationally or through regional agencies or arrangements, pursuant to this and other relevant resolutions;

18. *Requests* the Secretary-General and, as appropriate, the States concerned to report to the Council on a regular basis, the first such report to be made no later than fifteen days after the adoption of this resolution, on the implementation of this resolution and the attainment of the objective of establishing a secure environment so as to enable the Council to make the necessary decision for a prompt transition to continued peace-keeping operations;

19. *Requests* the Secretary-General to submit a plan to the Council initially within fifteen days after the adoption of this resolution to ensure that UNOSOM will be able to fulfil its mandate upon the withdrawal of the unified command;

20. *Invites* the Secretary-General and his Special Representative to continue their efforts to achieve a political settlement in Somalia;

21. *Decides* to remain actively seized of the matter.

B. SEVEN POINT AGREEMENT SIGNED BY MOHAMED FARAH AIDEED AND ALI MAHDI MOHAMED
DECEMBER 11, 1992

1. Immediate and total cessation of hostilities and restoration of unity of the U.S.C.

2. Immediate and total cessation of all negative propaganda.

3. To break the artificial lines in the capital city of Mogadishu.

4. All the forces and their technicals should report to their respective designated locations outside the city within the next 48 hours, and be controlled by the joint committee.

5. The already established reconciliation committee of the U.S.C. should convene their meetings within the next 24 hours.

6. We call upon all Somalis throughout the country to seriously engage on cessation of all hostilities and join with us for peace and unity of Somalia.

7. We express our deep appreciation to the international community for its efforts to assist Somalia and appeal to it to extend and expand its assistance including not only humanitarian relief aid but also reconstruction and rehabilitation as well as a national reconciliation conference.

Source: John Hirsch

C. GENERAL AGREEMENT AND SUPPLEMENT SIGNED IN ADDIS ABABA
JANUARY 8, 1993

We, the undersigned Somali political leaders, meeting at the United Nations Economic Commission for Africa in Addis Ababa, Ethiopia, from 4 January 1993 at the Informal Preparatory Meeting on National Reconciliation in Somalia, would like to thank the Secretary-General of the United Nations, H.E. Boutros Boutros-Ghali, who has facilitated this meeting in collaboration with the Organization of African Unity, the League of Arab States, the Organization of the Islamic Conference and the Standing Committee of the Horn of Africa.

We also thank the Government and people of Ethiopia, and H.E. Meles Zenawi, President of the Transitional Government of Ethiopia, for hosting the Meeting and for his personal engagement in assisting our efforts to reach the following agreement.

We, the Somali participants, further express our deep appreciation for the concern of the international community for the humanitarian crisis in our country and recognize their wish for us to reach a peaceful solution to our country's severe problems.

After discussing our problems and considering all options, we have agreed on the following points:

1. The convening of a National Reconciliation Conference in Addis Ababa on 15 March 1993;

2. The declaration of an immediate and binding cease-fire in all parts of the country under the control of the concerned warring factions, subject to paragraph (a) below;

3. The immediate cessation of all hostile propaganda against each other and the creation of an atmosphere conducive to reconciliation and peace;

4. The United Nations Operation in Somalia (UNOSOM), in consultation with the relevant regional and subregional organizations, will be responsible for the logistical preparations of the National Reconciliation Conference;

5. The establishment of further mechanisms for the continuation of free dialogue amongst all political factions and leaders in Somalia in preparation for the National Reconciliation Conference;

6. To continue and enhance our full and unrestrained cooperation with all international organizations working inside and outside Somalia to distribute humanitarian relief to our people;

7. To commit ourselves, without reservation, to facilitating the free movement of Somali people throughout the entire country as a measure of confidence-building before the National Reconciliation Conference.

This agreement shall be valid upon completion and adoption by consensus on the following three points, and a separate communiqué will be issued before leaving Addis Ababa:

a. The establishment of the modalities for implementing the cease-fire amongst all warring parties and the creation of a mechanism for disarmament;

b. The agenda of the National Reconciliation Conference;

c. The criteria for participation in the National Reconciliation Conference.

This agreement, signed in Addis Ababa on 8 January 1993, shall be considered binding on all the undersigned parties henceforth and all signatories shall be obliged to secure the support and implementation of this agreement amongst their movements and followers.

1. Somali Africans Muki Organization (SAMO)
   Mr. Mohamed Ramadan Arbow, Chairman

2. Somali Democratic Alliance (SDA)
   Mr. Mohamed Farah Abdullahi, Chairman

3. Somali Democratic Movement (SDM)
   Abdi Muse Mayo, Chairman
   Col. Mohamed Nur Aliyou, Chairman (SNA)

4. Somali National Democratic Union (SNDU)
   Mr. Ali Ismael Abdi, Chairman

5. Somali National Front (SNF)
   General Omar Hagi Mohamed Hersi, Chairman

6. Somali National Union (SNU)
   Dr. Mohamed Ragis Mohamed, Chairman

7. Somali Patriotic Movement (SPM)
   General Aden Abdillahi Noor, Chairman

8. Somali Patriotic Movement (SPM) (sna)
   Col. Ahmed Omar Jess, Chairman

9. Somali Salvation Democratic Front (SSDF)
   General Mohamed Abshir Musse, Chairman

10. Southern Somali National Movement (SSNM) (sna)
    Col. Abdi Warsame Isaaq, Chairman

11. United Somali Congress (USC) (sna)
    General Mohamed Farah Aidid, Chairman

12. United Somali Congress (USC)
    Mr. Mohamed Qanyare Afrah, Chairman

13. United Somali Front (USF)
    Mr. Abdurahman Dualeh Ali, Chairman

14. United Somali Party (USP)
    Mohamed Abdi Hashi, Chairman

January 8, 1993
Addis Ababa, ETHIOPIA
Africa Hall

Agreement on Implementing the Cease-fire and on Modalities of Disarmament

(Supplement to the General Agreement signed in Addis Ababa on 8 January 1993)

We, the undersigned Somali political leaders, meeting in Addis Ababa, Ethiopia, on 15 January 1993 at the Informal Preparatory Meeting on National Reconciliation in Somalia;

Having agreed on the need for a viable and verifiable cease-fire to promote the peace process in Somalia;

Recognizing that such cease-fire is intricately linked to questions of disarmament;

Further recognizing that disarmament cannot be accomplished in a single event but through a sustained process;
Hereby agree on the following:

I. Disarmament

1.1 All heavy weaponry under the control of political movements shall be handed over to a cease-fire monitoring group for safekeeping until such time as a legitimate Somali Government can take them over. This process shall commence immediately and be completed in March 1993.

1.2 The militias of all political movements shall be encamped in appropriate areas outside major towns where the encampment will not pose difficulties for peace. The encamped militias shall be disarmed following a process which will commence as soon as possible. This action shall be carried out simultaneously throughout Somalia. The international community will be requested to provide the encamped militias with upkeep.

1.3 The future status of the encamped militia shall be decided at the time of the final political settlement in Somalia. Meanwhile, the international community will be requested to assist in training them for civilian skills in preparation for possible demobilization.

1.4 All other armed elements, including bandits, shall be disarmed immediately and assisted through rehabilitation and integration into civil society.

II. Cease-fire monitoring group

2.1 A cease-fire monitoring group comprising UNITAF/United Nations troops shall be established immediately. There shall also be a committee composed of representatives of the warring factions to interlocute with the monitoring group and observe the implementation of the agreement by UNITAF/United Nations troops.

III. All sides agree in principle that properties unlawfully taken during the fighting shall be returned to the lawful owners. This shall be implemented as and when the situation allows.

IV. All POWs shall be freed and handed over to the International Committee of the Red Cross and/or UNITAF. This process shall commence immediately and be completed by 1 March 1993.

The present agreement shall enter into effect on 15 January 1993.

Source: "The Situation in Somalia," UN Document S/25168, January 26, 1993

D. ADDIS ABABA AGREEMENT OF THE FIRST SESSION OF THE CONFERENCE ON NATIONAL RECONCILIATION IN SOMALIA

MARCH 27, 1993

After long and costly years of civil war that ravaged our country, plunged it into famine, and caused acute suffering and loss of life among our people, there is the light of hope at last: progress has been made towards the restoration of peace, security and reconciliation in Somalia.

We, the Somali political leaders, recognize how vital it is that this process continue. It has our full commitment.

By our attendance at this historic Conference, we have resolved to put an end to armed conflict and to reconcile our differences through peaceful means. We pledge to consolidate and carry forward advances in peace, security and dialogue made since the beginning of this year. National reconciliation is now the most fervent wish of the Somali people.

We commit ourselves to continuing the peace process under the auspices of the United Nations and in cooperation with the Regional Organizations and the Standing Committee of the Horn as well as with our neighbors in the Horn of Africa.

After an era of pain, destruction and bloodshed that turned Somalis against Somalis, we have confronted our responsibility. We now pledge to work toward the rebirth of Somalia, to restore its dignity as a country and rightful place in the community of nations. At the close of the Holy Month of Ramadan, we believe this is the most precious gift we can give to our people.

The serenity and shade of a tree, which according to our Somali tradition is a place of reverence and rapprochement, has been replaced by the conference hall. Yet the promises made here are no less sacred or binding.

Therefore, we, the undersigned Somali political leaders, meeting at Africa Hall in Addis Ababa, Ethiopia between 15 and 27 March 1993, hereby reaffirm our commitment to the agreements signed during the Informal Preparatory Meeting on National Reconciliation in January 1993.

In concord to end hostilities, and to build on the foundation of peace for reconstruction and rehabilitation in Somalia, we agree to proceed within the framework of the following provisions and decisions:

I. *Disarmament and Security*

1. *Affirm* that uprooting of banditry and crime is necessary for peace, stability, security, reconciliation, reconstruction and development in Somalia;

2. *Further affirm* that disarmament must and shall be comprehensive, impartial and transparent;

3. *Commit* ourselves to complete, and simultaneous disarmament throughout the entire country in accordance with the disarmament concept and timeframe set by the Cease-fire Agreement of January 1993; and request that UNITAF/UNOSOM assist these efforts so as to achieve a substantial completion of the disarmament within 90 days;

4. *Further reiterate* our commitment to the strict, effective and expeditious implementation of the Cease-fire/Disarmament Agreement signed on 8 and 15 January 1993;

5. *Reaffirm* our commitment to comply with the requirements of the Cease-fire Agreement signed in January of 1993, including the total and complete handover of weapons to UNITAF/UNOSOM;

6. *Urge* UNITAF/UNOSOM to apply strong and effective sanctions against those responsible for any violation of the Cease-fire Agreement of January 1993;

7. *Stress* the need for the air, sea and land borders of Somalia to be closely guarded by UNITAF/UNOSOM in order to prevent any flow of arms into the country and to prevent violation of the territorial waters of Somalia;

8. *Further stress* the need for maximum cooperation by neighboring countries to assure that their common borders with Somalia are

not used for the movement of weapons in Somalia, in keeping with the United Nations arms embargo against Somalia;

9. *Agree* on the need to establish an impartial National and Regional Somali Police Force in all regions of the country on an urgent basis through the reinstatement of the former Somali Police force and recruitment and training of young Somalis from all regions, and request the assistance of the international community in this regard.

II. *Rehabilitation and Reconstruction*

1. *Affirm* the need to accelerate the supply and operation of relief, reconstruction and rehabilitation programs in Somalia;

2. *Welcome* the conclusion of the Third Coordination Meeting on Humanitarian Assistance to Somalia;

3. *Express* our appreciation to donor countries for their continued humanitarian assistance to Somalia and, in particular, for the generous pledge, made at the Third Coordination Meeting, to mobilize $142 million for relief and rehabilitation efforts in Somalia;

4. *Call upon* UNOSOM, aid agencies and donor countries to immediately assist in the rehabilitation of essential public and social services, and of necessary infrastructure, on a priority basis by the end of June 1993;

5. *Assure* the international community of the full desire of Somali leaders to reestablish, with the assistance of UNOSOM, a secure environment for relief, reconstruction and rehabilitation operations and the protection of relief and rehabilitation workers and supplies;

6. *Condemn* the acts of violence committed against relief workers and all forms of extortion regarding humanitarian operations;

7. *Urge* the organizations within the UN system and NGOs to effectively utilize Somali human resources in the rehabilitation and reconstruction process in Somalia.

III. *Restoration of Property and Settlement of Disputes*

1. *Affirm* that all disputes must henceforth be settled by dialogue, negotiations and other peaceful and legal means;

2. *Further affirm* that all private or public properties that were illegally confiscated, robbed, stolen, seized, embezzled or taken by other fraudulent means must be returned to their rightful owners;

3. *Decide* to deal with this matter within the framework specified in the report of the committee on the peaceful settlement of disputes.

IV. *Transitional Mechanisms*

The Somali people believe that there is concurrence among the people of Somalia that Somalia must retain its rightful place in the community of nations and that they must express their political views and make the decisions that affect them. This is an essential component of the search for peace.

To achieve this, political and administrative structures in Somalia need to be rebuilt to provide the people as a whole with an opportunity to participate in shaping the future of the country.

In this context, the establishment of transitional mechanisms which prepare the country for a stable and democratic future is absolutely essential. During the transitional period, which will last for a period of *two years* effective from the date of signature to this agreement, the emphasis will be upon the provision of essential services, complete disarmament, restoration of peace and domestic tranquility and on the attainment of the reconciliation and reconstruction of basic infrastructure and on the building of democratic institutions. All of this will prepare the country to enter a constitutional phase in which the institutions of democratic governance, rule of law, decentralization of power, protection of human rights and individual liberties, and the safeguarding of the integrity of the Somali Republic are all in place.

Therefore, we have agreed to a broad outline of a framework for a transitional system of governance to allow for the provision of essential services, the creation of a basis for long-term planning, and for the resumption of greater administrative responsibility by Somalis. In general terms, this system will be composed of four basic administrative components that will be mandated to function during the transitional period.

Taking into account the reality of the situation in Somalia today and the need for stability, we hereby agree to the establishment of the following four basic transitional organs of authority:

1. *The Transitional National Council (TNC)*

The TNC will:

a) be the repository of Somali sovereignty;

b) be the prime political authority having legislative functions during the period in question;

c) interact, as appropriate, with the international community, including UNOSOM;

d) appoint various committees, including the Transitional Charter Drafting Committee, as required;

e) appoint Officers for its various functions;

f) appoint the heads of administrative departments;

g) oversee the performance of the departments created; and

h) establish an independent Judiciary.

The TNC shall be composed of:

a) Three representatives from each of the 18 regions currently recognized, including one woman from each region;

b) Five additional seats for Mogadishu;

c) One nominee from each of the political factions currently participating in the First session of the National Reconciliation Conference;

2. *The Central Administrative Departments (CADs)*

The TNC will appoint the heads of the Central Administrative Departments, whose prime function will be to re-establish and operate the departments of civil administration, social affairs, economic affairs and humanitarian affairs, paving the way for the re-establishment and operation of a formal government. The CADs shall comprise skilled professionals having the ability to reinstate, gradually, the administrative functions of national public administration. The performance of these departments will be overseen by the TNC.

3. *Regional Council (RCs)*

Regional Councils shall be established in all the existing 18 regions of Somalia. The present 18 regions shall be maintained during the transitional period. The Regional Councils shall be entrusted primarily with the task of implementing humanitarian, social and economic programs in coordination with the TNC and will also assist in the conducting of the internationally-supervised census. The Regional Councils will liaise with UNOSOM II, UN specialized agencies, NGOs and other relevant organizations directly and through the Central Administrative Departments and Transitional National Council. The Regional Councils shall also be responsible for law and order at the regional level. In this regard, the law enforcement institution will be a regional police force and a regional judiciary. The District Councils (see below) in each region shall send representatives who will constitute the Regional Councils.

4. *District Councils*

District Councils shall be established in the present districts in every region. District council members shall be appointed through election or through consensus-based selection in accordance with Somali traditions. The District Councils shall be responsible for managing the affairs of the district including public safety, health, education and reconstruction.

V. *Conclusion*

The Conference agreed on the appointment, by the TNC, of a Transitional Charter Drafting Committee referred to in section IV 1 (d) above. In drafting the Transitional Charter, the Committee shall be guided by the basic principles of the Universal Declaration of Human Rights and by the Somali traditional ethics.

The Conference agreed that the TNC shall appoint a "Peace Delegation" composed of political movements and other social elements to travel to all parts of the country for the purpose of advancing the peace and reconciliation process as well as to explain the agreements reached in Addis Ababa.

We further agree that the TNC shall appoint a National Committee to bring about reconciliation and seek solutions to outstanding political problems with the SNM.

The Conference also calls upon the international community and in particular on the neighboring states to facilitate the noble effort at reconciliation by providing moral and material support.

In conclusion, we the undersigned, in agreeing to the above, resolve that never again will Somalia suffer the tragedy of the recent past. Emerging from the darkness of catastrophe and war, we Somalis herald the beginning of a new era of peace, of healing and rebuilding, in which cooperation and trust will overcome hatred and suspicion. It is a message we must pass on to our children and our grandchildren so that the proud Somali family, as we knew it, can once again become whole.

We, the undersigned, hereby pledge to abandon the logic of force for the ethic of dialogue. We will pursue the process of national reconciliation with vigor and sincerity, in accordance with this declaration and with the cooperation of the people of Somalia as a whole.

Recognizing the tragic and painful recent history of problems in our country, we pledge to achieve comprehensive national reconciliation through peaceful means. We also pledge to adopt, in all parts of Somalia, transitional measures that will contribute to harmony and healing of wounds among all the people of Somalia.

We invite the Secretary-General of the United Nations and his Special Representative in Somalia, in accordance with the mandate entrusted to them by the UN Security Council, to extend all necessary assistance to the people of Somali for the implementation of this agreement.

Signatories

1.	SAMO	Mohamed R. Arbow
2.	SDA	Mohamed F. Abdullahi
3.	SDM	Abdi Musse Mayow
4.	SDM (SNA)	Mohamed Nur Alio
5.	SNDU	Ali Ismail Abdi
6.	SNF	Gen. Omar Haji Mohamed
7.	SNU	Mohamed Rajis Mohamed
8.	SPM	Gen. Aden Abdullahi Nur
9.	SPM (SNA)	Ahmed Hashi Mahmmud
10.	SSDF	Gen. Mohammed Abshir Mussa
11.	SSNM	Abdi Warsame Isaq
12.	USC (SNA)	Gen. Mohammed Farah H. Aidid
13.	USC	Mohammed Qanyare Afrah
14.	USF	Abdurahman Dualeh Ali
15.	USP	Mohamed Abdi Hashi

Before proceeding to adopt the Agreement, I should also like to confirm to the participants that at the meeting of all political leaders yesterday, prior to the signing of the document they unanimously agreed that one of the five additional seats for Mogadishu would be reserved for Mr. Ali Mahdi. This understanding remains binding.

Source: John Hirsch

E. UN SECURITY COUNCIL RESOLUTION 814
MARCH 26, 1993

The Security Council,

Reaffirming its resolutions 733 (1992) of 23 January 1992, 746 (1992) of 17 March 1992, 751 (1992) of 24 April 1992, 767 (1992) of 27 July 1992, 775 (1992) of 28 August 1992 and 794 (1992) of 3 December 1992,

Bearing in mind General Assembly resolution 47/167 of 18 December 1992,

Commending the efforts of Member States acting pursuant to resolution 794 (1992) to establish a secure environment for humanitarian relief operations in Somalia,

Acknowledging the need for a prompt, smooth and phased transition from the Unified Task Force (UNITAF) to the expanded United Nations Operation in Somalia (UNOSOM II),

Regretting the continuing incidents of violence in Somalia and the threat they pose to the reconciliation process,

Deploring the acts of violence against persons engaging in humanitarian efforts on behalf of the United Nations, States, and non-governmental organizations,

Noting with deep regret and concern the continuing reports of widespread violations of international humanitarian law and the general absence of the rule of law in Somalia,

Recognizing that the people of Somalia bear the ultimate responsibility for national reconciliation and reconstruction of their own country,

Acknowledging the fundamental importance of a comprehensive and effective programme for disarming Somali parties, including movements and factions,

Noting the need for continued humanitarian relief assistance and for the rehabilitation of Somalia's political institutions and economy,

Concerned that the crippling famine and drought in Somalia, compounded by the civil strife, have caused massive destruction to the means of production and the natural and human resources of that country,

Expressing its appreciation to the Organization of African Unity, the League of Arab States, the Organization of the Islamic Conference and the Non-Aligned Movement for their cooperation with, and support of, the efforts of the United Nations in Somalia,

Further expressing its appreciation to all Member States which have made contributions to the Fund established pursuant to paragraph 11 of resolution 794 (1992) and to all those who have provided humanitarian assistance to Somalia,

Commending the efforts, in difficult circumstances, of the initial United Nations Operation in Somalia (UNOSOM) established pursuant to resolution 751 (1992),

Expressing its appreciation for the invaluable assistance the neighbouring countries have been providing to the international community in its efforts to restore peace and security in Somalia and to host large numbers of refugees displaced by the conflict and *taking note* of the difficulties caused to them due to the presence of refugees in their territories,

Convinced that the restoration of law and order throughout Somalia would contribute to humanitarian relief operations, reconciliation and political settlement, as well as to the rehabilitation of Somalia's political institutions and economy,

Convinced also of the need for broad-based consultations and deliberations to achieve reconciliation, agreement on the setting up of transitional government institutions and consensus on basic principles and steps leading to the establishment of representative democratic institutions,

Recognizing that the re-establishment of local and regional administrative institutions is essential to the restoration of domestic tranquillity,

Encouraging the Secretary-General and his Special Representative to continue and intensify their work at the national, regional and local levels, including and encouraging broad participation by all sectors of Somali society, to promote the process of political settlement and national reconciliation and to assist the people of Somalia in rehabilitating their political institutions and economy,

Expressing its readiness to assist the people of Somalia, as appropriate, on a local, regional or national level, to participate in free and fair elections, with a view towards achieving and implementing a political settlement,

Welcoming the progress made at the United Nations-sponsored Informal Preparatory Meeting on Somali Political Reconciliation in Addis Ababa from 4 to 15 January 1993, in particular the conclusion at that meeting of three agreements by the Somali parties, including movements and factions, and *welcoming also* any progress made at the Conference on National Reconciliation which began in Addis Ababa on 15 March 1993,

Emphasizing the need for the Somali people, including movements and factions, to show the political will to achieve security, reconciliation and peace,

Noting the reports of States concerned of 17 December 1992 (S/24976) and 19 January 1993 (S25126) and of the Secretary-General of 19 December 1992 (S/24992) and 26 January 1993 (S25168) on the implementation of resolution 794 (1992),

Having examined the report of the Secretary-General of 3 March 1993 (S/25354 and Add.1 and 2),

Welcoming the intention of the Secretary-General to seek maximum economy and efficiency and to keep the size of the United Nations presence, both military and civilian, to the minimum necessary to fulfil its mandate,

Determining that the situation in Somalia continues to threaten peace and security in the region,

A

1. *Approves* the report of the Secretary-General of 3 March 1993;

2. *Expresses* its appreciation to the Secretary-General for convening the Conference on National Reconciliation for Somalia in accordance with the agreements reached during the Informal Preparatory Meeting on Somalia Political Reconciliation in Addis Ababa in January 1993 and for the progress achieved towards political reconciliation in Somalia, and also for his efforts to ensure that, as appropriate, all Somalis, including movements, factions, community leaders, women, professionals, intellectuals, elders and other representative groups are suitably represented at such conferences;

3. *Welcomes* the convening of the Third United Nations Coordination Meeting for Humanitarian Assistance for Somalia in Addis Ababa from 11 to 13 March 1993 and the willingness expressed by Governments through this process to contribute to relief and rehabilitation efforts in Somalia, where and when possible;

4. *Requests* the Secretary-General, through his Special Representative and with assistance, as appropriate, from all relevant United Nations entities, offices and specialized agencies, to provide humanitarian and other assistance to the people of Somalia in rehabilitating their political institutions and economy and promoting political settlement and national reconciliation, in accordance with the recommendations contained in his report of 3 March 1993, including in particular:

   (a) To assist in the provision of relief and in the economic rehabilitation of Somalia, based on an assessment of clear, prioritized needs, and taking into account, as appropriate, the 1993 Relief and Rehabilitation Programme for Somalia prepared by the United Nations Department of Humanitarian Affairs;

   (b) To assist in the repatriation of refugees and displaced persons within Somalia;

   (c) To assist the people of Somalia to promote and advance political reconciliation, through broad participation by all sectors of Somali society, and the re-establishment of national and regional institutions and civil administration in the entire country;

   (d) To assist in the re-establishment of Somali police, as appropriate at the local, regional or national level, to assist in the restoration and maintenance of peace, stability and law and order, including in the investigation and facilitating the prosecution of serious violations of international humanitarian law;

(e) To assist the people of Somalia in the development of a coherent and integrated programme for the removal of mines throughout Somalia;

(f) To develop appropriate public information activities in support of the United Nations activities in Somalia;

(g) To create conditions under which Somali civil society may have a role, at every level, in the process of political reconciliation and in the formulation and realization of rehabilitation and reconstruction programmes;

B

Acting under Chapter VII of the Charter of the United Nations,

5. *Decides* to expand the size of the UNOSOM force and its mandate in accordance with the recommendations contained in paragraphs 56-88 of the report of the Secretary-General of 3 March 1993, and the provisions of this resolution;

6. *Authorizes* the mandate for the expanded UNOSOM (UNOSOM II) for an initial period through 31 October 1993, unless previously renewed by the Security Council;

7. *Emphasizes* the crucial importance of disarmament and the urgent need to build on the efforts of UNITAF in accordance with paragraphs 59-69 of the report of the Secretary-General of 3 March 1993;

8. *Demands* that all Somali parties, including movements and factions, comply fully with the commitments they have undertaken in the agreements they concluded at the Informal Preparatory Meeting on Somali Political Reconciliation in Addis Ababa, and in particular with their Agreement on Implementing the Cease-fire and on Modalities of Disarmament (S/25168, annex III);

9. *Further demands* that all Somali parties, including movements and factions, take all measures to ensure the safety of the personnel of the United Nations and its agencies as well as the staff of the International Committee of the Red Cross (ICRC), intergovernmental organizations and non-governmental organizations engaged in providing humanitarian and other assistance to the people of Somalia in rehabilitating their political institutions and economy and promoting political settlement and national reconciliation;

10. *Requests* the Secretary-General to support from within Somalia the implementation of the arms embargo established by resolution 733 (1992), utilizing as available and appropriate the UNOSOM II forces authorized by this resolution, and to report on this subject, with any recommendations regarding more effective measures if necessary, to the Security Council;

11. *Calls upon* all States, in particular neighbouring States, to cooperate in the implementation of the arms embargo established by resolution 733 (1992);

12. *Requests* the Secretary-General to provide security as appropriate to assist in the repatriation of refugees and the assisted resettlement of displaced persons, utilizing UNOSOM II forces, paying particular attention to those areas where major instability continues to threaten peace and security in the region;

13. *Reiterates its demand* that all Somali parties, including movements and factions, immediately cease and desist from all breaches of international humanitarian law and *reaffirms* that those responsible for such acts be held individually accountable;

14. *Requests* the Secretary-General, through his Special Representative, to direct the Force Commander of UNOSOM II to assume responsibility for the consolidation, expansion and maintenance of a secure environment throughout Somalia, taking account of the particular circumstances in each locality, on an expedited basis in accordance with the recommendations contained in his report of 3 March 1993, and in this regard to organize a prompt, smooth and phased transition from UNITAF to UNOSOM II;

15. *Requests* the Secretary-General to maintain the fund established pursuant to resolution 794 (1992) for the additional purpose of receiving contributions for maintenance of UNOSOM II forces following the departure of UNITAF forces and for the establishment of Somali police, and calls on Member States to make contributions to this fund, in addition to their assessed contributions;

16. *Expresses appreciation* to the United Nations agencies, intergovernmental and non-governmental organizations and the ICRC for their contributions and assistance and *requests* the Secretary-General to ask them to continue to extend financial, material and technical support to the Somali people in all regions of the country;

17. *Requests* the Secretary-General to seek, as appropriate, pledges and contributions from States and others to assist in financing the rehabilitation of the political institutions and economy of Somalia;

18. *Requests* the Secretary-General to keep the Security Council fully informed on action taken to implement the present resolution, in particular to submit as soon as possible a report to the Council containing recommendations for establishment of Somali police forces and thereafter to report no later than every ninety days on the progress achieved in accomplishing the objectives set out in the present resolution;

19. *Decides* to conduct a formal review of the progress towards accomplishing the purposes of the present resolution no later than 31 October 1993;

20. *Decides* to remain actively seized of the matter.

F. UN SECURITY COUNCIL RESOLUTION 837

JUNE 6, 1993

The Security Council,

Reaffirming its resolutions 733 (1992) of 23 January 1992, 746 (1992) of 17 March 1992, 751 (1992) of 24 April 1992, 767 (1992) of 27 July 1992, 775 (1992) of 28 August 1992, 794 (1992) of 3 December 1992 and 814 (1993) of 26 March 1993,

Bearing in mind General Assembly resolution 47/167 of 18 December 1992,

Gravely alarmed at the premeditated armed attacks launched by forces apparently belonging to the United Somali Congress (USC/SNA) against the personnel of the United Nations Operation in Somalia (UNOSOM II) on 5 June 1993,

Strongly condemning such actions, which directly undermine international efforts aimed at the restoration of peace and normalcy in Somalia,

Expressing outrage at the loss of life as a result of these criminal attacks,

Reaffirming its commitment to assist the people of Somalia in reestablishing conditions of normal life,

Stressing that the international community is involved in Somalia in order to help the people of Somalia who have suffered untold miseries due to years of civil strife in that country,

Acknowledging the fundamental importance of completing the comprehensive and effective programme for disarming all Somali parties, including movements and factions,

Convinced that the restoration of law and order throughout Somali would contribute to humanitarian relief operations, reconciliation and political settlement, as well as to the rehabilitation of Somalia's political institutions and economy,

Condemning strongly the use of radio broadcasts, in particular by the USC/SNA, to incite attacks against United Nations personnel,

Recalling the statements made by its president on 31 March 1993 (S/25493) concerning the safety of United Nations forces and personnel deployed in conditions of strife and *committed* to consider promptly measures appropriate to the particular circumstances to ensure that persons responsible for attacks and other acts of violence against United Nations forces and personnel are held to account for their actions,

Taking note of the information provided to the Council by the Secretary-General on 6 June 1993,

Determining that the situation in Somalia continues to threaten peace and security in the region,

Acting under Chapter VII of the Charter of the United Nations,

1. *Strongly condemns* the unprovoked armed attacks against the personnel of UNOSOM II on 5 June 1993, which appear to have been part of a calculated and premeditated series of cease-fire violations to prevent by intimidation UNOSOM II from carrying out its mandate as provided for in resolution 814 (1993);

2. *Expresses* its condolences to the Government and people of Pakistan and the families of the UNOSOM II personnel who have lost their lives;

3. *Reemphasizes* the crucial importance of the early implementation of the disarmament of all Somali parties, including movements and factions, in accordance with paragraphs 56-69 of the report of the Secretary-General of 3 March 1993 (S/25354), and of neutralizing radio broadcasting systems that contribute to the violence and attacks directed against UNOSOM II;

4. *Demand once again* that all Somali parties, including movements and factions, comply fully with the commitments they have undertaken in the agreements they concluded at the informal Preparatory Meeting on Somali Political Reconciliation in Addis Ababa, and in par-

ticular with their Agreement on Implementing the Cease-fire and on Modalities of Disarmament (S/25168, annex III);

5. *Reaffirms* that the Secretary-General is authorized under resolution 814 (1993) to take all necessary measures against all those responsible for the armed attacks referred to in paragraph 1 above, including against those responsible for publicly inciting such attacks, to establish the effective authority of UNOSOM II throughout Somalia, including to secure the investigation of their actions and their arrest and detention for prosecution, trial and punishment;

6. *Requests* the Secretary-General urgently to inquire into the incident, with particular emphasis on the role of those factional leaders involved;

7. *Encourages* the rapid and accelerated deployment of all UNOSOM II contingents to meet the full requirements of 28,000 men, all ranks, as well as equipment, as indicated in the Secretary-General's report of 3 March 1993 (S/25354);

8. *Urges* Member States to contribute, on an emergency basis, military support and transportation, including armored personnel carriers, tanks and attack helicopters, to provide UNOSOM II the capability appropriately to confront and deter armed attacks directed against it in the accomplishment of its mandate;

9. *Further requests* the Secretary-General to submit a report to the Council on the implementation of the present resolution, if possible within seven days from the date of its adoption;

10. *Decides* to remain actively seized of the matter.